Programming the Boundary Element Method

Programming the Boundary Element Method
An Introduction for Engineers

Gernot Beer
Institute for Structural Analysis
University of Technology Graz, Austria

JOHN WILEY & SONS, LTD

Chichester • New York • Weinheim • Brisbane • Singapore • Toronto

Other Wiley Editorial Offices

John Wiley & Sons, Inc., 605 Third Avenue,
New York, NY 10158-0012, USA

WILEY-VCH Verlag GmbH
Pappelallee 3, D-69469 Weinheim, Germany

John Wiley & Sons, Australia, Ltd
33 Park Road, Milton, Queensland 4064, Australia

John Wiley & Sons (Canada) Ltd, 22 Worcester Road
Rexdale, Ontario, M9W 1L1, Canada

John Wiley & Sons (Asia) Pte Ltd, 2 Clementi Loop #02-01,
Jin Xing Distripark, Singapore 129809

Library in Congress Cataloging-in-Publication Data

Beer, Gernot.
   Programming the boundary element method / Gernot Beer.
     p. cm.
   Includes bibliographical references and index.
   ISBN 0-471-85722-X (alk. Paper)
    1. Boundary element methods—Data processing. 2. Computer programming. 3.
   FORTRAN (Computer program language) I. Title.

   TA347.B69 B44 2001
   620'.001'515355—dc21

                                          00-054558

British Library Cataloguing in Publication Data

A catalogue record for this book is available from the British Library

ISBN 0-471-85722-X (cloth)
ISBN 0-471-86333-5 (paper)

Produced from pdf files supplied by the author.

CONTENTS

Preface

In 1909, Ritz proposed a method for the approximate solution of differential equations. The proposition was that a set of arbitrary functions which only satisfy boundary conditions could be used to approximate the exact solution. This paper subsequently provided the basis for the well known finite element method (FEM).

In 1926, Trefftz published a paper entitled *Ein Gegenstück zum Ritzschen Verfahren* (An alternative to the Ritz method) where he suggested that, instead of functions which satisfy boundary conditions, those satisfying only the governing differential equations could be used. This paper then supplied the basic idea for the lesser known boundary element method (BEM).

The development of both methods, which started almost simultaneously when digital computers became available, has been quite different. Whereas the FEM is very well known and widely used, the BEM has become a sort of 'Cinderella' of numerical methods, that is, one the beauty of which is being kept hidden away and not fully appreciated.

At the very early stage in the development of the FEM, the first edition of the book *The Finite Element Method in Engineering Science*, by O. C. Zienkiewicz, was published. This was a book that engineers could understand and, more importantly, one which showed how the method could be implemented and used. Furthermore, the undying enthusiasm and drive of Oleg Zeinkiewicz convinced a myriad of people, including myself, to work with the FEM.

In contrast, early texts on the BEM concentrated on the mathematics of the method and in most cases ignored the original contribution of Trefftz. Nearly all earlier texts used tensor notation. Engineers at that time were not used to this notation and this would have prevented many from working in this field or using the method. All this gave the impression that the method was difficult to understand and to program, an opinion which, unfortunately, seems to be still very much prevalent today.

The motivation for writing this book was to show that the numerical implementation of the BEM is not significantly more difficult than that of the FEM, and that, if the theory is presented in engineering terms, not more difficult to understand. With this book I hope to make more engineers aware that for certain applications the method can not only mean a substantial saving in effort but also result in greater accuracy. In contrast to other books on the BEM my aim was to present the topic in a very practical way and this is demonstrated by devoting a whole chapter on industrial applications.

I make no secret of the fact that this book was inspired by the very successful text by Smith and Griffiths *Programming the Finite Element Method*, published by Wiley, now in its third edition. Indeed, on the cover of both books there is an example of a numerical simulation of the rock caverns, that house the scientific equipment of the CERN particle accelerator. By comparing both pictures the subtle differences between the FEM and BEM may be clearly seen.

In a sense, the present book is a companion to my earlier one, co-authored by J. Watson, *Introduction to Finite and Boundary Element Methods for Engineers,* published by Wiley in 1992, which goes into much greater detail with respect to the mathematical treatment of the BEM and concentrates on the coupling with the FEM.

There is considerably less mathematical rigour in the present book as compared with others, and I have often explained some of the more difficult aspects of the method in engineering terms. No apologies are made for this, since one of the main aims of this book is to increase the use and popularity of the BEM in the engineering community. Excellent books dealing with the mathematical treatment of the method are readily available and may be consulted.

However, this book is not just about learning how to program the BEM, but about understanding the method, not through complicated mathematical derivations but through numerical experiments. Throughout the book the reader will be able to try and test the method on small examples by using a series of programs, and will thus be able to learn intuitively by experience.

It is hoped this book will help to bring the numerical Cinderella out into the open, so that her beauty can be seen and appreciated by more engineers.

COMPUTER PROGRAMS

All software (libraries and programs) can be downloaded free of charge from the website http://www.ifb.tu-graz.ac.at/publications

G. Beer
Graz, Austria, January 2001

Acknowledgements

A number of people have had considerable influence on my scientific career path and have in a way contributed to the fact that this book was written. My father Hermann Beer, who in his lectures (that I had the privilege to attend) inspired the clear and simple way in which difficult topics could be explained. Oleg Zienkiewicz, who inspired me, through his early books on the finite element method and his friendship, to work in this area. I still remember vividly the late evening in the *Traminer Weinstube* many years ago in my hometown Graz where, drawing on napkins stained with the light red wine of the region, he first introduced me to the Trefftz method and to his idea of coupling finite and boundary elements. I strongly believe that this gave me the impetus to extend my research interest into the boundary element method. John Watson, with whom I wrote my first book on finite and boundary elements, helped me in the understanding of the method, especially with respect to numerical implementation of isoparametric boundary elements, where he did some pioneering work.

Writing a book is a task that cannot be achieved without a 'support group'. In the case of the present book, which was written in a very short time, this was paramount. Of course, this support includes one's wife and children, who are affected by the long working hours. However, my wife Sylvia deserves special thanks beyond the one usually offered for wifely support, because she actively helped me with the book. Using her experience in desktop publishing she provided valuable assistance in editing it, checking the text, layout and compiling the index. Without her help this book would certainly not have been finished so quickly.

Christian Dünser, one of my PhD students at the time of writing, provided considerable support by checking and modifying the computer programs. My undergraduate students, Christian Linder and Werner Sachs, did most of the work for Chapter 9 and Appendix B. Special thanks are also due to students of the course I presented at Helsinki University while the book was in preparation, for handing in the exercises and their suggestions. I have used some diagrams provided by Hans Grippenberg in Chapter 5.

Thanks are also due to to Fulvio Tonon, for the use of the tunnel analysis examples from his thesis and to Harald Golser, for supplying the example of multi-region analysis in slope stability presented in Chapter 15.

Numerical analysis would largely remain academic if it were not for the end users, the consultants, who use it to solve real life problems and for the clients who need solutions. In this respect, I would like to acknowledge Lahmeyer International, in particular Peter Zacher, who supplied the application in dam engineering, and Geoconsult ZT, in the person of Gernot Jedlitschka, who provided the material on the CERN caverns. Schoeller Bleckmann Edelstahlrohre, Austria, supplied the interesting application in mechanical engineering, and their permission to use this work as an example of industrial application is appreciated. I would also like to acknowledge all those consultants who have supported the use of the BEM in consulting, especially Golder/Grundteknik in Stockholm.

Last but not least, I would like to gratefully acknowledge the help of Trevor Davies, who looked through the draft of the manuscript and made many helpful suggestions. As it happened, we were both in the finishing stages of book writing when he visited the TU Graz for a one month stay. I very much appreciated his discussions on various topics which prompted me to think of better ways to explain certain difficult aspects.

1

Preliminaries

*A journey of a million miles
starts with the first step*

Confucius

1.1   INTRODUCTION

There are many textbooks which describe the mathematical background of boundary element methods (the interested reader may refer to an excellent compilation of literature on the BEM in the BENET homepage *http://www.ce.udel.edu/faculty/cheng/benet*). Unfortunately, most texts to date have been written by mathematicians or engineers with a strong background in mathematics and, therefore, they tend to dwell on the theoretical treatment of the method rather than concentrating on the physical meaning and implementation. Furthermore, although many books include simple programs in the appendix few have examples on how the theory explained is translated into a computer program. Since it is obvious that the methods would not be useful without computers, the lack of emphasis on computer implementation is surprising. In contrast to most mathematicians, who are happy just to prove the existence of a solution and error bounds, the engineer is interested in the application of the method in solving real problems. Invariably this means that either an existing program must be used or a program developed for solving a specific problem. Either way, a user of boundary element software must have a basic understanding about how the theory is implemented, as it is important that one be aware of the limitations of the program, possible pitfalls to look out for, etc. So in addition to understanding the theoretical assumptions made to obtain an approximate solution of the problem and the theoretical error bounds, such as they may be derived for certain examples, the user must be aware of the errors introduced by the numerical implementation, for example, the numerical integration and boundary interpolation.

Most readers of this book will be familiar with the finite element method[1]. In the most common version of this method in elasticity, we subdivide the domain into elements and approximate the displacement field with functions which are defined at element level, i.e. are piecewise continuous. The parameters of these functions, which are the displacements at the nodes where elements are connected to each other, are determined by a suitable general equilibrium condition, such as the minimum of potential energy. The functions themselves generally do not satisfy the governing differential equations. Therefore, for elasticity problems, we may have overall equilibrium satisfied, but a local violation of equilibrium exists, for example at element boundaries. It can be proven that the method converges, i.e. the exact solution is approached as the element size approaches zero.

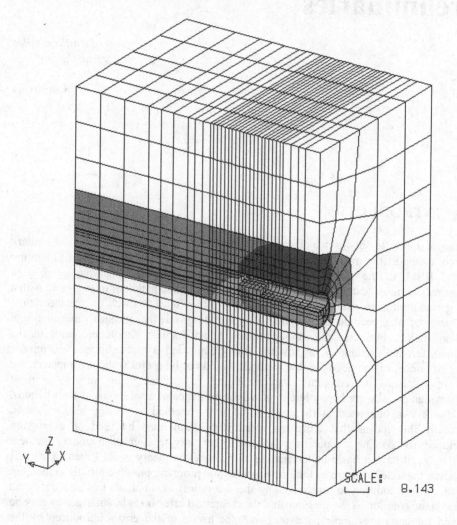

Figure 1.1   Finite element mesh for the analysis of tunnel excavation with contours of z-displacements on elements near tunnel

Figure 1.1 shows an example of a finite element mesh for the three-dimensional analysis of sequential excavation and construction of a tunnel. A plane of symmetry is applied so that only half of the tunnel is discretised. Note that to model the rock mass through which the tunnel is driven, which for all practical purposes can be assumed to be infinite, we must make a 'box' of solid elements. At the outer boundaries of this box we either set the displacements to zero or apply stress boundary conditions which represent the *in situ* stress. The mesh shown here has approximately 100 000 degrees of freedom, and a solution took several hours on a PC.

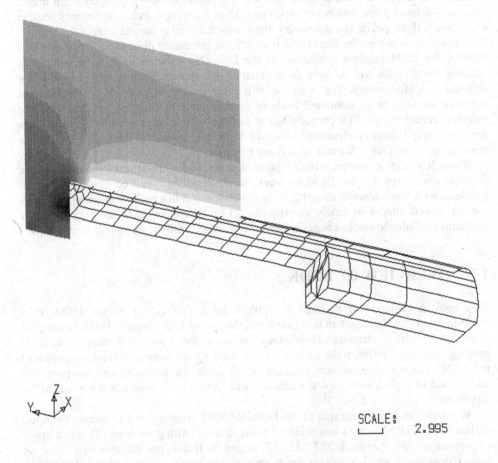

Figure 1.2   Boundary element mesh for the simulation of tunnel excavation with contours of σ_z on a result plane

In contrast, the BEM does not require the subdivision of the domain into elements because the functions used for approximating the solution inside the domain are chosen to be only those which satisfy the governing differential equations exactly. Consequently, only boundary conditions have to be satisfied. This can be achieved in an approximate way by subdividing the boundary into elements over which the values (for example, tractions or displacements in the case of elasticity) are interpolated, much in the same way as with the FEM. The advantage of the method is obvious: the dimensionality of the problem is reduced by one order, that is, only a surface instead of a volume discretisation is required. This means that the number of unknowns is reduced dramatically, especially for three-dimensional problems, because unknowns occur only on the problem boundary. Another advantage is that equilibrium is satisfied exactly everywhere in the domain, and that it is easy to deal with an infinite continuum.

As an example, Figure 1.2 shows the boundary element mesh for the same tunnel as analysed by the FEM in Figure 1.1. This mesh has approximately 1000 degrees of freedom and takes 3 minutes to solve on a PC. The stress contours, computed and drawn on a user-defined plane inside the rock mass, show no jumps, as are sometimes seen in FEM results if no special stress recovery algorithms are used to smooth results.

Since functions must be found which satisfy the governing differential equation (DE) exactly, the BEM requires a solution of the DE. This solution must be as simple as possible because, as will be seen in the chapter on implementation, this is crucial for efficiency. Unfortunately, the simplest solutions which we can find (fundamental solutions) are due to concentrated loads or sources and are singular, i.e., have infinite values at certain points. This property has to be taken into account when integrating these functions over boundary elements. This will make the numerical integration procedure more complicated than is the case with finite elements.

There has been a general misconception that because a fundamental solution of the problem must exist for the BEM to work, the method can only be applied to linear problems with homogeneous material. As will be shown in this book, non-linear problems can be solved almost as easily as with the FEM by the repeated solution of linear problems and inhomogeneous material can be considered with the multi-region method.

1.2   OVERVIEW OF BOOK

This book is designed to be used as a basis for a first course on the BEM. It is recommended that the chapters be read consecutively, as later chapters build on material discussed in earlier chapters. Throughout the book the reader will build a suite of subprograms which perform the various tasks needed for the numerical implementation of the BEM. Various exercises are included which allow the reader to test the programs written and to experience how the method works. Answers to exercises are given in the Appendix.

We start with an introduction to the FORTRAN 95 programming language, the latest dialect of FORTRAN which was released at the time of writing the book. This is a major improvement over the old FORTRAN 77 variant in that it has extensive facilities for matrix operations which result in much shorter and more readable code. FORTRAN, which stands for FORmula TRANslation, is the most widely used language for programming engineering applications and is easier to learn than other higher level languages such as C++. However, there is no reason why the procedures outlined in some detail in this book could not be implemented in another language.

The next chapter deals with the way in which we can describe the geometrical boundary of a problem and the boundary conditions in a numerical way. This is done by subdividing the surface into small elements and by interpolating between nodal values. This is essential for the later treatment of integral equations. With the aid of the examples we cannot only test the subroutines developed, but also get an understanding of the error introduced by the approximations used to describe boundaries.

Another fundamental building block is the description of the material response. In Chapter 4 we introduce basic concepts of elasticity and potential flow and develop fundamental solutions, that is, simple solutions which satisfy the governing differential equations. These will be central to our subsequent deliberations.

We introduce the concepts of boundary element methods next, using the method originally proposed by Trefftz. Although this very simple method cannot be used for general purpose programs, it serves very well to explain the fundamental ideas of the method. A small computer program can be developed to solve some simple problems. Again, this will serve as a tool for learning by experience.

The direct boundary element method used in the majority of BEM software is introduced next. Here we will use the reciprocal theorem by Betti, which is well known to engineers, to obtain an integral equation where the integrals contain singular functions, i.e. ones which have infinite value at certain points. The major task in the implementation is to solve the integral equations numerically.

The next chapter on numerical implementation therefore deals with the evaluation of integrals using numerical integration. Those familiar with isoparametric finite elements will recognise the Guass Quadrature method used. However, in contrast to its use in the FEM, one must be very careful to select the number of integration points, as they are dependent on how close the singularity is to the integration region. This is the most difficult and crucial part in the implementation of the BEM. The integration over the boundary surface is carried out over a boundary element and the contributions of all elements which describe a boundary are then added. We will see that this is very similar to the assembly procedure in the FEM.

After the numerical treatment of the integral equations, we end up with a system of equations. In contrast to the FEM, the coefficient matrix is fully populated and unsymmetrical. Standard Gauss elimination can be used but, for large systems, the storage requirement and the computation times may be reduced considerably by iterative solvers, such as conjugate gradient methods.

The results which we obtain after the solution are values of displacement or traction at the boundary depending on the boundary condition specified. In contrast to the FEM, values are not obtained immediately in the interior of the domain, but are computed by post-processing. In Chapter 8 it is explained how the stresses at the boundary and in the interior can be obtained from boundary displacements and tractions. This is an advantage of the method because the user has a free choice where results are obtained.

We now have all the building blocks together, and are able to compile a computer program which is able to solve two and three-dimensional problems in elasticity and potential flow, depending on which fundamental solution is used. In Chapter 9 we therefore apply the program developed to test examples to find out what level of accuracy can be obtained in comparison with the FEM.

For the solution of inhomogeneous domains, where we cannot obtain a fundamental solution, we introduce the concept of multiple regions, where the domain is subdivided into subregions, much in the same way as with the FEM. There is an additional advantage in this concept because sparseness is introduced in the system of equations. We will also find out in a later chapter that the multi-region method allows contact problems to be solved.

If we use the multi-region concept, we find that there is a possibility that the traction vector can be discontinuous if there is a corner in the boundary of a subregion. We find that this causes some difficulty, since we have more unknown than available equations to solve for them. Since in nearly all textbooks on the BEM this problem is not discussed in any detail, we devote the whole of Chapter 11 to the treatment of corners and edges.

Because there are no elements in the interior of the domain, as is the case with the FEM, the BEM has difficulty dealing with problems where 'body forces' have to be applied. In the case of gravity or centrifugal forces, which are constant, the body forces can be treated by a surface integral. If the body forces are not constant throughout the domain, a volume integral will have to be evaluated. This can be done by supplying internal cells which look like finite elements, but do not involve any additional degrees of freedom, as they are only used for integration. The implementation of this procedure, discussed in Chapter 12, however, allows the solution of problems in elastoplasticity and viscoplasticity.

In Chapter 13 we show that the solution of non-linear problems follows the same procedures as in the FEM, so the general solution algorithm is the same. Here we discuss two types of non-linear problems in more detail: plasticity and contact problems. We find that the analysis of problems in plasticity involves the consideration of residual stresses as 'body forces'. We also find that the method is well suited to model contact and crack propagation problems.

It is possible to couple the BEM with the FEM thus getting the 'best of both worlds'. In Chapter 14, methods of coupling are presented. Basically, a stiffness matrix of the BE region is obtained and assembled with the FEM stiffness matrices. Since many general purpose programs allow the input of a user defined element stiffness matrix, this may be used to extend the capabilities of a reader's FEM code.

To demonstrate that the method also works for large scale industrial problems, Chapter 15 shows some applications of the boundary and coupled methods in engineering. The purpose of this chapter is twofold: firstly, it shows how complex problems, as they invariably occur in real life, can be simplified and how a boundary element mesh can be obtained. Secondly, it shows the range of problems that can be solved.

By the end of this book, the reader should have an understanding of how the method works and how it can be implemented into a computer program.

1.3   MATHEMATICAL PRELIMINARIES

A good consistent notation is essential to any textbook. For developing and explaining numerical methods two notations are used by engineers: matrix and tensor notation. Traditionally, textbooks on the BEM have used tensors, whereas those about the FEM have used matrix notation, although many engineers working in solid mechanics are using tensors nowadays because of the ease with which non-linear problems can be treated. For this book we have chosen to use matrix/vector notation.

There are two reasons for this: firstly, the book which is probably still the most widely read on numerical modelling, *The Finite Element Method*, by O. C. Zienkiewicz and R.L Taylor, now in its fifth edition, uses matrix notation throughout. Since we hope to attract more engineers to the BEM, this was one motivation. The other reason is that even classical books on the BEM that use tensor notation have to revert to matrix at some stage, for example when discussing the assembly of the system of equations.

As will be shown for the (mostly) linear problems discussed here, there is not so much difference between the two notations. Indeed, the only time one notices a significant difference is when deriving fundamental solutions, which are much simpler using tensor indicial notation. The reader will see that some compromise solution has been reached in this book for this case.

In the following we discuss some basic mathematics which will be used in this book and also attempt a comparison of matrix and tensor notation. A good introduction into tensor algebra can be found in the book by Holzapfel[2].

1.3.1 Vector algebra

Vectors are used to represent a vectorial quantity such as displacement or force. They can also be used to define the position of a point relative to a set of Cartesian axes. By contrast, a column vector is simply a matrix with one column.

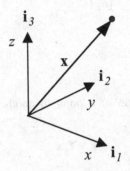

Figure 1.3   Position vector **x** defining a point in space

We define the position of a point in 3-D space with respect to Cartesian axes x, y, z (Figure 1.3) as

$$\mathbf{x} = \begin{Bmatrix} x \\ y \\ z \end{Bmatrix} \qquad (1.1)$$

Alternatively, we may represent the point in terms of Cartesian coordinates x_i, where $i = 1,2,3$ (this is referred to as *range*).

The Cartesian coordinates are specified with respect to a set of orthogonal coordinate axes, which are defined by base vectors of unit length, $\mathbf{i}_i$ and which have the property

$$\mathbf{i}_i \bullet \mathbf{i}_j = \delta_{ij} \quad \begin{Bmatrix} 1 \text{ for } i = j \\ 0 \text{ for } i \neq j \end{Bmatrix} \qquad (1.2)$$

where $(\ )\bullet(\ )$ denotes the scalar (dot) product, as explained on page 17 and δ_{ij} is known as the Kronekker delta.

Vector **x** may then be represented in indicial notation as

$$\mathbf{x} = \sum_{i=1}^{3} x_i \mathbf{i}_i = x_i \mathbf{i}_i \tag{1.3}$$

where the summation convention has been used for the last expression. This convention specifies that for any index which is repeated, a summation of all terms within the *range* is implied.

Another vector quantity is the displacement which can be written either as

$$\mathbf{u} = \begin{Bmatrix} u_x \\ u_y \\ u_z \end{Bmatrix} \quad \text{or} \quad \mathbf{u} = u_i \mathbf{i}_i \tag{1.4}$$

in indicial notation.

Coordinate transformation

If we want to express the location of a point in another orthogonal system with the directions $\mathbf{v}_1$, $\mathbf{v}_2$, $\mathbf{v}_3$ then we have

$$\mathbf{x}' = \mathbf{T}_g \mathbf{x} \tag{1.5}$$

where the transformation matrix is defined as

$$\mathbf{T}_g = \begin{bmatrix} \mathbf{v}_1 & \mathbf{v}_2 & \mathbf{v}_3 \end{bmatrix} \tag{1.6}$$

Alternatively, we may write in indicial notation

$$x_i' = \Lambda_{ij} x_j \tag{1.7}$$

where

$$\Lambda_{ij} = \mathbf{v}_i \bullet \mathbf{i}_j \tag{1.8}$$

Projection of one vector onto another

If we want to compute the projection of a vector onto a direction specified by a unit vector **v**, then it is very convenient to use the dot product. For example, the component, u' of the displacement **u** in the direction specified by **v** is given by (Figure 1.4)

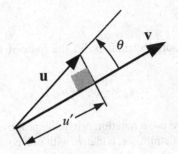

Figure 1.4   Projection of vector

$$u' = \mathbf{u} \bullet \mathbf{v} \qquad (1.9)$$

The angle θ between the two vectors is computed by

$$cos\theta = \frac{1}{|\mathbf{u}|}\mathbf{u} \bullet \mathbf{v} \qquad (1.10)$$

where the length of vector $\mathbf{u}$ is given by

$$|\mathbf{u}| = \sqrt{u_x^2 + u_y^2 + u_z^2} \qquad (1.11)$$

Derivatives of vectors

The derivatives of the displacement vector may be written as

$$\frac{\partial \mathbf{u}}{\partial x} = \begin{Bmatrix} \dfrac{\partial u_x}{\partial x} \\ \dfrac{\partial u_y}{\partial x} \\ \dfrac{\partial u_z}{\partial x} \end{Bmatrix}; \quad \frac{\partial \mathbf{u}}{\partial y} = \begin{Bmatrix} \dfrac{\partial u_x}{\partial y} \\ \dfrac{\partial u_y}{\partial y} \\ \dfrac{\partial u_z}{\partial y} \end{Bmatrix}; \quad \frac{\partial \mathbf{u}}{\partial z} = \begin{Bmatrix} \dfrac{\partial u_x}{\partial z} \\ \dfrac{\partial u_y}{\partial z} \\ \dfrac{\partial u_z}{\partial z} \end{Bmatrix} \qquad (1.12)$$

In indicial notation we simply write

$$\frac{\partial u_i}{\partial x_j} = u_{i,j} \qquad (1.13)$$

1.3.2   Stress and strain

Stresses and strains are tensorial quantities. In the indicial notation the strain tensor is defined by

$$\varepsilon_{ij} = \frac{1}{2}\left(u_{i,j} + u_{j,i}\right)$$

(1.14)

In this book, however, we use a notation originally proposed by Timoshenko[3]. We define a *pseudo-vector* of strain, i.e. a matrix with one column:

$$\boldsymbol{\varepsilon} = \left\{ \begin{array}{c} \varepsilon_x \\ \varepsilon_y \\ \varepsilon_z \\ \gamma_{xy} \\ \gamma_{yz} \\ \gamma_{xz} \end{array} \right\} = \left\{ \begin{array}{c} \dfrac{\partial u_x}{\partial x} \\[2mm] \dfrac{\partial u_y}{\partial y} \\[2mm] \dfrac{\partial u_z}{\partial z} \\[2mm] \dfrac{\partial u_x}{\partial y} + \dfrac{\partial u_y}{\partial x} \\[2mm] \dfrac{\partial u_y}{\partial z} + \dfrac{\partial u_z}{\partial y} \\[2mm] \dfrac{\partial u_x}{\partial z} + \dfrac{\partial u_z}{\partial x} \end{array} \right\}$$

(1.15)

Note that in the *pseudo-vector* notation we only have six strain components, whereas the symmetric strain tensor has nine. Also note that the ½ term is missing for the shear strains in order to achieve consistency between the tensor and matrix operations. The index number of the location of the strain or stress components for matrix notation and tensor notation is given in Table 1.1.

Table 1.1 Index numbering for strain and stress

Notation	Index number					
Matrix	1	2	3	4	5	6
Tensor	11	22	33	12 and 21	23 and 32	31&13
	xx	yy	zz	xy and yx	yz and zy	zx and xz

Similarly, the stress tensor σ_{ij} can be written as a *pseudo-vector*

$$\boldsymbol{\sigma} = \begin{bmatrix} \sigma_x & \sigma_y & \sigma_z & \tau_{xy} & \tau_{yz} & \tau_{xz} \end{bmatrix}^{T}$$

(1.16)

1.4   CONCLUSIONS

At the beginning of this chapter we have shown on a geomechanics example that substantial gains can be made with the BEM, in terms of mesh generation and solution times. These gains are most pronounced for problems which involve infinite or semi-infinite domains. Other examples where the BEM seems to be superior to the FEM is for problems where boundary stresses are important, for example, in mechanical engineering. Examples of this will be shown later.

The purpose of this book is to encourage the use of the method. The simple computer programs included here contain all the necessary building blocks necessary for building more advanced and more specific computer programs for research or industrial applications. For example, by simply substituting different fundamental solutions one can solve problems other than elasticity and potential flow.

The reader will note that although the size of this book is similar to *Programming the Finite Element Method* [4], the range of applications that have been discussed here is less. For example, here we do not mention transient and dynamic problems. The reason for this is twofold: firstly, the subject matter is more complex and an attempt has been made to gently lead the reader into it. The integration over elements to obtain the stiffness matrix in the FEM can be explained, for example, in a few pages, whereas due to the singularity of the fundamental solutions a whole chapter has to be dedicated to integration in the BEM. Secondly, some problems simply do not arise at all in the FEM. The treatment of edges and corners causes no problems in the FEM because boundary tractions are not referred to at all.

In conclusion, the reader should see this book as an introduction to the BEM, containing some basic building blocks for computer programming. Armed with this knowledge, the readers interested in more advanced topics may find it easier to understand more advanced text books and publications on this subject.

1.5   REFERENCES

1. Zienkiewicz, O.C. and Taylor, R.L. (2000) *The Finite Element Method,* (5th ed.). Butterworth-Heinemann, Oxford.
2. Holzapfel, G.A. (2000) *Non-linear Solid Mechanics.* Wiley, Chichester.
3. Timoshenko, S.P. and Goodier, J.N. (1970) *Theory of Elasticity.* McGraw-Hill, London.
4. Smith, I.M. and Griffiths, D.V. (1998) *Programming the Finite Element Method,* (3rd ed.). Wiley, Chichester.

2
Programming

*Art is only pleasing if it
has the character of lightness*

J.W. von Goethe

2.1   STRATEGIES

Although the first idea which provided the background for the boundary element method dates back to the early 1900s, the method only emerged when digital computers became available. This is because, except for the simplest problems, the number of computations required is too large for 'hand calculation'.

The implementation into a computer application basically consists of giving the processor a series of instructions or tasks to perform. In the early days these instructions had to be given in complicated machine code and writing them was mainly the domain of specialised programmers. However, higher level languages were soon developed which made the programming task easier and this had the additional advantage that code developed could run on any hardware. One of these languages, especially developed for scientists and engineers, was FORTRAN. In the past decades, the language has undergone tremendous development. Whereas with FORTRAN IV the writing of programs was rather lengthy and tedious and the code difficult to read, the new facilities of FORTRAN 90/95 make it suitable for writing short, readable code. This has mainly to do with features that do away with the need to use statement numbers, and the availability of powerful array and matrix manipulation tools. Today, any engineer should be able to write a program in a reasonably short time.

When developing a relatively large program, such as will be attempted in this book, it is important to use the concept of modular programming. This means that the task has to be divided into many subtasks. Therefore, we will develop a library of procedures which perform certain tasks, for example, computing the value or the derivative of a function at a certain point. The subprocedures or functions can be called as needed from the main program or from other procedures.

In the following, we will give a short introduction into some new features of FORTRAN 90/95 that will be used in this book. Here, it will be assumed that readers already have some knowledge of FORTRAN. A more detailed description of FORTRAN 90 is given by Smith[1].

In this chapter we will also introduce the notation used in this book, especially with respect to vectors and matrices. A short introduction to matrix algebra and vectors will also be given.

2.2   FORTRAN 90/95 FEATURES

2.2.1   Representation of numbers

Numbers are stored in the computer in binary form. Real numbers are stored in two parts: one consists of the digits that make up the number, the other of the exponent. The exact way in which a number is stored depends on the hardware used. For real numbers, either 4 or 8 bytes could be allocated for storage in FORTRAN 77 (F77) by declaring the variable REAL*4 (single precision) or REAL*8 (double precision). However, since the storage is machine dependent, we do not know exactly how many digits can be stored in either mode. FORTRAN 95 provides a facility for specifying the precision in number of digits with the KIND statement.

First, one must interrogate the processor as to which value of KIND provides a certain number of digits. For example

IWP= **SELECTED_REAL_KIND**(P=16)

would assign to the variable IWP the KIND number which would give 16 significant digits. The declaration of the real variable would then be

REAL(KIND=IWP) :: A

instead of the F77 statement

REAL*8 A

The precision in which the numbers are stored is significant because numerical round-off may occur if two numbers, which differ greatly in magnitude, are subtracted from each other. If a great number of such operations are carried out, the error produced by the round-off may become significant. For example, in the solution of a large system of equations by Gauss elimination, there are many subtractions, and therefore a double precision storage is necessary, whereas other parts of the program may not be so sensitive. For simplicity in all the programs developed in this book we use REAL for single precision and REAL (KIND=8) for double precision. The software user, however, must be aware that if a particular code is run on two different types of processors, very small differences may occur in the results because of the different way numbers are stored.

2.2.2 Arrays

FORTRAN 90/95 (F90) has powerful features to handle arrays. An array can be of rank 1,2,3, etc. A rank 1 array is a vector, a rank 2 array a matrix. In this book, we will distinguish between real vectors (identified by a lowercase bold letter) and matrices with one column, or pseudo-vectors (identified by lowercase bold Greek letters).

The *shape* of an array indicates the number of elements in each dimension. For example the vector

$$\mathbf{v} = \begin{Bmatrix} v_x \\ v_y \\ v_z \end{Bmatrix} \tag{2.1}$$

may be seen also as a matrix of *shape* (3,1), whereas the matrix

$$\mathbf{A} = \begin{bmatrix} a_{11} & a_{12} \\ a_{21} & a_{22} \\ a_{31} & a_{32} \end{bmatrix} \tag{2.2}$$

would be of *shape* (3,2).

The declaration of the two arrays in F90 would be

REAL :: V(3),A(3,2)

One of the most important new features of F90, however, is that arrays may be declared dynamically, that is, the programmer does not need to know the dimensions of an array when writing the program, but these can be read or calculated at run-time. Since array dimensions in the BEM will depend on number of nodes and/or number of degrees of freedom, this is a particularly useful feature. To declare an array **A**, whose dimensions are known at run-time only, we write:

REAL, ALLOCATABLE :: A(:,:)

In the program, we can then allocate the dimensions of the array with computed values of dimensions I,J by:

ALLOCATE(A(I,J))

When the array is no longer used then the space in memory can be freed by

DEALLOCATE(A)

An array may be assigned initial values by the statement

$$\textbf{REAL} :: V(3)=(/\ 1.0,\ 0.0,\ 0.0\ /)$$

2.2.3 Array operations

FORTRAN 90 has features for array and vector operations which simplify the manipulation of arrays from the programmer's point of view. The operations include matrix/vector additions and subtractions, multiplication and vector product. They also include operations on part of the arrays, examining arrays, determining max/min values, the gathering of submatrices, etc.

Matrix addition: If all the coefficients of A are to be added, then one can simply write

$$A= A+B$$

Multiplication by a scalar: If all coefficients of A are to be multiplied by a scalar (say 3.0), then this would translate into

$$A= A*3.0$$

Operation on selected coefficients of a matrix: For example, to add the first 3 coefficients of the second column of array A to array B and store it in A, one needs a single statement instead of a DO loop as in FORTRAN 77:

$$A(1:3,2)= A(1:3,2) + B(1:3,2)$$

Transpose of a matrix: The transpose of a matrix means exchanging rows and columns. For example, the transpose of vector **v** defined above is given by:

$$\mathbf{v}^{\mathrm{T}} = \begin{bmatrix} v_x & v_y & v_z \end{bmatrix} \qquad (2.3)$$

this translates into

$$VT= \textbf{TRANSPOSE}(V)$$

the resulting vector $\mathbf{v}^{\mathrm{T}}$ would be of shape (1,3).

Matrix multiplication: The multiplication of two matrices of shapes (1,3) and (3,2) gives a result of shape (1,2). For example

$$\mathbf{B} = \mathbf{v}^T \mathbf{A} = \begin{bmatrix} v_x & v_y & v_z \end{bmatrix} \begin{bmatrix} a_{11} & a_{12} \\ a_{21} & a_{22} \\ a_{31} & a_{32} \end{bmatrix} = \begin{bmatrix} v_x a_{11} + v_y a_{21} + v_z a_{31} & v_x a_{12} + v_x a_{22} + v_x a_{32} \end{bmatrix} \quad (2.4)$$

translates into the FORTRAN 90 statement

$$B= \textbf{MATMUL}(VT,A)$$

It is obvious that for matrix multiplication to be possible, the shapes of the matrices to be multiplied have to obey certain rules.

Scalar (dot) product: The scalar dot product of two vectors is defined as:

$$\mathbf{v}_1 = \begin{Bmatrix} v_{1x} \\ v_{1y} \\ v_{1z} \end{Bmatrix}; \quad \mathbf{v}_2 = \begin{Bmatrix} v_{2x} \\ v_{2y} \\ v_{2z} \end{Bmatrix}; \quad x = \mathbf{v}_1 \bullet \mathbf{v}_2 = v_{1x}v_{2x} + v_{1y}v_{2y} + v_{1z}v_{2z} \qquad (2.5)$$

this translates into

$$X= \textbf{DOT\_PRODUCT}(V1,V2)$$

Maximum value in an array: To find the element of array **A** which has the maximum value one writes

$$AMAX= \textbf{MAXVAL}(A)$$

Location of maximum value: To find the location of the maximum element of array **A**, NMAX, execute the statement

$$NMAX= \textbf{MAXLOC}(A)$$

Upper bound of an array: Sometimes it is useful for the program to find out what shape an array was assigned. This will be used extensively in SUBROUTINES in order to reduce parameter lists. The statement

$$N= \textbf{UBOUND}(A,1)$$

will return the number of the last row of **A** which is 3, whereas

$$M= \textbf{UBOUND}(A,2)$$

will return the number of the last column of **A**, which is 2.

Check on array elements: Another useful function is one which checks if all elements of an array fulfil a certain condition. For example

$$\textbf{ALL}(A >0)$$

will return a logical .TRUE. if all elements of **A** are greater than zero.

Sum of array elements: In F77, summing of the coefficients of an array required at least one **DO** loop. With the intrinsic function **SUM** we may calculate the sum of all coefficients of an array by simply writing

$$C= \textbf{SUM}(A)$$

Masking can be used to sum only coefficients which fulfil certain conditions. For example,

$$C= \textbf{SUM}(A, \textbf{MASK}=A>0.0)$$

would sum only coefficients of A which are greater than zero.

Gathering and scattering: A new feature in F90 makes the 'gathering' and 'scattering' of values, which we will need later, very simple. To explain these operations consider the two-dimensional mesh of boundary elements in Figure 2.1.

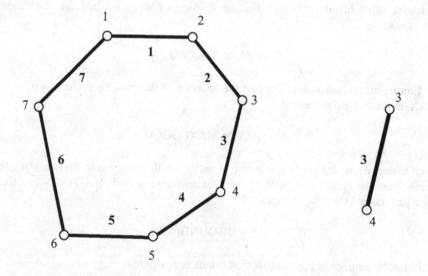

Figure 2.1   Two-dimensional boundary element mesh and connectivity of element 3

The nodes of the mesh where elements are connected with each other can be numbered in two different ways: locally and globally. When referring, for example, to the unknown u (e.g. temperature in the case of heat conduction problems) one has two vectors, a global one

$$\mathbf{u} = \begin{Bmatrix} u_1 \\ u_2 \\ u_3 \\ u_4 \\ u_5 \\ u_6 \\ u_7 \\ u_8 \end{Bmatrix} \tag{2.6}$$

and a local one defined at element level, for example, for the two nodes of element 3:

$$\mathbf{u}^3 = \begin{Bmatrix} u_1 \\ u_2 \end{Bmatrix}^3 \tag{2.7}$$

For element 3 we may define a 'connectivity vector' of dimension 2, which contains the global node numbers of the two nodes of the element

CONNECTIVITY=(/3,4/)

The 'scatter' operation is where the locally defined unknowns are put into the global vector

U_GLOBAL(CONNECTIVITY)= U_LOCAL

This statement would put u_1^3 and u_2^3 of element 3 into locations 3 and 4 of the global vector **u**.

The 'gather' operation would do the opposite, i.e.

U_LOCAL= U_GLOBAL(CONNECTIVITY)

would put the global values of u_3 and u_4 into the local positions 1 and 2 of $\mathbf{u}^3$.

2.2.4 Control

Various features which can be used to control the flow of the program have been improved and new ones added. With these new features it should no longer be necessary to have **GOTO** statements and statement numbers, features which made programs sometimes very difficult to read. The new feature is the **SELECT CASE**, which replaces the computed **GOTO**. This feature allows us to control which parts of the code are executed under certain conditions.

For example, the coding:

```
SELECT CASE(NUMBER_OF_FREEDOMS)
        CASE(1)
                Coding for one degree of freedom
        CASE(2)
                Coding for two degrees of freedom
        CASE DEFAULT
                Error message
END SELECT
```

would execute two different types of instructions, depending on the degrees of freedom per node (i.e. potential vs. elasticity problems) and would issue an error message if another value is encountered.

The IF statement is also useful in controlling execution. This has been improved in that symbols which are familiar to engineers can be used.

For example, the old F77 operators:

$$\begin{array}{lll} .NE. & \text{can be written as} & /= \\ .EQ. & \text{can be written as} & == \\ .GT. & \text{can be written as} & > \\ .GE. & \text{can be written as} & >= \\ .LT. & \text{can be written as} & < \\ .LE. & \text{can be written as} & <= \end{array}$$

The DO loop has also been improved. Now it is possible to give each DO loop a name which enhances readability. Also, there is an easier way of exiting a loop when a certain condition is reached. For example, in an iteration loop the condition for exiting may be that a convergence has been achieved. The code

```
Iteration_loop:        &
DO ITER=1,NITERS
        Statements
        IF(CONVERGED) EXIT
        Statements
END DO &
Iteration_loop
```

would exit the loop if the value of CONVERGED is **.TRUE.**

Another nice feature is the CYCLE. For example, the coding

```
Element_loop:        &
DO NEL=1,NELEM
        Statements 1
        IF(NEL >= NELB) CYCLE
        Statements 2
END DO &
Element_loop
```

would skip Statements 2 if NEL becomes greater than or equal to NELB.

2.2.5 Subroutines and functions

Subroutines and functions perform frequently used tasks and split a complex problem into smaller ones. For example, to normalise a vector we may define a Subroutine Vector_norm as

```
SUBROUTINE VECTOR_NORM(V,VLEN)
!---------------------------------------
! Normalise vector
!---------------------------------------
REAL, INTENT(INOUT) :: V(:)        !   Vector to be normalised
REAL, INTENT(OUT)    :: VLEN       !   Length of vector
VLEN= SQRT( SUM(V*V))
IF(VLEN == 0.) RETURN
V= V/VLEN
RETURN
END SUBROUTINE VECTOR_NORM
```

Two things are different from F77 here. Firstly, in the declaration of variables in the parameter list we may specify if a parameter is to be used for input (IN), output (OUT) or input and output (INOUT). This not only helps to clarify the readability of the code but also protects variables from being changed accidentally in the subprogram. Secondly, we do not need to specify the dimension of vector V, since this will be determined in the program calling the subroutine. For example, the calling program will have

```
REAL :: V(3)
.
.
CALL VECTOR_NORM(V,VLEN)
```

Another very useful feature which we will use in the book is that a function can also return an array. For example, we may write a function for determining the vector ex-product of two vectors as:

```
FUNCTION VECTOR_EX(V1,V2)
!---------------------------------------
! Returns vector x-product v1xv2
! where v1 and v2 are dimension 3
!---------------------------------------
REAL, INTENT(IN) :: V1(3),V2(3)         !  Input
REAL             :: VECTOR_EX(3)        !  Result
VECTOR_EX(1)=V1(2)*V2(3)-V2(2)*V1(3)
VECTOR_EX(2)=V1(3)*V2(1)-V1(1)*V2(3)
VECTOR_EX(3)=V1(1)*V2(2)-V1(2)*V2(1)
RETURN
END FUNCTION VECTOR_EX
```

In the calling program we use this function in the following way

REAL :: V1(3),V2(3),V3(3)
.
.
.
V3= VECTOR_EX(V1,V2)

2.2.6 Subprogram libraries and common variables

As indicated previously for developing large programs, it is convenient to subdivide the big task into small ones. This means that a library of subroutines will be developed. There are basically two ways in which these subroutines were able to communicate with each other in F77: via parameter lists or via COMMON blocks. FORTRAN 95 has replaced the somewhat tedious COMMON block structure by the **MODULE** and **USE** statements. A **MODULE** is simply a set of declarations and/or subroutines. If a program or subprogram wants to use the declarations and subroutines, it simply has a **USE** statement at the beginning. For example, to define some variables which are used by subprograms we specify:

MODULE Common_Variables
    REAL :: A, B
    REAL, ALLOCATABLE :: C(:)
END MODULE Common_Variables

This replaces the old COMMON statements. Any program or subprogram that uses the common declarations has a **USE** statement such as:

PROGRAM TEST
    USE Common_Variables
    -
    -
    -
END PROGRAM TEST

To help with the management of large programs it is convenient to group the subroutines which are used by programs into different files. The MODULE facility can be used for this purpose. For example, we may group all the subroutines which have to do with the geometrical description of boundary elements into a module Geometry_lib

MODULE Geometry_lib
    REAL :: Pi= 3.149
    CONTAINS
      SUBROUTINE Shape ...
      END SUBROUTINE Shape
    ...
END MODULE Geometry_lib

2.3   CHARTS AND PSEUDO-CODE

Even though the new features in FORTRAN 95 have made the programs much more readable, there is still a need to show the general layout of the program in a simple way. The use of flowcharts, as they were used in the early days of programming, has not been found to be very useful.

Instead, structured charts and pseudo-code are used, i.e. a FORTRAN-like code which gives a general description of what to do. Since nested DO LOOPS are complicated to read in a FORTRAN code they can be better explained in a structure chart. For instance, the chart for the example of two-dimensional numerical integration discussed in the next chapter is shown in Figure 2.2.

The advantage of the structure chart is that the structure of the nested DO loop can be clearly seen. Another feature where structure charts may be useful is in IF statements, especially when they are complicated. For example, if we wish to check all diagonal elements of the coefficient matrix and take appropriate action if they are negative, zero or positive, the structure chart in Figure 2.3 can be used.

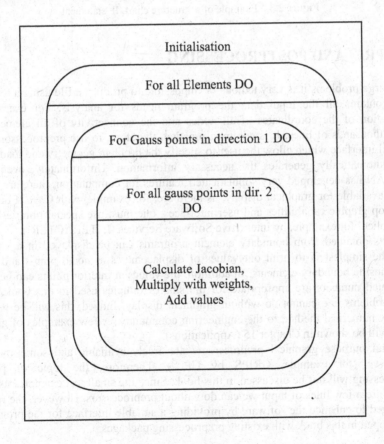

Figure 2.2   Example of a structure chart, nested DO loop

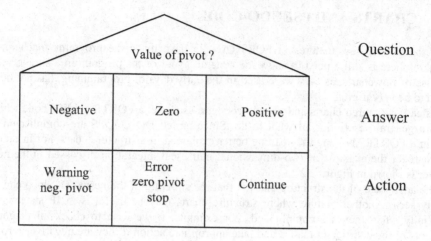

Figure 2.3   Example of a structure chart, IF statement

2.4   PRE- AND POSTPROCESSING

For large problems it is very tedious or impossible to produce a file using a text editor which contains all the input data the program needs for analysis. For example, the specification of the coordinates of all nodes and the connectivity of all elements may involve thousands of lines. It is common practice, therefore, to use preprocessors with a graphical interface which allow the user to specify the problem geometry and loading, and which automatically generates the necessary information. Unfortunately, even though FORTRAN has developed very sophisticated features for computation, there are no built in tools available for graphical display, as there are, for example, with C++. If one wants to develop graphic capabilities and user interfaces, one must use special libraries such as that supplied, for example, by Interactive Software Services (INTERACTER[2]).

Results obtained from boundary element programs can be displayed in a variety of ways. The simplest is to print out values of displacements at nodal points and surface stresses inside boundary elements. In addition, the values at interior points can be printed out. Printed numbers are appropriate for the small examples used in this book, but for larger problems one cannot do without graphical display. Indeed, this will be what will 'sell' any numerical method to the engineering community. A few examples of graphical display will be shown in Chapter 15 (Applications).

General purpose graphical preprocessors are freely available and sometimes quite inexpensive (for example, GRIPS by UPC[3]). Therefore, the topic of pre and postprocessing will not be discussed in this book. Since the small test examples used here only require a few lines of input we can do without preprocessors. However, the reader is encouraged to enhance the software by providing a suitable interface for the programs to be developed in this book with existing preprocessing packages.

2.5   CONCLUSIONS

In this chapter we have given a short overview of some of the new features of FORTRAN 95, the latest dialect of FORTRAN which we are going to use in the book. There are as many programming styles as they are programmers and each programmer will no doubt claim that his/hers is best.

The aim in good programming should be to produce efficient, readable and easy to check code. The last is a very stringent requirement in quality control. Easy to read programs sometimes also tend to be efficient, however a small gain in efficiency should not be made if clarity is sacrificed. For example, sometimes it is clearer and more efficient not to use a DO loop if less than four cycles will occur. If permutations of indices have to be made, such as in the fundamental solutions shown later, it is often better to generate all of the coefficients using the editors copy and paste facility, since the code can be checked visually much faster.

In the past, subprograms had either many COMMON blocks or huge parameter lists. These were needed to pass variables between SUBROUTINES and the main program. Fortunately, FORTRAN 95 has done away with COMMON blocks and the number of parameters for SUBROUTINES can be further reduced by the dynamic array allocation, the use of UBOUND and the USE statement. However, one must consider very carefully which variables should be declared in the Common Module, as explained previously, and which should be declared in each subroutine.

Regarding the programs presented in this book, we claim neither that they are very efficient nor that this is the only way that the procedures outlined may be implemented. Indeed, we encourage the reader to think of different ways in which the theory can be converted efficiently and elegantly into code. In the programs that we present here we have placed our emphasis on readability. Otherwise there would be no point in including the code in the text. In many cases we have sacrificed efficiency and limited ourselves to solving small problems, because we do not use direct access files for storing values, but assume instead that all data required fit into the RAM. With the dramatic increase in the amount of RAM available on standard PCs, however, this is not likely to become a main issue during the lifetime of this book. With regard to efficiency some rearranging of DO loops would be necessary so that computations which only need to be carried out once are not unnecessarily carried out many times. This especially occurs in the subroutines for integration of kernel-shape function products. However, such rearrangement would have made the programs more difficult to follow, and therefore was not implemented.

The programs were developed on a Visual Fortran[4] compiler. However, since only standard FORTRAN 90 features have been used, the source codes can be compiled with any FORTRAN 90 or FORTRAN 95 compiler.

2.6   EXERCISES

Exercise 2.1

Two integer arrays of rank one named Inci1, Inci2 contain element node numbers. Write a **LOGICAL FUNCTION Match**(Inci1,Inci2) which returns .TRUE. if all the numbers of Inci2 match all the numbers in Inci1. Note that the sequence of the numbers in Inci1 and Inci2 will in general not be the same. The dimension of both arrays will be declared in the calling program.

Exercise 2.2
Write a **REAL FUNCTION DETERMINANT**(Array) which computes the determinant of the matrix Array which can be of shape (2,2) or (3,3).

Exercise 2.3
A submatrix **A** of shape (2,2) is given together with a matrix **B** whose shape is declared in the calling program. Write a **SUBROUTINE ASSEMBLE**(A,B,I,J) which assembles the submatrix **A** into the matrix **B** at location i,j (see the figure below).

$$
\begin{array}{c}
\quad\quad\quad i \\
j \begin{bmatrix} \cdots & \cdots & \cdots & \cdots \\ \cdots & A_{11} & A_{12} & \\ \cdots & A_{21} & A_{22} & \\ \cdots & & & \end{bmatrix}
\end{array}
$$

2.7 REFERENCES

1. Smith, I.M. (1995) *Programming in FORTRAN90*. Wiley, Chichester.
2. Interactive Software Services (1999) *INTERACTER Library Subroutine Reference*. Huntington, U.K.
3. http://gid.cimne.upc.es
4. Digital Equipment Corporation (1997) *Digital FORTRAN Language Reference Manual*. Maynard, Mass., USA.

3

Discretisation and Interpolation

*Nature is indifferent
towards the difficulties it
causes to a mathematician*

Fourier

3.1   INTRODUCTION

One of the fundamental requirements for numerical modelling is a description of the problem, its boundaries, boundary conditions and material properties, in a mathematical way. The exact definition of the complex shape of the boundary of a particular problem to be analysed would require the specification of the location (relative to the origin of a set of axes) of an infinite number of points on the surface. In order to be able to model such a problem with a reasonable amount of input data, only a limited number of points may be defined and the shape between the points approximated by functions. This is known as solid modelling[1] and is a technique which is not only restricted to the boundary element method. Solid modelling is being used, for example, to describe the shape of car bodies in mechanical engineering and ore bodies in mining, for the purpose of generating displays on computer graphics terminals. Thus, a new form of car body can be visualised in perspective from various angles, even before a scale model is built and the location and grade of ore bodies can be displayed for optimising excavation strategies in mine planning.

In the following, we will discuss one and two-dimensional elements as defined by the number of intrinsic (element) coordinate directions. One-dimensional elements exist in two-dimensional Cartesian space and two-dimensional elements in three-dimensional space. Thus, in this chapter, we consider discretisation methods used in the boundary method and start building the library of subroutines which are needed later.

3.2 ONE-DIMENSIONAL ELEMENTS

One-dimensional elements are used for the description of the boundary of plane sections of a problem. In the simplest case we have linear elements which connect two nodes i and j, the positions of which are defined by Cartesian coordinates, as shown in Figure 3.1. For each element, we now define an element or intrinsic coordinate ξ which follows the direction of the element, equals zero at the centre and ± 1 at the ends (Figure 3.2).

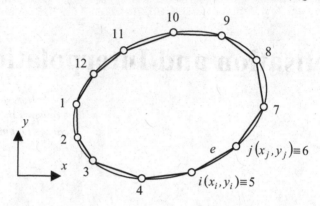

Figure 3.1   Plane domain, boundary approximated by linear elements

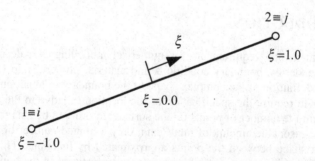

Figure 3.2   Linear element with local numbering and coordinate system

It can easily be verified that the Cartesian coordinates of a point on element e with the intrinsic coordinate ξ are given by

$$x(\xi) = \frac{x_i + x_j}{2} + \frac{x_j - x_i}{2}\xi$$
$$y(\xi) = \frac{y_i + y_j}{2} + \frac{y_j - y_i}{2}\xi$$

(3.1)

This equation can be checked by substituting $\xi = -1$ and $\xi = +1$ to obtain the coordinates of nodes i and j.

It is now convenient to substitute for the global coordinates:

$$\left.\begin{array}{l} x_i = x_1^{\,e} \\[4pt] y_i = y_1^{\,e} \end{array}\right\} \text{ first node of element } e$$

$$\left.\begin{array}{l} x_j = x_2^{\,e} \\[4pt] y_j = y_2^{\,e} \end{array}\right\} \text{ second node of element } e$$

In this way we establish a link between local and global numbering of nodes.

The global numbers of nodes which belong to the element are referred to as *element incidences* or *element connectivity*. For example, the connectivity of element e is i,j where the numbers *5,6* may be substituted in the example in Figure 3.1. The sequence in which the element node numbers are entered will be significant later, as it will affect the direction of the *outward normal*. From now on we will work with the local numbering system and use the element incidences to 'gather' coordinates from the global values, as explained in the previous chapter.

We can rewrite equation (3.1) as

$$x(\xi) = \frac{1}{2}(1-\xi)x_1^e + \frac{1}{2}(1+\xi)x_2^e \tag{3.2}$$

$$y(\xi) = \frac{1}{2}(1-\xi)y_1^e + \frac{1}{2}(1+\xi)y_2^e$$

or in abbreviated form as

$$\begin{Bmatrix} x \\ y \end{Bmatrix} = \sum_{n=1}^{L} N_n(\xi) \begin{Bmatrix} x_n^{\,e} \\ y_n^{\,e} \end{Bmatrix} \tag{3.3}$$

where L is the number of element nodes and N_n are element 'shape' functions. Equation (3.3) can be written in matrix notation

$$\mathbf{x} = \sum N_n(\xi)\mathbf{x}_n^e \tag{3.4}$$

where $\mathbf{x}$ is a vector containing coordinates of a point on element e and $\mathbf{x}_n^e$ is a vector of coordinates of the nth node of element e.

For the two-node element just derived, the shape functions are (Figure 3.3)

$$N_1 = \frac{1}{2}(1-\xi)$$

$$N_2 = \frac{1}{2}(1+\xi) \tag{3.5}$$

or in short

$$N_n = \frac{1}{2}\left(1+\xi_n\xi\right)$$
(3.6)

where the local coordinates of the nodes are given by

$$n = 1 \quad \xi_n = -1$$
$$n = 2 \quad \xi_n = 1$$
(3.7)

   Complicated shapes can be more accurately described by a smaller number of elements with three nodes and quadratic shape functions (Figure 3.4). The coordinate ξ now follows the element shape, i.e. is curvilinear and the third node is placed at $\xi = 0$. The shape function associated with the mid-side node is a parabola which has unit value at the third node and zero value at the other nodes, that is,

$$N_3(\xi) = 1 - \xi^2$$
(3.8)

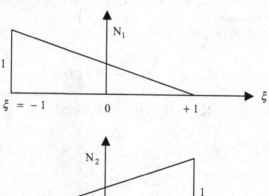

Figure 3.3   Linear shape functions

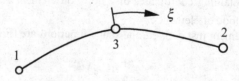

Figure 3.4   Quadratic element

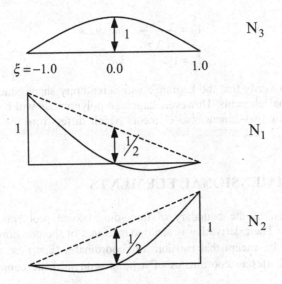

Figure 3.5   Quadratic shape functions

The end node shape functions can be obtained by subtracting half of the centre node function from each of the linear shape functions (Figure 3.5)

$$N_n = \frac{1}{2}(1 + \xi_n \xi) - \frac{1}{2}N_3 \qquad n = 1,2 \tag{3.9}$$

The shape functions presented here have so far not been derived mathematically, but written down intuitively. Shape functions derived this way have been called *Serendipity* functions. It can be seen that the shape functions derived so far have had the following properties:

$$N_n(\xi_n) = 1$$
$$N_n(\xi_i) = 0 \quad for \quad i \neq n \tag{3.10}$$
$$\sum N_n(\xi) = 1$$

These are indeed the properties all shape functions must have.

The mathematical derivation of functions which satisfy conditions (3.10) is possible using Lagrange polynomials. For the parabolic elements the Lagrange shape functions are defined as

$$L_i(\xi) = A_{i1} A_{i2} A_{i3} \tag{3.11}$$

where

$$A_{il} = \frac{\xi - \xi_l}{\xi_i - \xi_l} \quad for \quad i \neq l$$

$$A_{il} = 1 \quad\quad for \quad i = l \tag{3.12}$$

The reader can verify that the Lagrange and Serendipity shape functions are identical for one-dimensional elements. However, Lagrange polynomials will be used to construct shape functions for two-dimensional elements which differ from the Serendipity shape functions.

3.3   TWO-DIMENSIONAL ELEMENTS

For the description of the boundary of three-dimensional problems, two-dimensional elements are used. Their derivation is analogous to that of the one-dimensional elements described previously, except that two intrinsic coordinates (ξ, η) are used, as shown in Figure 3.6. The Cartesian coordinates of a point with intrinsic coordinates (ξ, η) are obtained by

$$\mathbf{x} = \sum_{n=1}^{L} N_n(\xi, \eta) \mathbf{x}_n^e \tag{3.13}$$

Bilinear shape functions are used for the quadrilateral element in Figure 3.6:

$$N_n = \frac{1}{2}(1 + \xi_n \xi)\frac{1}{2}(1 + \eta_n \eta) \tag{3.14}$$

The shape function N_1 is shown in Figure 3.7. It describes a curved surface consisting of straight lines in the ξ, η directions.

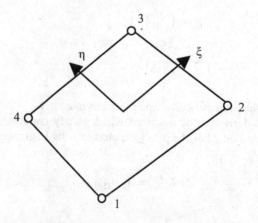

Figure 3.6   Quadrilateral element

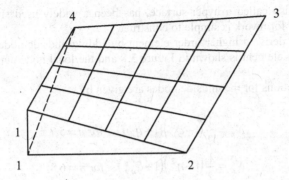

Figure 3.7   Bilinear shape function N_1

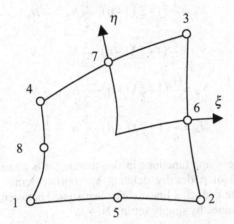

Figure 3.8   Quadratic Serendipity element

Table 3.1   Intrinsic coordinates of nodes

n	ξ_n	η_n
1	-1.0	-1.0
2	1.0	-1.0
3	1.0	1.0
4	-1.0	1.0
5	0.0	-1.0
6	1.0	0.0
7	0.0	1.0
8	-1.0	0.0

The surface, also called a hyper-surface, has been a widely used shape for concrete shells because the formwork is simple to construct.

Again, we can derive a higher order element by adding midside nodes on the element sides. A quadratic element is shown in Figure 3.8 and the local node numbering is shown in Table 3.1.

The shape functions for the midside nodes are given by

$$N_n = \frac{1}{2}\left(1-\xi^2\right)\left(1+\eta_n\eta\right) \quad for\ n = 5,7$$

$$N_n = \frac{1}{2}\left(1-\eta^2\right)\left(1+\xi_n\xi\right) \quad for\ n = 6,8$$

(3.15)

The corner node functions are constructed in a similar way as for the one-dimensional element (Figure 3.9):

$$N_1 = \frac{1}{4}(1-\xi)(1-\eta) - \frac{1}{2}N_5 - \frac{1}{2}N_8$$

$$N_2 = \frac{1}{4}(1+\xi)(1-\eta) - \frac{1}{2}N_5 - \frac{1}{2}N_6$$

$$N_3 = \frac{1}{4}(1+\xi)(1+\eta) - \frac{1}{2}N_6 - \frac{1}{2}N_7$$

$$N_4 = \frac{1}{4}(1-\xi)(1+\eta) - \frac{1}{2}N_7 - \frac{1}{2}N_8$$

(3.16)

By writing down the shape functions in this manner, it is possible to derive elements with variable numbers of nodes by deleting appropriate terms. For example, for an element with no midside node 5, a linear shape is assumed between nodes 1 and 2 and the shape functions are obtained by simply setting $N_5 = 0$.

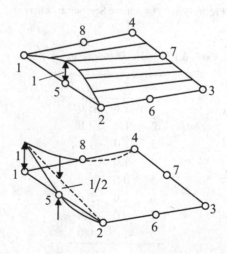

Figure 3.9   Shape functions for midside and corner nodes

If the element shape functions for the quadratic element are derived from Lagrange polynomials, then there is an additional node at the centre of the element (Figure 3.10). The shape functions are given by

$$L_{n(i,j)} = A_{i1} A_{i2} A_{i3} B_{j1} B_{j2} B_{j3} \tag{3.17}$$

where $A_{i,l}$ is defined in equation (3.12) and

$$B_{jl} = \frac{\eta - \eta_l}{\eta_j - \eta_l} \quad \text{if } j \neq l \tag{3.18}$$
$$B_{jl} = 1 \qquad \text{if } j = l$$

where i and j are the column and row numbers of the nodes. This numbering is defined in Figure 3.10. The node numbers are defined as

$$\begin{array}{lll}
n\,(1,1) = 1 & n\,(2,1) = 2 & n\,(3,1) = 5 \\
n\,(1,2) = 4 & n\,(2,2) = 3 & n\,(3,2) = 7 \\
n\,(1,3) = 8 & n\,(2,3) = 6 & n\,(3,3) = 9
\end{array}$$

The Serendipity and Lagrange shape functions are compared in Figure 3.11. The Lagrange element has an additional 'bubble mode' and is, therefore, able to describe complicated shapes more accurately.

Triangular elements can be formed from quadrilateral elements by assigning the same global node number and coordinates to two or three nodes. Such 'degenerate' elements are shown in Figure 3.12. Alternatively, triangular elements can be derived using area coordinates[4] instead of curvilinear ones. A convenient set of coordinates L_1, L_2 L_3 is defined as (Figure 3.13).

$$L_1 = \frac{Area\,P23}{Area\,123} \; , \; L_2 = \frac{Area\,P13}{Area\,123} \; , \; L_3 = 1 - L_1 - L_2 \tag{3.19}$$

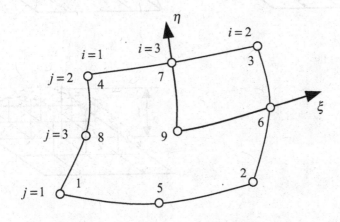

Figure 3.10   Quadratic Lagrange element

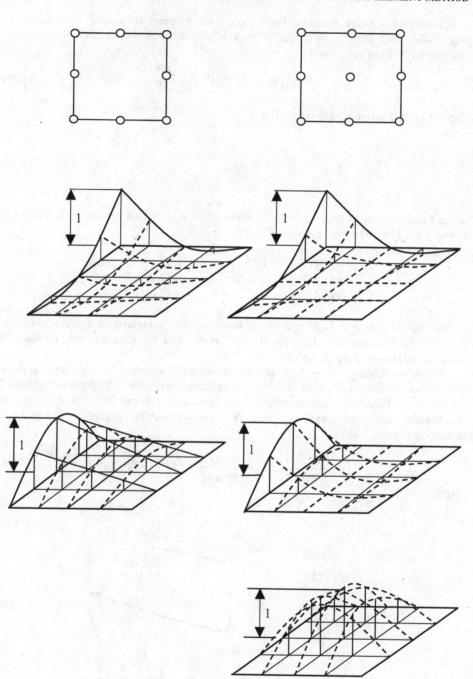

Figure 3.11   Serendipity and Lagrange shape functions

The coordinates of point P for the linear element are given by

$$\mathbf{x} = \sum_{n=1}^{3} L_n \mathbf{x}_n^e$$ (3.20)

It is also possible to derive quadratic and higher order element shape functions using area coordinates.

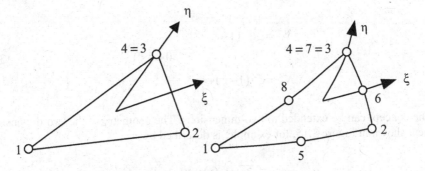

Figure 3.12   Linear and parabolic degenerate elements

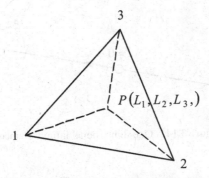

Figure 3.13   Triangular linear element

3.4   ELEMENTS OF INFINITE EXTENT

It is sometimes necessary to describe surfaces of infinite extent. Examples are found in geomechanics, in which either the surface of the ground extends to infinity or a tunnel can be assumed to be infinitely long. To describe the geometry of an element of infinite extent in one intrinsic coordinate direction, say ξ, we may use special shape functions which tend to infinity as ξ tends to +1. For the one-dimensional element shown in Figure 3.14, the coordinate transformation

$$\mathbf{x}(\xi) = \sum_{n=1}^{3} N_n^{\infty}(\xi) \mathbf{x}_n^e$$ (3.21)

gives infinite Cartesian coordinates at $\xi = 1$ if the shape functions are taken to vary as follows:

$$N_1^\infty = -\xi$$

$$-1 < \xi < 0 \tag{3.22}$$

$$N_2^\infty = \xi + 1$$

and

$$N_1^\infty = -\xi /(1-\xi)$$

$$0 < \xi < 1 \tag{3.23}$$

$$N_2^\infty = \xi /(1-\xi)+1$$

The concept can be extended to two-dimensions. The geometry of the two-dimensional element shown in Figure 3.15, for example, is described by

$$x = \sum_{n=1}^{4} N_n(\xi) N_n^\infty(\eta) x_n^e \tag{3.24}$$

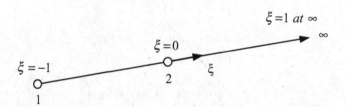

Figure 3.14   One-dimensional infinite element

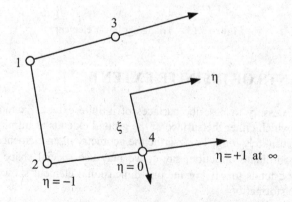

Figure 3.15   Two-dimensional infinite element

where $N_n(\xi)$ are Serendipity shape functions, as discussed previously, and N_n^∞ are defined by

$$N_1^\infty = \eta$$
$$N_2^\infty = \eta$$
$$N_3^\infty = 1 + \eta \qquad -1 < \eta \le 0 \tag{3.25}$$
$$N_4^\infty = 1 + \eta$$

and

$$N_1^\infty = -\eta/(1-\eta)$$
$$N_2^\infty = -\eta/(1-\eta)$$
$$N_3^\infty = \eta/(1-\eta) + 1 \qquad 0 < \eta \le 1 \tag{3.26}$$
$$N_4^\infty = \eta/(1-\eta) + 1$$

3.5 SUBROUTINES FOR SHAPE FUNCTIONS

Here we start building our library of subroutines for future use. We create routines for the calculation of Serendipity, infinite and Lagrange shape functions. Only the listings for the first one is shown here.

As explained in Chapter 2, some variables will be defined as global, that is, as accessible to all the subroutines in a **MODULE** and all programs which use them via the **USE** statement. The dimension for the array Ni, which contains the shape functions, depends on the type of element and will be set by the main program.

```
SUBROUTINE Serendip_func(Ni,xsi,eta,ldim,nodes,inci)
!------------------------------------
! Computes Serendipity shape functions Ni(xsi,eta)
! for one and two-dimensional (linear/parabolic) finite
! boundary elements
!------------------------------------
REAL,INTENT(OUT)   :: Ni(:)  ! Array with shape function
REAL, INTENT(IN)   :: xsi,eta! intrinsic coordinates
INTEGER,INTENT(IN) :: ldim   ! element dimension
INTEGER,INTENT(IN) :: nodes  ! number of nodes
INTEGER,INTENT(IN) :: inci(:)! element incidences
REAL:: mxs,pxs,met,pet       ! temporary variables
SELECT CASE (ldim)
CASE(1)! one-dimensional element
   Ni(1)= 0.5*(1.0 - xsi); Ni(2)= 0.5*(1.0 + xsi)
   IF(nodes == 2) RETURN     ! linear element finished
   Ni(3)=  1.0 - xsi*xsi
   Ni(1)= Ni(1) - 0.5*Ni(3); Ni(2)= Ni(2) 0.5*Ni(3)
CASE(2)! two-dimensional element
```

```
    mxs=1.0-xsi; pxs=1.0+xsi; met=1.0-eta; pet=1.0+eta
    Ni(1)= 0.25*mxs*met ; Ni(2)= 0.25*pxs*met
    Ni(3)= 0.25*pxs*pet ; Ni(4)= 0.25*mxs*pet
    IF(nodes == 4) RETURN !  linear element finished
    IF(Inci(5) > 0) THEN  !  zero node = node missing
      Ni(5)= 0.5*(1.0 -xsi*xsi)*met
      Ni(1)= Ni(1) - 0.5*Ni(5) ;  Ni(2)= Ni(2)0.5*Ni(5)
    END IF
    IF(Inci(6) > 0) THEN
      Ni(6)= 0.5*(1.0 -eta*eta)*pxs
      Ni(2)= Ni(2) - 0.5*Ni(6) ;  Ni(3)= Ni(3)0.5*Ni(6)
    END IF
    IF(Inci(7) > 0) THEN
      Ni(7)= 0.5*(1.0 -xsi*xsi)*pet
      Ni(3)= Ni(3) - 0.5*Ni(7) ; Ni(4)= Ni(4)0.5*Ni(7)
    END IF
    IF(Inci(8) > 0) THEN
      Ni(8)= 0.5*(1.0 -eta*eta)*mxs
      Ni(4)= Ni(4) - 0.5*Ni(8) ; Ni(1)= Ni(1)0.5*Ni(8)
    END IF
CASE DEFAULT !   error message
   CALL Error_message('Element dimension not 1 or 2')
END SELECT
RETURN
END SUBROUTINE Serendip_func
```

3.6   DESCRIPTION OF PHYSICAL QUANTITIES

In addition to defining the shape of the solid to be modelled, we can also specify the variation of physical quantities (displacements, temperatures, tractions, etc.) on the boundaries. Here we can use exactly the same method as has been used for the modelling of the geometrical shape. The physical quantities are specified at nodal points of an element and interpolation functions are used to calculate values between the nodes. The value at a point inside an element e can therefore be written as

$$q = \sum N_n q_n^e \qquad (3.27)$$

where q_n^e is the value of the function at the nth node of element e and N_n are interpolation functions identical to the shape functions discussed previously.

If, for a particular element, the same functions are used for the element shape and for the interpolations of physical quantities inside the element, then the element is called 'isoparametric' (i.e., same number of parameters).

An example of an isoparametric boundary element is shown in Figure 3.16, in which the geometrical shape and variation of q are described using the same shape functions.

The interpolations are given by:

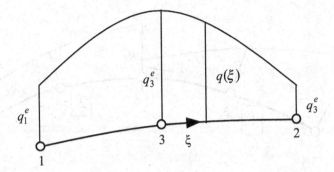

Figure 3.16 Variation of q along quadratic boundary element

Geometry:

$$\mathbf{x} = \sum N_n(\xi)\mathbf{x}_n^e$$

Physical quantity: $q(\xi) = \sum N_n(\xi)q_n^e$

(3.28)

The variation of physical quantities on the surface of two-dimensional elements can be described (Figure 3.17) in a similar way:

$$q(\xi,\eta) = \sum N_n(\xi,\eta)q_n^e$$

(3.29)

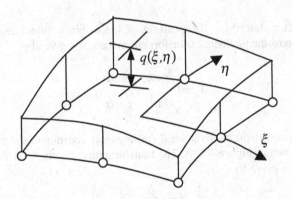

Figure 3.17 Variation of q over quadratic surface element

Note that q may be a scalar or a vector. The physical quantities are defined for each element separately, so for the elements considered here they can be discontinuous at nodes shared by two elements: that is they can have 'jumps', as shown in Figure 3.18. If Serendipity or Lagrange shape functions are used, C^0 continuity can be enforced between elements by specifying the same function value for each element at a shared node.

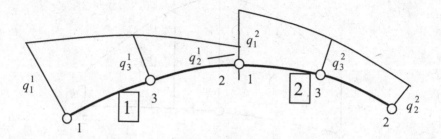

Figure 3.18 Distribution with 'jumps'

3.7 COORDINATE TRANSFORMATION

When defining the coordinates of a node it might be convenient to define them in a local Cartesian coordinate system. A local coordinate system is defined by the location of its origin, x_0 and the direction of the axes. In two dimensions we define the direction with an angle α, as shown in Figure 3.19a.

The global coordinates defined by local coordinates x' are given by

$$x = x_0 + T_g x' \tag{3.30}$$

where x_0 is a vector describing the position of the origin of the local axes. For two-dimensional problems the geometric transformation matrix is given by

$$T_g = \begin{bmatrix} \cos\alpha & -\sin\alpha \\ \sin\alpha & \cos\alpha \end{bmatrix} \tag{3.31}$$

For three-dimensional problems a local (orthogonal) coordinate system is defined by unit vectors, as shown in Figure 3.19b. The transformation matrix for a three-dimensional coordinate system is given by

$$T_g = \begin{bmatrix} v_{1x} & v_{2x} & v_{3x} \\ v_{1y} & v_{2y} & v_{3y} \\ v_{1z} & v_{2z} & v_{3z} \end{bmatrix} \tag{3.32}$$

The inverse relationship between local and global coordinates is given by

$$x' = T_g^T (x - x_0) \tag{3.33}$$

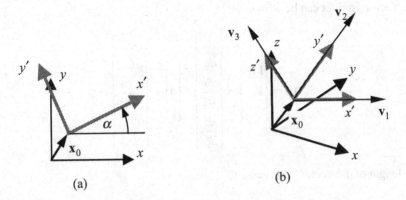

(a)   (b)

Figure 3.19   Local coordinate systems

3.8   DIFFERENTIAL GEOMETRY

In the boundary element method it will be necessary to work out the direction normal to a line or surface element. The best way to determine these directions is by using vector algebra.

Consider a one-dimensional element (Figure 3.20). The vector in the direction of ξ is

$$V_\xi = \frac{\partial}{\partial \xi} x \tag{3.34}$$

where x has two Cartesian components. By the differentiation of equation (3.4) we obtain the tangential vector as

$$V_\xi = \frac{\partial}{\partial \xi} x = \sum \frac{\partial N_n}{\partial \xi} x_n^e \tag{3.35}$$

A vector normal to the line element, V_3, may then be computed by taking the cross-product of V_ξ with a unit vector in the z-direction (v_z):

$$V_3 = V_\xi \times v_z \tag{3.36}$$

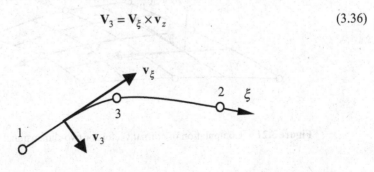

Figure 3.20   Normal and tangential vectors for one-dimensional element

This vector product can be written as:

$$\mathbf{V}_3 = \begin{Bmatrix} V_{3x} \\ V_{3y} \\ V_{3z} \end{Bmatrix} = \begin{Bmatrix} \dfrac{dx}{d\xi} \\ \dfrac{dy}{d\xi} \\ \dfrac{dz}{d\xi} \end{Bmatrix} \times \begin{Bmatrix} 0 \\ 0 \\ 1 \end{Bmatrix} = \begin{Bmatrix} \dfrac{dy}{d\xi} \\ -\dfrac{dx}{d\xi} \\ 0 \end{Bmatrix} \tag{3.37}$$

The length of the vector $\mathbf{V}_3$ is equal to

$$V_3 = |\mathbf{V}_3| = \sqrt{\frac{dy}{d\xi}^2 + \frac{dx}{d\xi}^2} \tag{3.38}$$

and therefore the unit vector in the direction normal to a line element is given by

$$\mathbf{v}_3 = \mathbf{V}_3 \frac{1}{V_3} \tag{3.39}$$

It can be shown that V_3 is also the length of $d\xi$ on the element and hence the Jacobian of the transformation.

For two-dimensional surface elements (Figure 3.21) there are two tangential vectors, $\mathbf{V}_\xi$ in the ξ-direction and

$$\mathbf{V}_\eta = \frac{d}{d\eta}\mathbf{x} \tag{3.40}$$

in the η-direction, where

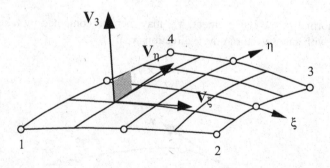

Figure 3.21   Computation of normal vector for two-dimensional elements

$$\frac{\partial}{\partial \eta}\mathbf{x} = \sum \frac{\partial N_n}{\partial \eta}\mathbf{x}_n^e \tag{3.41}$$

The vector normal to the surface may be computed by taking the cross-product of $\mathbf{V}_\xi$ and $\mathbf{V}_\eta$:

$$\mathbf{V}_3 = \mathbf{V}_\xi \times \mathbf{V}_\eta \tag{3.42}$$

that is

$$\mathbf{V}_3 = \begin{Bmatrix} \dfrac{dx}{d\xi} \\[6pt] \dfrac{dy}{d\xi} \\[6pt] \dfrac{dz}{d\xi} \end{Bmatrix} \times \begin{Bmatrix} \dfrac{dx}{d\eta} \\[6pt] \dfrac{dy}{d\eta} \\[6pt] \dfrac{dz}{d\eta} \end{Bmatrix} = \begin{Bmatrix} \dfrac{dy}{d\xi}\dfrac{dz}{d\eta} - \dfrac{dy}{d\eta}\dfrac{dz}{d\xi} \\[6pt] \dfrac{dz}{d\xi}\dfrac{dx}{d\eta} - \dfrac{dz}{d\eta}\dfrac{dx}{d\xi} \\[6pt] \dfrac{dx}{d\xi}\dfrac{dy}{d\eta} - \dfrac{dx}{d\eta}\dfrac{dy}{d\xi} \end{Bmatrix} \tag{3.43}$$

The unit normal vector $\mathbf{v}_3$ is obtained as indicated previously, by first computing the length of the vector:

$$V_3 = \sqrt{V_{3x}^2 + V_{3y}^2 + V_{3z}^2} \tag{3.44}$$

and then by normalising

$$\mathbf{v}_3 = \mathbf{V}_3 \frac{1}{V_3} \tag{3.45}$$

To determine orthogonal directions tangential to the surface we may quite arbitrarily choose the tangential direction 1, $\mathbf{v}_1$, to be defined by the cross-product of the vector $\mathbf{v}_3$ and the unit vector in the y-direction, $\mathbf{v}_y$:

$$\mathbf{v}_1 = \mathbf{v}_3 \times \mathbf{v}_y \tag{3.46}$$

To construct an orthogonal system of axes it follows that the tangential direction 2, $\mathbf{v}_2$ is obtained by

$$\mathbf{v}_2 = \mathbf{v}_1 \times \mathbf{v}_3 \tag{3.47}$$

A special case might exist in which $\mathbf{v}_3$ is exactly in the y-direction. In this case the vector $\mathbf{v}_1$ is obtained by taking the cross-product with a vector in the x-direction. The computation of the normal vector requires the derivatives of the shape functions. These are computed by SUBROUTINE Serendip_deriv shown below.

```fortran
SUBROUTINE Serendip_deriv(DNi,xsi,eta,ldim,nodes,inci)
!-----------------------------------
! Computes Derivatives of Serendipity shape functions
! for one and two-dimensional (linear/parabolic)
! finite boundary elements
!-----------------------------------
REAL,INTENT(OUT)    :: DNi(:,:) ! Array, shape function
                               ! deriv.
REAL, INTENT(IN)  :: xsi,eta  ! intrinsic coordinates
INTEGER,INTENT(IN):: ldim     ! element dimension
INTEGER,INTENT(IN):: nodes    ! number of nodes
INTEGER,INTENT(IN):: inci(:)  ! element incidences
REAL:: mxs,pxs,met,pet        ! temporary variables
SELECT CASE (ldim)
   CASE(1)                         ! one-dimensional element
      DNi(1,1)= -0.5
      DNi(2,1)= 0.5
      IF(nodes == 2)RETURN    ! linear element finished
      DNi(3,1)=  -2.0*xsi
      DNi(1,1)= DNi(1,1) - 0.5*DNi(3,1)
      DNi(2,1)= DNi(2,1) - 0.5*DNi(3,1)
   CASE(2)                         ! two-dimensional element
      mxs= 1.0-xsi
      pxs= 1.0+xsi
      met= 1.0-eta
      pet= 1.0+eta
      DNi(1,1)= -0.25*met
      DNi(1,2)= -0.25*mxs
      DNi(2,1)=  0.25*met
      DNi(2,2)= -0.25*pxs
      DNi(3,1)=  0.25*pet
      DNi(3,2)=  0.25*pxs
      DNi(4,1)= -0.25*pet
      DNi(4,2)=  0.25*mxs
      IF(nodes == 4) RETURN   ! linear element finshed
      IF(Inci(5) > 0) THEN   ! zero node = node  missing
         DNi(5,1)= -xsi*met
         DNi(5,2)= -0.5*(1.0 -xsi*xsi)
         DNi(1,1)= DNi(1,1) - 0.5*DNi(5,1)
         DNi(1,2)= DNi(1,2) - 0.5*DNi(5,2)
         DNi(2,1)= DNi(2,1) - 0.5*DNi(5,1)
         DNi(2,2)= DNi(2,2) - 0.5*DNi(5,2)
      END IF
      IF(Inci(6) > 0) THEN
         DNi(6,1)= 0.5*(1.0 -eta*eta)
         DNi(6,2)= -eta*pxs
         DNi(2,1)= DNi(2,1) - 0.5*DNi(6,1)
         DNi(2,2)= DNi(2,2) - 0.5*DNi(6,2)
         DNi(3,1)= DNi(3,1) - 0.5*DNi(6,1)
         DNi(3,2)= DNi(3,2) - 0.5*DNi(6,2)
      END IF
```

```
      IF(Inci(7) > 0) THEN
        DNi(7,1)= -xsi*pet
        DNi(7,2)= 0.5*(1.0 -xsi*xsi)
        DNi(3,1)= DNi(3,1) - 0.5*DNi(7,1)
        DNi(3,2)= DNi(3,2) - 0.5*DNi(7,2)
        DNi(4,1)= DNi(4,1) - 0.5*DNi(7,1)
        DNi(4,2)= DNi(4,2) - 0.5*DNi(7,2)
      END IF
      IF(Inci(8) > 0) THEN
        DNi(8,1)= -0.5*(1.0-eta*eta)
        DNi(8,2)= -eta*mxs
        DNi(4,1)= DNi(4,1) - 0.5*DNi(8,1)
        DNi(4,2)= DNi(4,2) - 0.5*DNi(8,2)
        DNi(1,1)= DNi(1,1) - 0.5*DNi(8,1)
        DNi(1,2)= DNi(1,2) - 0.5*DNi(8,2)
      END IF
  CASE DEFAULT                    ! error message
      CALL Error_message('Element dimension not 1 or 2' )
END SELECT
RETURN
END SUBROUTINE Serendip_deriv
```

The computation of the vector normal to the surface and of the Jacobian is combined in one SUBROUTINE Normal_Jac.

```
SUBROUTINE &
Normal_Jac(v3,Jac,xsi,eta,ldim,nodes,inci,coords)
!-------------------------------------------------------
! Computes normal vector and Jacobian
! at point with local coordinates xsi,eta
!-------------------------------------------------------
REAL,INTENT(OUT)   :: v3(:)        ! Vector normal to point
REAL,INTENT(OUT)   :: Jac          ! Jacobian
REAL, INTENT(IN)   :: xsi,eta      ! intrinsic coords of point
INTEGER,INTENT(IN) :: ldim         ! element dimension
INTEGER,INTENT(IN) :: nodes        ! number of nodes
INTEGER,INTENT(IN) :: inci(:)      ! element incidences
REAL,  INTENT(IN)  :: coords(:,:)! node coordinates
REAL,ALLOCATABLE   :: DNi(:,:)     ! Derivatives of Ni
REAL,ALLOCATABLE   :: v1(:),v2(:)! Vectors in xsi,eta dir
INTEGER :: Cdim     !    Cartesian dimension
!Cartesian dimension:
Cdim= ldim+1
!Allocate temporary arrays
ALLOCATE (DNi(nodes,Cdim),V1(Cdim),V2(Cdim))
!Compute derivatives of shape function
CALL Serendip_deriv(DNi,xsi,eta,ldim,nodes,inci)
!Compute vectors in xsi (eta) direction(s)
DO I=1,Cdim
```

```
   V1(I)= DOT_PRODUCT(DNi(:,1),COORDS(I,:))
   IF(ldim == 2) THEN
     V2(I)= DOT_PRODUCT(DNi(:,2),COORDS(I,:))
   END IF
END DO
!Compute normal vector
IF(ldim == 1) THEN
   v3(1)= V1(2)
   v3(2)= -v1(1)
ELSE
   V3= Vector_ex(v1,v2)
END IF
! Normalise
CAll Vector_norm(V3,Jac)
DEALLOCATE (DNi,V1,V2)
RETURN
END SUBROUTINE Normal_Jac
```

3.9   INTEGRATION OVER ELEMENTS

The main reason for selecting a range of +1 to −1 for the intrinsic coordinates is to enable the use of numerical (Gauss Quadrature) integration over the elements. In the discussion of the implementation of boundary element methods we will find that this integration plays an important role. The functions to be integrated over elements will be quite complex and therefore require numerical treatment. One prerequisite for numerical integration is an expression for rate of change of length, area or volume with respect to intrinsic coordinates. For example, for a one-dimensional element the total element length is given by

$$S^e = \int_{-1}^{+1} J \, d\xi \tag{3.48}$$

where the Jacobian J is given by equation (3.38).

Similarly, the area of a two-dimensional boundary element is given by

$$A^e = \int_{-1}^{+1} \int_{-1}^{+1} J \, d\xi \, d\eta \tag{3.49}$$

where the Jacobian J is given by equation (3.43).

3.9.1   Numerical integration

In the numerical treatment we approximately evaluate the integral by replacing it by a sum, i.e.

$$I = \int_{-1}^{+1} f(\xi) d\xi \approx \sum_{i=1}^{N} W_i f(\xi_i) \tag{3.50}$$

In the above, W_i are weights and ξ_i are the intrinsic coordinates of the integration (*sampling*) points. If the well known trapezoidal rule is used, for example, then N=2, the weights are 1 and the *sampling* points are at +1 and –1. That is:

$$I = \int_{-1}^{+1} f(\xi)d\xi \approx f(-1) + f(1) \tag{3.51}$$

However, the trapezoidal rule is much too inaccurate for the functions that we will attempt to integrate. The Gauss Quadrature with a variable number of integration points can be used to integrate more accurately. In this method, the function to be integrated is replaced by a polynomial of the form:

$$f(\xi) = a_0 + a_1\xi + a_2\,\xi^2 + \ldots + a_p\,\xi^p \tag{3.52}$$

where the coefficients are adjusted in such a way as to give the best fit to $f(\xi)$. We determine the number and location of the sampling points or *Gauss* points and the weights by the condition that the polynomial is integrated exactly. We find that with increasing degree of polynomial p we need an increasing number of Gauss points. Table 3.2 gives an overview of the number of Gauss points N needed to integrate a polynomial of degree p up to $p= 5$. The computed location of the sampling points and the weights are given in Table 3.2 for one to three Gauss points (information for up to 8 Gauss points is given in the program listing).

Table 3.2   Gauss point and degree of polynomial

No. of Gauss Points N	Degree of Polynomial p
1	1 (linear)
2	3 (cubic)
3	5 (quintic)

Table 3.3   Gauss point coordinates and weights

N	ξ_i	W_i
1	0.0	2.0
2	0.57735 , -0.57735	1.0,1.0
3	0.77459, 0.0 , -0.77459	0.55555, 0.88888, 0.55555

If we apply the numerical integration to two-dimensional elements, then a double sum has to be specified:

$$I = \int_{-1}^{+1}\int_{-1}^{+1} f(\xi,\eta)d\xi d\eta \approx \sum_{i=1}^{N}\sum_{j=1}^{M} W_i W_j f(\xi_i,\eta_j) \tag{3.53}$$

The Gauss integration points for a two-dimensional element are shown in Figure 3.22.

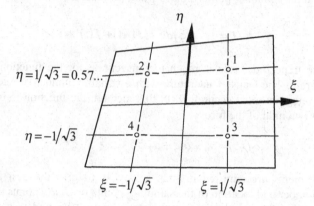

Figure 3.22 Gauss integration points for a two-dimensional element

A subroutine can be written which returns the Gauss point coordinates and weights depending on the number of Gauss points for an integration order of up to 8.

```
SUBROUTINE Gauss_coor(Cor,Wi,Intord)
!-----------------------------------------
! Returns Gauss coords and Weights for up to 8 Gauss points
!-----------------------------------------
REAL, INTENT(OUT)   :: Cor(8)   !   Gauss point coordinates
REAL, INTENT(OUT)   :: Wi(8)    !   weigths
INTEGER,INTENT(IN)  :: Intord   !   integration order
SELECT CASE (Intord)
CASE(1)
   Cor(1)= 0.
   Wi(1)  = 2.0
CASE(2)
   Cor(1)= .577350269  ; Cor(2)= -Cor(1)
   Wi(1)  = 1.0 ;  Wi(2)  = Wi(1)
CASE(3)
   Cor(1)= .774596669  ; Cor(2)= 0.0 ; Cor(3)= -Cor(1)
   Wi(1)  = .555555555  ; Wi(2)  = .888888888 ; Wi(3)  = Wi(1)
CASE(4)
   Cor(1)= .861136311 ; Cor(2)= .339981043 ; Cor(3)= -Cor(2)
   Cor(4)= -Cor(1)
   Wi(1)  = .347854845 ; Wi(2)  = .652145154 ; Wi(3)  = Wi(2)
   Wi(4)  = Wi(1)
CASE(5)
   Cor(1)= .906179845 ; Cor(2)= .538469310 ; Cor(3)= .0
   Cor(4)= -Cor(2) ; Cor(5)= -Cor(1)
```

```
      Wi(1)= .236926885 ; Wi(2)= .478628670 ; Wi(3)= .568888888
      Wi(4) = Wi(2) ; Wi(5) = Wi(1)
   CASE(6)
      Cor(1)=.932469514 ; Cor(2)=.661209386 ; Cor(3)=.238619186
      Cor(4)= -Cor(3) ;  Cor(5)= -Cor(2) ; Cor(6)= -Cor(1)
      Wi(1)= .171324492 ; Wi(2)= .360761573 ; Wi(3)= .467913934
      Wi(4) = Wi(3) ; Wi(5) = Wi(2) ; Wi(6) = Wi(1)
   CASE(7)
      Cor(1)=.949107912 ; Cor(2)=.741531185 ; Cor(3)=.405845151
      Cor(4)= 0.
      Cor(5)= -Cor(3) ;Cor(6)= -Cor(2) ;Cor(7)= -Cor(1)
      Wi(1)= .129484966 ; Wi(2)= .279705391 ; Wi(3)= .381830050
      Wi(4) = .417959183
      Wi(5) = Wi(3) ; Wi(6) = Wi(2) ; Wi(7) = Wi(1)
   CASE(8)
      Cor(1)=.960289856 ; Cor(2)=.796666477 ; Cor(3)=.525532409
      Cor(4)= .183434642
      Cor(5)= -Cor(4) ; Cor(6)= -Cor(3) ; Cor(7)= -Cor(2)
      Cor(8)= -Cor(1)
      Wi(1)= .101228536 ; Wi(2)= .222381034 ; Wi(3)= .313706645
      Wi(4) = .362683783
      Wi(5)= Wi(4) ; Wi(6)= Wi(3) ; Wi(7)= Wi(2) ; Wi(8)= Wi(1)
   CASE DEFAULT
      CALL Error_Message('Gauss points not in range 1-8')
   END SELECT
   END SUBROUTINE Gauss_coor
```

3.10 PROGRAM 3.1: CALCULATION OF SURFACE AREA

We now have developed sufficient library subroutines for writing our first program. The program is intended to calculate the length or surface area of a boundary described or discretised by boundary elements. First we define the libraries of subroutines to be used. The names after the **USE** statement refer to the **MODULE** names in the listing in the Appendix. There are three types of libraries:

- The Geometry_lib which contains all the shape functions, derivative of the shape functions and the routines to compute the Jacobian and the outward normal.

- The Utility_lib which contains utility subroutines for computing, for example, vector ex-products, normalising vectors and printing error messages.

- The Integration_lib which contains Gauss point coordinates and weights.

We define allocable arrays for storing the incidences of all elements, the incidences of one element, the coordinates of all node points, the coordinates of all nodes of one element and the vector normal to the surface. The dimensions of these arrays depend on the element dimension (one-dimensional, two-dimensional), the number of element nodes (linear/parabolic shape function) and the number of elements and nodes. The dimension of these arrays will be allocated once this information is known.

The first executable statements read the information necessary to allocate the dynamic arrays and the integration order that is to be used for the example. Here we use two files INPUT.DAT and OUTPUT.DAT for input and output. The input file has to be created by the user before the program can be run. The FORMAT of inputting data is free-field, that is, numbers can be separated by blanks. After reading the general information the incidences or the connectivity is read for all elements and stored in the array InciG. While reading this information we find out the largest node number, information which we need for allocating the dimension for the array containing node coordinates which we do before reading the node coordinates. We make use of the new feature in FORTRAN 95 that allows the subscripts of an array to start with zero, because a transition element that has the midside node missing will have a node number of 0 in the incidences. We assign zero coordinates to node number 0.

We loop over all elements which describe the boundary. For each element we get the Gauss point coordinates and weightings by a call to Gaus_coor with the corresponding interegration order as specified by intord. We then add all the Gauss point contributions, i.e. the Jacobians computed (by a call to **Normal_jac**) for each Gauss point multiplied by the weighting. Note that there are two cases to be considered: for a one-dimensional case, that is, if we work out a length of a curve, only one DO loop is required. For the two-dimensional case, that is, when we work out a surface area, two nested DO loops are required (see equation 3.52).

```fortran
PROGRAM Compute_Area
!---------------------------------------------
!    Program to compute the length/surface area
!    of a line/surface modelled by boundary elements
!---------------------------------------------
USE Geometry_lib ; USE Utility_lib ; USE Integration_lib
IMPLICIT NONE
INTEGER :: ldim,noelem,nelem,lnodes,maxnod,node,Cdim
INTEGER,ALLOCATABLE :: inciG(:,:)! Incidences
INTEGER,ALLOCATABLE :: inci(:)! Incidences one element
REAL,ALLOCATABLE :: corG(:,:) ! Coordinates (all nodes)
REAL,ALLOCATABLE :: cor(:,:)  ! Coordinates one element
REAL,ALLOCATABLE :: v3(:)      ! Normal vector
REAL :: Gcor(8),Wi(8) ! Gauss point coords and weights
REAL :: Jac, xsi, eta, Area
OPEN(UNIT=10,FILE='INPUT.DAT',STATUS='OLD')
OPEN(UNIT=11,FILE='OUTPUT.DAT',STATUS='UNKNOWN')
READ(10,*) ldim,lnodes,noelem,intord
WRITE(11,*) ' Element dimension=',ldim
WRITE(11,*) ' No. of elem.nodes=',lnodes
WRITE(11,*) ' Number of elements=',noelem
WRITE(11,*) ' Integration order =',intord
Cdim= ldim+1    !Cartesian dimension
ALLOCATE(v3(Cdim))
ALLOCATE(inciG(8,noelem))! Allocate global incid. Array
DO nelem=1,noelem
   READ(10,*) (inciG(n,nelem),n=1,lnodes)
END DO
```

```
maxnod= MAXVAL(inciG)
ALLOCATE(corG(Cdim,0:maxnod)!Allocate array for coords
corG(:,0)= 0.0!Node # 0 = node is missing
DO node=1,maxnod
   READ(10,*) (corG(i,node),i=1,Cdim)
END DO
ALLOCATE(inci(lnodes),cor(Cdim,lnodes))
CALL Gauss_coor(Gcor,Wi,Intord)!Gauss coor and wgths
Area= 0.0 !   Start sum for area/length
Element_loop: &
DO nelem=1,noelem
   inci=  inciG(:,nelem)!   Store incidences locally
   cor= corG(:,inci)!   gather element coordinates
   SELECT CASE (ldim)
   CASE (1)!   One-dim. problem determine length
     Gauss_loop:&
     DO I=1,INTORD
        xsi= Gcor(i)
        CALL Normal_Jac(v3,Jac,xsi,eta,ldim,lnodes,inci,cor)
        Area= Area + Jac*Wi(i)
     END DO &
     Gauss_loop
   CASE (2)!   Two-dim. problem determine area
     Gauss_loop1:&
     DO I=1,INTORD
        DO j=1,INTORD
           xsi= Gcor(i)
           eta= Gcor(j)
           CALL Normal_Jac(v3,Jac,xsi,eta,ldim,lnodes,inci,cor)
           Area= Area + Jac*Wi(i)*Wi(j)
        END DO
     END DO &
     Gauss_loop1
   CASE DEFAULT
   END SELECT
END DO &
Element_loop
IF(ldim == 1) THEN
   WRITE(11,*) ' Length =',Area
ELSE
   WRITE(11,*) ' Area =',Area
END IF
END PROGRAM Compute_Area
```

3.11   CONCLUSIONS

In this chapter we have dealt with methods for describing the geometry of a problem. Here we have obviously concentrated on describing problem boundaries. The method consists of subdividing the boundary curve or surface into small elements, and is commonly known as *discretisation*. The concept of isoparametric elements was introduced, where we use interpolation functions not only to describe the boundary surface in terms of nodal values, but also the variation of known or unknown values. We have laid the foundation for Chapter 6 (Boundary Element Methods) here, where we will use the concepts which have been described. We find that, once we use this advanced discretisation method in the BEM, the analytical integration is no longer possible. Therefore, we have also introduced the method of numerical integration, which is most commonly used in numerical work, the Gauss Quadrature. For general purpose programs using the isoparametric concept, the accuracy of the numerical integration will be crucial. We have started our process for building a subroutine library which will be needed later. A small program has been written which we can use to test the subroutines and to do numerical experiments.

3.12 EXERCISES

Exercise 3.1
Using program Compute_area, calculate the length of a quarter circle using:

(a) one linear element
(b) two linear element
(c) one quadratic element.

Determine the discretisation error. Use 2x2 integration.

Exercise 3.2
Using program Compute_area, calculate the area of a quarter circle using:

(a) the discretisation into one element, as shown in Figure 3.23 (a)
(b) the discretisation three elements, as shown in Figure 3.23 (b).

Plot the variation of the Jacobian over the element using the Gauss point values. Determine the discretisation error.

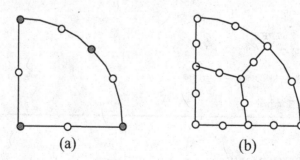

(a)   (b)

Figure 3.23   Discretisations for determining the area of a quarter circle

Exercise 3.3
Using program Compute_area, calculate the area of 1/8 of a sphere using

(a) the discretisation into one element, as shown in Figure 3.24 (a)
(b) the discretisation into three elements as shown in Figure 3.24 (b)

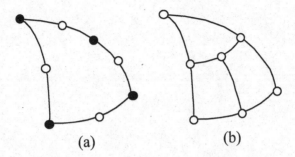

(a)   (b)

Figure 3.24   Discretisations for determining the surface area of 1/8 sphere (for mesh (a) the black dots mark the corner nodes)

Plot the variation of the Jacobian over the element using the Gauss point values. Determine the discretisation error.

3.13   REFERENCES

1. Encarnacao, J. and Schlechtendahl, J. (1983) *Computer Aided Design*. Springer Verlag, Berlin.
2. Irons, B. M. (1966) Engineering applications of numerical integration in stiffness method, *J.A.I.A.A.*, **14**, 2035-7.
3. Zienkiewicz, O. C., Irons, B. M., Ergatoudis, J. G., Ahmad, S. and Scott, F. C. (1969) Iso-parametric and associate element families for two- and three-dimensional analysis. *Finite Element Methods in Stress Analysis* (eds. I. Holland and K. Bell). Tapir Press, Norway, Chap. 14.
4. Fraejis de Veubeke, B. (1965) Displacement and equilibrium models in the finite element method. *Stress Analysis* (eds. O.C. Zienkiewicz and G.S. Holister). John Wiley & Sons, Chichester, Chap. 9.

4

Material Modelling and Fundamental Solutions

*If you can measure what
you are speaking about, and
express it in numbers, you
know something about it.*

Lord Kelvin

4.1 INTRODUCTION

In addition to specifying the geometry of the problem, it is necessary to describe the physical response of the material in a mathematical way. This is done by defining the response characteristics of an infinitesimally small portion of the solid. The *constitutive law* establishes a relationship between heat flow and the temperature gradient or between strain and stress. The constants in such relationships are characteristic values or properties of the material. We distinguish between material properties which are direction independent (*isotropic* material), and those which are dependent on direction (*anisotropic* material). Furthermore, there are materials which have the same properties everywhere (*homogeneous* materials), and those where the properties change from location to location (*non-homogeneous* materials).

In the material response we distinguish between linear and non-linear behaviour. For linear materials we can establish a unique (linear) relationship between stress/strain, temperature/heat flow or potential/fluid flow. For non-linear materials this relationship changes with time. The results for these problems are therefore dependent on the deformation (thermal) history.

As outlined previously, solutions of the governing equation have to be obtained which are as simple as possible. In the literature, these solutions are referred to as fundamental solutions, Green's functions or Kernels. Obviously, these solutions can only be found for a linear and homogeneous continuum.

The equations which have to be satisfied by the fundamental solutions are:

- Constitutive law
- Equilibrium / conservation of energy
- Compatibility / continuity.

The last condition will be automatically satisfied for solutions which are continuous in the domain. In the following, we will first derive the governing differential equations using the first two conditions, and then present the fundamental solutions for potential problems (heat flow and seepage) and for elasticity problems in two and three dimensions.

There has been a misconception that, because fundamental solutions can only be obtained for homogeneous domains and linear elastic materials, the BEM is only suitable for analysing these types of problems. As will be shown in this book, this is not the case. Inhomogeneous materials can be analysed by the multi-region approach and non-linear problems with an internal cell subdivision.

4.2   STEADY STATE POTENTIAL PROBLEMS

Heat conduction in solids and flow in porous media (seepage) are diffusion problems and can be treated concurrently because they are governed by the same differential equation (Laplace).

Steady state heat flux or fluid flow $\mathbf{q}$ per unit area (also referred to as flow velocity) is related to temperature or potential u by Fourier's / Darcy's law:

$$\mathbf{q} = -\mathbf{D}\nabla u$$

(4.1)

where the negative sign is due to the fact that the flow is always from higher to lower temperature/potential. The flow vector is defined as:

$$\mathbf{q} = \begin{Bmatrix} q_x \\ q_y \\ q_z \end{Bmatrix}$$

(4.2)

The conductivity/pemeabilty matrix $\mathbf{D}$ is given by

$$\mathbf{D} = \begin{bmatrix} k_{xx} & k_{xy} & k_{xz} \\ k_{yx} & k_{yy} & k_{yz} \\ k_{zx} & k_{zy} & k_{zz} \end{bmatrix}$$

(4.3)

where k_{xx}, etc., are conductivities measured in [W/m°K] in the case of thermal problems and permeabilities measured in [m/sec] in the case of seepage problems. The coefficients in $\mathbf{D}$ represent flow values for unit values of temperature or potential gradient. It can be shown that $\mathbf{D}$ has to be symmetric and positive definite.

The differential operator ∇ for three-dimensional problems is defined as

$$\nabla = \left\{ \begin{array}{c} \dfrac{\partial}{\partial x} \\[2mm] \dfrac{\partial}{\partial y} \\[2mm] \dfrac{\partial}{\partial z} \end{array} \right\} \tag{4.4}$$

and for two-dimensional problems

$$\nabla = \left\{ \begin{array}{c} \dfrac{\partial}{\partial x} \\[2mm] \dfrac{\partial}{\partial y} \end{array} \right\} \tag{4.5}$$

The conservation of energy condition states that for heat flow problems the outflow must be equal to the inflow plus the heat generated inside, $\hat{Q}$.

For the infinitesimal cube in Figure 4.1 this gives the following

$$\left(q_x + \frac{\partial q_x}{\partial x} dx \right) dydz + \left(q_y + \frac{\partial q_y}{\partial y} dy \right) dxdz + \left(q_z + \frac{\partial q_z}{\partial z} dz \right) dxdy = \tag{4.6}$$

$$q_x dydz + q_y dxdz + q_z dxdy + \hat{Q} dxdydz$$

After cancelling terms we obtain

$$\frac{\partial q_x}{\partial x} + \frac{\partial q_y}{\partial y} + \frac{\partial q_z}{\partial z} - \hat{Q} = 0 \tag{4.7}$$

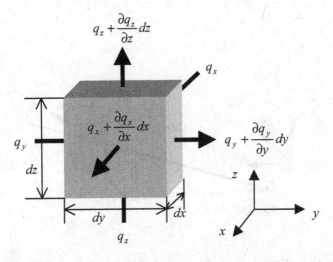

Figure 4.1   Heat flow in an infinitesimal cube

Substituting equation (4.1) for isotropic material (i.e. $k_{xx} = k_{yy} = k_{zz} = k$ and $k_{xy} = k_{xz} = k_{yz} = 0$), we obtain the governing differential equation for which we seek a fundamental solution:

$$k\left(\frac{\partial^2 u}{\partial x^2} + \frac{\partial^2 u}{\partial y^2} + \frac{\partial^2 u}{\partial z^2}\right) - \hat{Q} = 0 \tag{4.8}$$

The simplest solution we can find is that of a concentrated source at point P *(source point)* of magnitude one in an infinite homogeneous domain. This means that internal heat generation only occurs at one point in the domain and is zero elsewhere. The function describing this spatial distribution of $\hat{Q}$ is also referred to as a *Dirac Delta* function.

For this problem the temperature or potential at point Q *(field point)* can be written for the three-dimensional case as

$$U(P,Q) = \frac{1}{4\pi\, rk} \tag{4.9}$$

In our notation the meaning of the values in the parenthesis is defined as

$$U\ (source\ point,\ field\ point)$$

and $r = |\mathbf{r}|$ is the distance between source point and field point (Figure 4.2).

As we will see later, the flow in a direction normal to a boundary defined by a vector $\mathbf{n}$ is also required. For three-dimensional problems, the flow is computed by

$$T(P,Q) = -k\frac{\partial U}{\partial \mathbf{n}} = -k\left(n_x\frac{\partial U}{\partial x} + n_y\frac{\partial U}{\partial y} + n_z\frac{\partial U}{\partial z}\right) \tag{4.10}$$

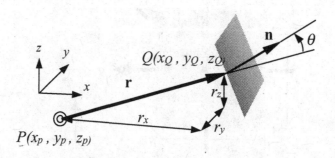

Figure 4.2   Notation for fundamental solution (three-dimensional potential problems)

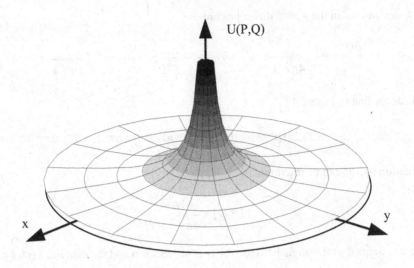

Figure 4.3   Variation of fundamental solution U (potential/temperature) for 3-D potential problems (source at origin of coordinate system)

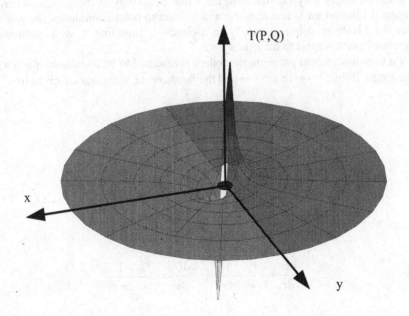

Figure 4.4   Variation of fundamental solution T for **n** = {1,0,0} (flow in x-direction) for 3-D potential problems

The derivatives in the global directions are

$$\frac{\partial U}{\partial x} = -\frac{r_x}{r}\frac{1}{4\pi r^2 k}; \quad \frac{\partial U}{\partial y} = -\frac{r_y}{r}\frac{1}{4\pi r^2 k}; \quad \frac{\partial U}{\partial z} = -\frac{r_z}{r}\frac{1}{4\pi r^2 k} \tag{4.11}$$

where, according to Figure 4.2,

$$r_x = x_Q - x_P; \quad r_y = y_Q - y_P; \quad r_z = z_Q - z_P \tag{4.12}$$

Equation (4.10) can be rewritten as

$$T(P,Q) = \frac{\cos\theta}{4\pi r^2} \tag{4.13}$$

where θ is defined as the angle between the normal vector $\mathbf{n}$ and the distance vector $\mathbf{r}$, i.e.

$$\cos\theta = \frac{1}{r}\mathbf{n}\bullet\mathbf{r} \quad with \quad \mathbf{n} = \{n_x, n_y, n_z\} \quad and \quad \mathbf{r} = \{r_x, r_y, r_z\} \tag{4.14}$$

The variation of kernels U and T is plotted in Figures 4.3 and 4.4. It can be seen that both solutions decay very rapidly from the value of infinity at the source. Whereas the fundamental solution for U is symmetric with respect to polar coordinates, the solution for T with the vector $\mathbf{n}$ pointing exactly in x-direction (meaning flow in x-direction) is antisymmetric with respect to the y-axis.

For a two-dimensional problem, the source is assumed to be distributed along a line of infinite length from $z = -\infty$ to $z = +\infty$ and the fundamental solutions are given by

$$U(P,Q) = \frac{1}{2\pi k} \ln\left(\frac{1}{r}\right) \tag{4.15}$$

and

$$T(P,Q) = -k\frac{\partial U}{\partial \mathbf{n}} = -k\left(n_x\frac{\partial U}{\partial x} + n_y\frac{\partial U}{\partial y}\right) \tag{4.16}$$

where

$$\frac{\partial U}{\partial x} = -\frac{r_x}{r}\frac{1}{2\pi rk}; \quad \frac{\partial U}{\partial y} = -\frac{r_y}{r}\frac{1}{2\pi rk} \tag{4.17}$$

Equation (4.16) can also be rewritten as

$$T(P,Q) = \frac{\cos\theta}{2\pi r} \tag{4.18}$$

where:

$$cos\theta = \frac{1}{r}\mathbf{n} \bullet \mathbf{r} \quad with \quad \mathbf{n} = \{n_x, n_y\} \quad and \quad \mathbf{r} = \{r_x, r_y\} \tag{4.19}$$

Subroutines for the isotropic solutions are presented below.

```fortran
MODULE Laplace_lib
REAL :: PI=3.14159265359
CONTAINS
  REAL FUNCTION U(r,k,Cdim)
    !-------------------------------
    !   Fundamental solution for Potential problems
    !   Temperature/Potential
    !-------------------------------
    REAL,INTENT(IN)      :: r  ! Distance source and field point
    REAL,INTENT(IN)      :: k  ! Conducivity
    INTEGER,INTENT(IN)   :: Cdim ! Cartesian dimension (2-D,3-D)
    SELECT CASE (CDIM)
      CASE (2)            ! Two-dimensional solution
        U= 1.0/(2.0*Pi*k)*LOG(1/r)
      CASE (3)            ! Three-dimensional solution
        U= 1.0/(4.0*Pi*r*k)
      CASE DEFAULT
        U=0.0
        CALL Error_Message('Cdim not equal 2 or 3 in Function U')
    END SELECT
  END FUNCTION U


  REAL FUNCTION T(r,dxr,Vnorm,Cdim)
    !-------------------------------
    !   Fundamental solution for Potential problems
    !   Flow
    !-------------------------------
    REAL,INTENT(IN)::       r  ! Distance source and field point
    REAL,INTENT(IN)::   dxr(:) ! rx/r , ry/r , rz/r
    REAL,INTENT(IN):: Vnorm(:) ! Normal vector
    INTEGER,INTENT(IN) :: Cdim ! Cartesian dimension
    SELECT CASE (Cdim)
      CASE (2)            ! Two-dimensional solution
        T= -DOT_PRODUCT (Vnorm,dxr)/(2.0*Pi*r)
      CASE (3)            ! Three-dimensional solution
        T= -DOT_PRODUCT (Vnorm,dxr)/(4.0*Pi*r*r)
      CASE DEFAULT
        T=0.0
        CALL Error_Message('Cdim not equal 2 or 3 in Function T')
    END SELECT
  END FUNCTION T
END MODULE Laplace_lib
```

For anisotropic problems, the fundamental solutions have been presented by Bonnet[5]. For example, the solution for temperature/potential is given by

$$U(P,Q) = \frac{1}{2\pi \bar{k}} ln \frac{1}{\bar{r}}$$

(4.20)

for two-dimensional problems and

$$U(P,Q) = \frac{1}{4\pi \bar{r}\bar{k}}$$

(4.21)

for three-dimensional problems, where the values in the denominator are defined as

$$\bar{k} = \sqrt{det\,\mathbf{D}} \quad and \quad \bar{r} = \mathbf{r}^T \mathbf{D}^{-1} \mathbf{r}$$

(4.22)

For general anisotropy in three dimensions, the D-matrix has 9 material parameters but, because of the property of symmetry, only 6 components need to be input. A special case of anisotropy exists where the material parameters are different in three orthogonal coordinate directions. This is known as *orthotropic* material. If these conductivities are defined in the direction of global coordinates, then all off-diagonal elements of **D** are zero. If we denote the conductivities in x,y and z-directions as k_1, k_2, k_3 then

$$\mathbf{D} = \begin{bmatrix} k_1 & 0 & 0 \\ 0 & k_2 & 0 \\ 0 & 0 & k_3 \end{bmatrix}$$

(4.23)

For this case the values in equation (4.22) are simply given by:

$$\bar{k} = \sqrt{k_1 k_2 k_3} \quad and \quad \bar{r} = \sqrt{r_x^2 \frac{1}{k_1} + r_y^2 \frac{1}{k_2} + r_z^2 \frac{1}{k_3}}$$

(4.24)

4.3   STATIC ELASTICITY PROBLEMS

In solid mechanics applications, a relationship between stress and strain must be established. Stresses are forces per unit area inside the solid. They can be visualised by cutting the solid on a particular plane (i.e. z-y, x-z, y-z). In Figure 4.5 the definitions of these stresses for an infinitesimally small cube are shown. We define three stress components normal to the cutting plane σ_x, σ_y, σ_z and three tangential to the cutting plane τ_{xy}, τ_{xz}, τ_{yz}. The stress *pseudo-vector* can be defined as

$$\sigma = \begin{Bmatrix} \sigma_x \\ \sigma_y \\ \sigma_z \\ \tau_{xy} \\ \tau_{yz} \\ \tau_{xz} \end{Bmatrix} \tag{4.25}$$

In *plane stress* problems, such as in thin plates subject to in-plane loading, all stresses associated with the z direction are assumed to be zero, i.e. $\sigma_z = \tau_{xz} = \tau_{yz} = 0$.

Another way of defining stresses is as distributed forces in any direction per unit area which act on a plane section, as shown in Figure 4.6. The distributed forces defined that way are also referred to later as boundary stresses or *tractions*. The relationship between the *tractions* and stresses in three dimensions is given by

$$t_x = n_x \sigma_x + n_y \tau_{xy} + n_z \tau_{xz}$$
$$t_y = n_y \sigma_y + n_x \tau_{xy} + n_z \tau_{yz} \tag{4.26}$$
$$t_z = n_z \sigma_z + n_x \tau_{zx} + n_y \tau_{zy}$$

where n_x, n_y and n_z are the x-, y and z- components of the vector normal to the cutting plane.

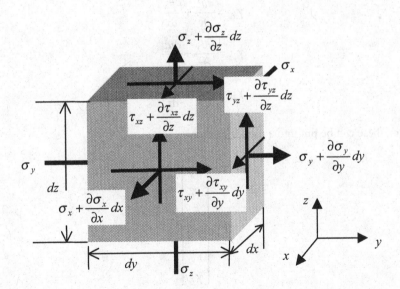

Figure 4.5   Stresses acting on an infinitesimal cube

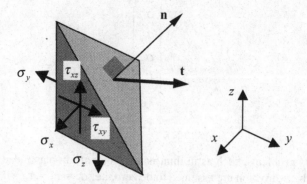

Figure 4.6   Relationship between tractions and stresses in three dimensions

Strains are defined in terms of displacement components in the x, y, z directions (u_x, u_y, u_z) by

$$\varepsilon_x = \frac{\partial u_x}{\partial x}$$

$$\varepsilon_y = \frac{\partial u_y}{\partial y}$$

$$\varepsilon_z = \frac{\partial u_z}{\partial z}$$

$$\gamma_{xy} = \frac{\partial u_x}{\partial y} + \frac{\partial u_y}{\partial x} \qquad (4.27)$$

$$\gamma_{yz} = \frac{\partial u_y}{\partial z} + \frac{\partial u_z}{\partial y}$$

$$\gamma_{zx} = \frac{\partial u_z}{\partial x} + \frac{\partial u_x}{\partial z}$$

Again, these can be put into a pseudo-vector

$$\boldsymbol{\varepsilon} = \left\{ \begin{array}{c} \varepsilon_x \\ \varepsilon_y \\ \varepsilon_z \\ \gamma_{xy} \\ \gamma_{yz} \\ \gamma_{zx} \end{array} \right\} \qquad (4.28)$$

In matrix form this can be written as

$$\boldsymbol{\varepsilon} = \mathbf{B}\,\mathbf{u} \qquad (4.29)$$

where **u** is a vector of displacements

$$\mathbf{u} = \begin{Bmatrix} u_x \\ u_y \\ u_z \end{Bmatrix} \tag{4.30}$$

and **B** is a differential operator matrix

$$\mathbf{B} = \begin{bmatrix} \dfrac{\partial}{\partial x} & 0 & 0 \\[6pt] 0 & \dfrac{\partial}{\partial y} & 0 \\[6pt] 0 & 0 & \dfrac{\partial}{\partial z} \\[6pt] \dfrac{\partial}{\partial y} & \dfrac{\partial}{\partial x} & 0 \\[6pt] 0 & \dfrac{\partial}{\partial z} & \dfrac{\partial}{\partial y} \\[6pt] \dfrac{\partial}{\partial z} & 0 & \dfrac{\partial}{\partial x} \end{bmatrix} \tag{4.31}$$

In some circumstances, simplifications can be made and certain strain components taken to be zero. A state of *plane strain* can be assumed if the solid extends a long distance in the z-direction, the loading is uniform in this direction and $u_z = 0$ everywhere. We then have $\varepsilon_z = \gamma_{xz} = \gamma_{yz} = 0$. Another special case is a state of complete plane strain, in which derivatives in the z direction of all displacements are taken to be zero, but u_z may be non-zero. This gives

$$\varepsilon_x = \frac{\partial u_x}{\partial x}$$

$$\varepsilon_y = \frac{\partial u_y}{\partial y} \tag{4.32}$$

$$\varepsilon_z = 0$$

$$\gamma_{xy} = \frac{\partial u_x}{\partial y} + \frac{\partial u_y}{\partial x}, \quad \gamma_{yz} = \frac{\partial u_z}{\partial y}, \quad \gamma_{xz} = \frac{\partial u_z}{\partial x}$$

Complete plane strain can be split into the plane strain case already discussed, and an antiplane strain or St Venant torsion component for which $\varepsilon_x = \varepsilon_y = \varepsilon_z = \gamma_{xy} = 0$ and

$$\gamma_{zy} = \frac{\partial u_z}{\partial y}$$

$$\gamma_{zx} = \frac{\partial u_z}{\partial x} \tag{4.33}$$

In complete plane strain it is possible to have shear strains and stresses acting in the z-direction.

Sometimes it is necessary to compute the components of stress or strain in directions which do not coincide with the global axes. In this case a *transformation* of stress or strain is necessary. The transformation of local stresses $\boldsymbol{\sigma}'$ acting on planes in the material parallel with the x', y' and z' axes to global stresses $\boldsymbol{\sigma}$ acting on cuts parallel with the x, y, z axes can be written as

$$\boldsymbol{\sigma} = \mathbf{T}_\sigma \boldsymbol{\sigma} \tag{4.34}$$

For the two-dimensional case, in which the local axes are defined by a rotation about the z-axis, $\mathbf{T}_\sigma$ is obtained by the two transformations shown in Figure 4.7

$$\mathbf{T}_\sigma = \begin{bmatrix} \cos^2\alpha & \sin^2\alpha & -2\cos\alpha\sin\alpha \\ \sin^2\alpha & \cos^2\alpha & 2\sin\alpha\cos\alpha \\ \cos\alpha\sin\alpha & \cos\alpha\sin\alpha & \cos^2\alpha - \sin^2\alpha \end{bmatrix} \tag{4.35}$$

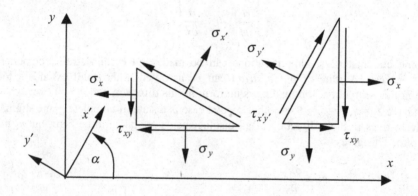

Figure 4.7   Transformation of stresses in two dimensions

For the stress transformation in three-dimensional space it is convenient to refer to the components of unit vectors in the directions of the local axes, Figure 4.5.

For example, we denote by

$$\mathbf{v}_1 = \begin{Bmatrix} v_{1x} \\ v_{1y} \\ v_{1z} \end{Bmatrix} \tag{4.36}$$

the unit vector in the direction of the x'- axis. Similarly, $\mathbf{v}_2$ and $\mathbf{v}_3$ are unit vectors along the y'- and z'- axes. In terms of these vector components, the matrix $\mathbf{T}_\sigma$ is written as:

$$\mathbf{T}_\sigma = \begin{bmatrix} \mathbf{T}_{\sigma 11} & \mathbf{T}_{\sigma 12} \\ \mathbf{T}_{\sigma 21} & \mathbf{T}_{\sigma 22} \end{bmatrix}$$

$$\mathbf{T}_{\sigma 11} = \begin{bmatrix} v_{1x}^2 & v_{2x}^2 & v_{3x}^2 \\ v_{1y}^2 & v_{2y}^2 & v_{3y}^2 \\ v_{1z}^2 & v_{2z}^2 & v_{3z}^2 \end{bmatrix}$$

$$\mathbf{T}_{\sigma 12} = \begin{bmatrix} 2v_{1y}v_{1x} & 2v_{2y}v_{2x} & 2v_{3y}v_{3x} \\ 2v_{1y}v_{1z} & 2v_{2y}v_{2z} & 2v_{3y}v_{3z} \\ 2v_{1x}v_{1z} & 2v_{2x}v_{2z} & 2v_{3x}v_{3z} \end{bmatrix}$$

$$(4.37)$$

$$\mathbf{T}_{\sigma 21} = \begin{bmatrix} v_{1x}v_{2x} & v_{2x}v_{3x} & v_{1x}v_{3x} \\ v_{1y}v_{2y} & v_{2y}v_{3y} & v_{1y}v_{3y} \\ v_{1z}v_{2z} & v_{2z}v_{3z} & v_{1z}v_{3z} \end{bmatrix}$$

$$\mathbf{T}_{\sigma 22} = \begin{bmatrix} v_{1x}v_{2y}+v_{1y}v_{2x} & v_{2x}v_{3y}+v_{2y}v_{3x} & v_{1x}v_{3y}+v_{1y}v_{3x} \\ v_{1y}v_{2z}+v_{1z}v_{2y} & v_{2y}v_{3z}+v_{2z}v_{3y} & v_{1y}v_{3z}+v_{1z}v_{3y} \\ v_{1x}v_{2z}+v_{1z}v_{2x} & v_{2x}v_{3z}+v_{2z}v_{3x} & v_{1x}v_{3z}+v_{1z}v_{3x} \end{bmatrix}$$

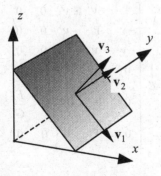

Figure 4.8   Transformation of stress in three dimensions

4.3.1 Constitutive equations

The elastic material response is governed by Hooke's law. For an isotropic material, this is in three dimensions

$$\varepsilon_x = \frac{1}{E}\left[\sigma_x - v\left(\sigma_y + \sigma_z\right)\right]$$

$$\varepsilon_y = \frac{1}{E}\left[\sigma_y - v\left(\sigma_x + \sigma_z\right)\right]$$

$$\varepsilon_z = \frac{1}{E}\left[\sigma_z - v\left(\sigma_x + \sigma_y\right)\right]$$

$$\gamma_{xy} = \frac{1}{G}\tau_{xy}, \quad \gamma_{yz} = \frac{1}{G}\tau_{yz}, \quad \gamma_{zx} = \frac{1}{G}\tau_{zx}$$

(4.38)

where E is the modulus of elasticity, v the Poisson's ratio and G the shear modulus, given by

$$G = \frac{E}{2(1+v)}$$

(4.39)

Equation (4.34) can be conveniently written in matrix form

$$\varepsilon = \mathbf{C}\sigma$$

(4.40)

where matrix $\mathbf{C}$ is defined as

$$\mathbf{C} = \frac{1}{E}\begin{bmatrix} 1 & -v & -v & 0 & 0 & 0 \\ -v & 1 & -v & 0 & 0 & 0 \\ -v & -v & 1 & 0 & 0 & 0 \\ 0 & 0 & 0 & \dfrac{E}{G} & 0 & 0 \\ 0 & 0 & 0 & 0 & \dfrac{E}{G} & 0 \\ 0 & 0 & 0 & 0 & 0 & \dfrac{E}{G} \end{bmatrix}$$

(4.41)

The inverse relationship can be defined by

$$\sigma = \mathbf{D}\varepsilon$$

(4.42)

where

$$\mathbf{D} = \mathbf{C}^{-1} = C_1 \begin{bmatrix} 1 & C_2 & C_2 & 0 & 0 & 0 \\ C_2 & 1 & C_2 & 0 & 0 & 0 \\ C_2 & C_2 & 1 & 0 & 0 & 0 \\ 0 & 0 & 0 & \dfrac{G}{C_1} & 0 & 0 \\ 0 & 0 & 0 & 0 & \dfrac{G}{C_1} & 0 \\ 0 & 0 & 0 & 0 & 0 & \dfrac{G}{C_1} \end{bmatrix} \qquad (4.43)$$

with

$$C_1 = E \frac{1-v}{(1+v)(1-2v)}; \quad C_2 = \frac{v}{(1-v)} \qquad (4.44)$$

A subroutine to compute the isotropic D-matrix is given below.

```fortran
SUBROUTINE D_mat(E,ny,D,Cdim)
!-----------------------------------
!   Computes isotropic D-matrix
!   Plane-strain (Cdim= 2)
!   or 3-D        (Cdim= 3)
!-----------------------------------
REAL, INTENT(IN)    :: E      !  Young's modulus
REAL, INTENT(IN)    :: ny     !  Poisson's ratio
INTEGER,INTENT(IN)  :: Cdim   !  Cartesian Dimension
REAL, INTENT(OUT)   :: D(:,:) !  D-matrix
REAL                :: c1,c2,G
c1= E*(1.0-ny)/( (1.0+ny)*(1.0-2.0*ny) )
c2= ny/(1.0-ny)
G = E/(2.0*(1.0+ny))
D = 0.0
SELECT CASE (Cdim)
CASE (2)
  D(1,1)= 1.0  ; D(2,2)= 1.0
  D(2,1)= c2   ; D(1,2)= c2
  D(3,3)= G/c1
CASE (3)            !     3-D
  D(1,1)= 1.0  ;  D(2,2)= 1.0  ;  D(3,3)= 1.0
  D(2,1)= c2   ;  D(1,3)= c2   ;  D(2,3)= c2
  D(1,2)= c2   ;  D(3,1)= c2   ;  D(3,2)= c2
  D(4,4)= G/c1 ;  D(5,5)= G/c1 ;  D(6,6)= G/c1
CASE DEFAULT
END SELECT
D= c1*D
RETURN
END SUBROUTINE D_mat
```

For general anisotropic material, 21 characteristic material properties are required. It is impractical to determine these. However, special types of anisotropy may exist for the case where the material properties are different in orthogonal directions. We may have a *laminate* or *stratified* material where E is the same in two orthogonal directions but different in the third orthogonal direction. Examples of this are a stratified rock mass or fibre reinforced plastics. If the orthogonal directions are in the coordinate directions, then the D-matrix is modified as follows:

$$\mathbf{D'} = \begin{bmatrix} C_1' & C_2' & C_3' & 0 & 0 & 0 \\ C_2' & C_1' & C_3' & 0 & 0 & 0 \\ C_3' & C_3' & C_4' & 0 & 0 & 0 \\ 0 & 0 & 0 & G_1 & 0 & 0 \\ 0 & 0 & 0 & 0 & G_2 & 0 \\ 0 & 0 & 0 & 0 & 0 & G_2 \end{bmatrix} \tag{4.45}$$

with

$$C_1' = n\left(1-v_2^{\,2}\right)C; \quad C_2' = n\left(v_1 + nv_2^{\,2}\right)C; \quad C_3' = nv_2\left(1+v_1\right)C$$

$$C_4' = \left(1+v_1^{\,2}\right)C; \quad C = \frac{E_2}{\left(1-v_1\right)\left(1-v_1-2nv_2^{\,2}\right)}; \quad n = \frac{E_1}{E_2} \quad and \quad G_1 = \frac{E_1}{2\left(1+v_1\right)} \tag{4.46}$$

The translation of the theory into a subroutine is shown below.

```
SUBROUTINE D_mat_anis(D,E1,G1,E2,G2,ny2,Cdim)
!---------------------------------
!Computes anisotropic D-matrix
!Plane-strain (Cdim= 2)
!or 3-D        (Cdim= 3)
!---------------------------------
REAL, INTENT(OUT)  :: D(:,:)! D-matrix
REAL, INTENT(IN)   :: E1    ! Young's modulus, dir 1
REAL, INTENT(IN)   :: G1    ! Shear modulus , dir 1
REAL, INTENT(IN)   :: E2    ! Young's modulus, dir 2
REAL, INTENT(IN)   :: G2    ! Shear Modulus , dir 2
REAL, INTENT(IN)   :: ny2   ! Poisson's ratio, dir 2
INTEGER,INTENT(IN):: Cdim   ! Cartesian Dimension
REAL              :: n      ! ratio E1/E2
REAL ::.cc,c1,c2,c3,c4,ny1 ! temps
ny1= 0.5*E1/G1 -1.0
n= E1/E2
cc= E2/(1.+ny1)/(1.-ny1-2.*n*ny2**2)
c1= n*(1.-n*ny2**2)*cc
c3= n*ny2*(1.0+ny1)*cc
c4= (1 - ny1**2)*cc
```

```
D= 0. !  only nonzero components of D are assigned
SELECT CASE (Cdim)
   CASE (2)              !     plane strain
            D(1,1)= c1 ; D(2,2)= c4
            D(1,2)= c3 ; D(2,1)= c3
            D(3,3)= G2
   CASE (3)              !       3-D
            c2= n*(ny1+n*ny2**2)*cc
            D(1,1)= C1 ; D(2,2)= c1 ; D(3,3)= c4
            D(1,2)= C2 ; D(1,3)= c3 ; D(2,3)= C3
            D(2,1)= C2 ; D(3,1)= c3 ; D(3,2)= C3
            D(4,4)= G1 ; D(5,5)= G2 ; D(6,6)= G2
   CASE DEFAULT
END SELECT
RETURN
END SUBROUTINE D_mat_anis
```

If the orthogonal directions are not in the directions of the Cartesian coordinates, then the following transformation of the D-matrix has to be made:

$$D = T_\sigma^T D' T_\sigma \qquad (4.47)$$

where D' is defined in local coordinates. This translates to one simple F95 statement

$$D= \text{MATMUL(MATMUL(TRANSPOSE(T),Dprime),T)}$$

4.3.2 Fundamental solutions

The governing differential equations are obtained from the condition of equilibrium. For two-dimensional problems the equilibrium conditions are:

$$\frac{\partial \sigma_x}{\partial x} + \frac{\partial \tau_{xy}}{\partial y} + b_x = 0$$

$$\frac{\partial \tau_{xy}}{\partial x} + \frac{\partial \sigma_y}{\partial y} + b_y = 0 \qquad (4.48)$$

where b_x and b_y are components of body force in the x and y directions.

Substitution of the equations for strain (4.24) and the Hooke's law for plane strain conditions gives

$$C_1 \frac{\partial^2 u_x}{\partial x^2} + C_2 \frac{\partial^2 u_x}{\partial y^2} + C_3 \frac{\partial^2 u_y}{\partial x \partial y} + C_2 \frac{\partial^2 u_y}{\partial x \partial y} + b_x = 0$$

$$C_3 \frac{\partial^2 u_x}{\partial x \partial y} + C_2 \frac{\partial^2 u_x}{\partial x \partial y} + C_2 \frac{\partial^2 u_y}{\partial x^2} + C_1 \frac{\partial^2 u_y}{\partial y^2} + b_y = 0 \qquad (4.49)$$

where

$$C_1 = E/(1-v^2); \quad C_2 = E/(1+v); \quad C_3 = Ev/(1-v^2) \tag{4.50}$$

For the plane strain problem, the fundamental solution is obtained for line loads in the x and y directions of magnitude one, which are of infinite extent in the $+z$ and $-z$ directions. The solution was first worked out by Lord Kelvin[1].

The solutions for the displacements in the x and y directions due to a unit load in the x-direction can be written as (Figure 4.9).

$$U_{xx}(P,Q) = C\left[C_1 \ln\left(\frac{1}{r}\right) + \left(\frac{r_x}{r}\right)^2\right]$$

$$U_{xy}(P,Q) = C\left(\frac{r_x}{r}\right)\left(\frac{r_y}{r}\right) \tag{4.51}$$

with

$$C = 1/(8\pi G(1-v)), \quad C_1 = 3 - 4v$$

The first subscript of U refers to the direction of the unit load, whereas the second relates to the direction of the displacement. We note that as the distance between source point P and field point Q is approaching infinity, the solution tends to negative infinity. This is due to the fact that the source is distributed along an infinite line. This does not present any difficulties, because as we will see later, scaling is introduced for the coordinates which limit the maximum distance to 1. The fundamental solution also has a positive singularity as P and Q coincide.

The reader may easily verify that for a unit load in the y-direction

$$U_{yy}(P,Q) = C\left[C_1 \ln\left(\frac{1}{r}\right) + \left(\frac{r_y}{r}\right)^2\right] \tag{4.52}$$

$$U_{yx}(P,Q) = U_{xy}(P,Q)$$

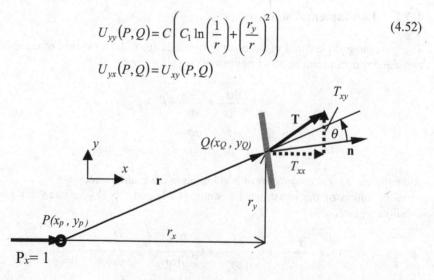

Figure 4.9   Notation for two-dimensional Kelvin solution (unit load in x-direction)

the last term indicating the symmetry of the solution. For the boundary element method we also need the solutions for the boundary stresses (tractions) acting on a surface with an outward normal direction of **n** (see Figure 4.9).

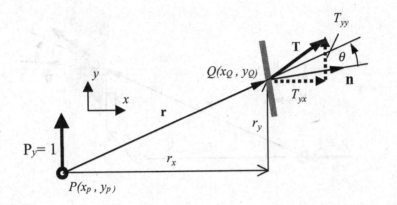

Figure 4.10   Notation for two-dimensional fundamental solution (unit load in y-direction)

The fundamental solution for the *tractions* due to a unit load at P in the x-direction are

$$T_{xx}(P,Q) = -\frac{C_2}{r}\left[C_3 + 2\left(\frac{r_x}{r}\right)^2\right]\cos\theta$$

$$T_{xy}(P,Q) = -\frac{C_2}{r}\left[2\left(\frac{r_x}{r}\right)\left(\frac{r_y}{r}\right)\cos\theta - C_3\left[n_y\left(\frac{r_x}{r}\right) - n_x\left(\frac{r_y}{r}\right)\right]\right] \qquad (4.53)$$

$$C_2 = 1/(4\pi(1-v)), \quad C_3 = 1-2v, \quad \cos\theta = \left(\frac{r_x}{r}\right)n_x + \left(\frac{r_y}{r}\right)n_y$$

where θ is defined in Figures 4.9 and 4.10.

For a unit load in the y-direction we have

$$T_{yy}(P,Q) = -\frac{C_2}{r}\left[C_3 + 2\left(\frac{r_y}{r}\right)^2\right]\cos\theta$$

$$\qquad (4.54)$$

$$T_{yx}(P,Q) = -\frac{C_2}{r}\left[2\left(\frac{r_x}{r}\right)\left(\frac{r_y}{r}\right)\cos\theta - C_3\left[n_x\left(\frac{r_y}{r}\right) - n_y\left(\frac{r_x}{r}\right)\right]\right]$$

We note that the first part of the solution is symmetrical (i.e. the first part of T_{xy} equals T_{yx}), but the second part is not.

For the three-dimensional problem, the fundamental solution is obtained for point loads in the x, y and z directions.

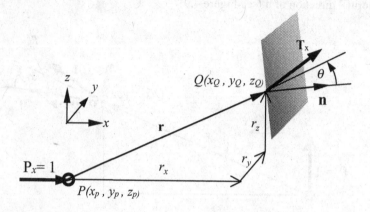

Figure 4.11   Notation for three-dimensional Kelvin solution (point load in x- direction)

The solutions for the displacements in the x,y and z directions due to a unit load in the x-direction can be written as

$$U_{xx}(P,Q) = \frac{C}{r}\left(C_1 + \left(\frac{r_x}{r}\right)^2\right)$$

$$U_{xy}(P,Q) = C\frac{1}{r}\left(\frac{r_x}{r}\right)\left(\frac{r_y}{r}\right)$$

$$U_{xz}(P,Q) = C\frac{1}{r}\left(\frac{r_x}{r}\right)\left(\frac{r_z}{r}\right)$$

(4.55)

with

$$C = 1/(16\pi G(1-v)), \quad C_1 = 3 - 4v$$

Now the solution approaches zero, as the distance between source point P and field point Q tends to infinity. However, both the two-dimensional and three-dimensional solution approach an infinite value, as r tends to zero. This fact will pose some problems with integrating the fundamental solutions which we will address later. The reader may easily verify that for a unit load in the y-direction

$$U_{yy}(P,Q) = \frac{C}{r}\left(C_1 + \left(\frac{r_y}{r}\right)^2\right)$$

$$U_{yx}(P,Q) = U_{xy}(P,Q)$$

(4.56)

$$U_{yz}(P,Q) = C\frac{1}{r}\left(\frac{r_y}{r}\right)\left(\frac{r_z}{r}\right)$$

For a unit load in the z-direction we have

$$U_{zz}(P,Q)=\frac{C}{r}\left(C_1+\left(\frac{r_z}{r}\right)^2\right)$$

$$U_{zx}(P,Q)=U_{xz}(P,Q)$$

$$U_{zy}(P,Q)=U_{yz}(P,Q)$$

(4.57)

The solutions for stresses acting on a boundary surface with an outward normal direction of $\mathbf{n}$ (see Figure 4.11) are presented next.

The fundamental solution for *tractions* due to a unit load at P in the x-direction, are

$$T_{xx}(P,Q)=-\frac{C_2}{r^2}\left(C_3+3\left(\frac{r_x}{r}\right)^2\right)\cos\theta$$

$$T_{xy}(P,Q)=-\frac{C_2}{r^2}\left[3\left(\frac{r_x}{r}\right)\left(\frac{r_y}{r}\right)\cos\theta-C_3\left[n_y\left(\frac{r_x}{r}\right)-n_x\left(\frac{r_y}{r}\right)\right]\right]$$

$$T_{xz}(P,Q)=-\frac{C_2}{r^2}\left[3\left(\frac{r_x}{r}\right)\left(\frac{r_z}{r}\right)\cos\theta-C_3\left[n_z\left(\frac{r_x}{r}\right)-n_x\left(\frac{r_z}{r}\right)\right]\right]$$

(4.58)

with

$$C_2=1/(8\pi(1-v)),\quad C_3=1-2v,$$

$$\cos\theta=\left(\frac{r_x}{r}\right)n_x+\left(\frac{r_y}{r}\right)n_y+\left(\frac{r_z}{r}\right)n_z$$

(4.59)

For a unit load in the y-direction we have

$$T_{yy}(P,Q)=-\frac{C_2}{r^2}\left(C_3+3\left(\frac{r_y}{r}\right)^2\right)\cos\theta$$

$$T_{yx}(P,Q)=-\frac{C_2}{r^2}\left[3\left(\frac{r_x}{r}\right)\left(\frac{r_y}{r}\right)\cos\theta-C_3\left[n_x\left(\frac{r_y}{r}\right)-n_y\left(\frac{r_x}{r}\right)\right]\right]$$

$$T_{yz}(P,Q)=-\frac{C_2}{r^2}\left[3\left(\frac{r_y}{r}\right)\left(\frac{r_z}{r}\right)\cos\theta-C_3\left[n_z\left(\frac{r_y}{r}\right)-n_y\left(\frac{r_z}{r}\right)\right]\right]$$

(4.60)

Finally, for a load in the z-direction we have

$$T_{zz}(P,Q) = -\frac{C_2}{r^2}\left(C_3 + 3\left(\frac{r_z}{r}\right)^2\right)\cos\theta$$

$$T_{zx}(P,Q) = -\frac{C_2}{r^2}\left[3\left(\frac{r_x}{r}\right)\left(\frac{r_z}{r}\right)\cos\theta - C_3\left[n_x\left(\frac{r_z}{r}\right) - n_z\left(\frac{r_x}{r}\right)\right]\right]$$

$$T_{zy}(P,Q) = -\frac{C_2}{r^2}\left[3\left(\frac{r_y}{r}\right)\left(\frac{r_z}{r}\right)\cos\theta - C_3\left[n_y\left(\frac{r_z}{r}\right) - n_z\left(\frac{r_y}{r}\right)\right]\right]$$

(4.61)

The Kelvin solutions for the displacements due to a unit load in x direction are plotted in Figures 4.12 and 4.13. A small circle of exclusion is used to avoid plotting very high values near the singularity. The variation of the x-component of the displacement shows symmetry about the x- and y-axes. The variation of the y-component of displacements shows antisymmetry about both axes. The influence of the Poisson's ratio on the displacements perpendicular to the load axis can be clearly seen in Figure 4.13. Note that the finite element method has difficulty dealing with a Poisson's ratio of 0.5 (incompressible material) because of the definition of C_1 in equation (4.44) which would give an infinite value for $v = 0.5$.

Figure 4.14 shows the variation of the fundamental solution for the x-component of boundary traction acting on a boundary defined by a normal vector, **n**, that points exactly in the x-direction. This means that this traction is the same as the stresscomponent σ_x. We can see clearly that the fundamental solution is antisymmetric about the y-axis and decays very rapidly from the singularity.

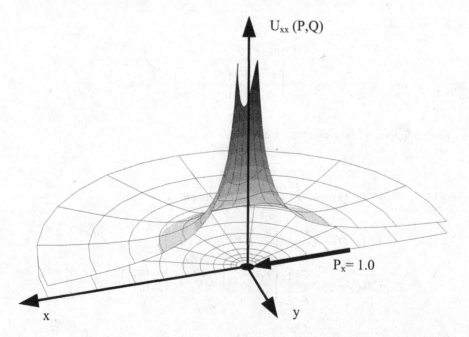

Figure 4.12   Three-dimensional Kelvin solution: variation of the x-component of displacement due to $P_x = 1.0$

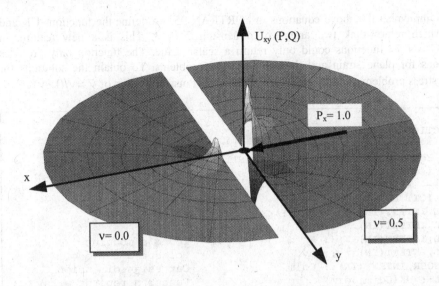

Figure 4.13   Three-dimensional Kelvin solution: variation of the y-component of displacement due to $P_x= 1.0$ for Poissons ratio of 0.0 (left figure) and 0.5 (right figure)

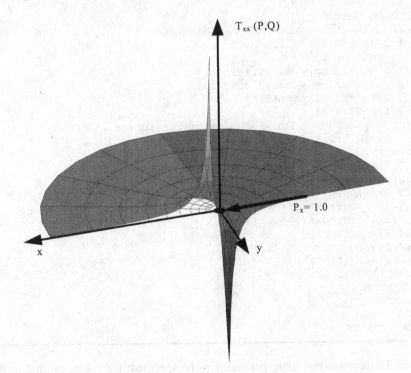

Figure 4.14   Three-dimensional Kelvin solution: variation of T_{xx} for $\mathbf{n} =\{1,0,0\}$. This is equivalent to σ_x

To implement the above equations in FORTRAN 95, we define the functions UK and TK which return rank two arrays of dimension 2 or 3. This is a new feature; in FORTRAN 77 functions could only return a scalar value. The function only provides solutions for plane strain and three-dimensional problems. To obtain the solutions for plane stress problems, simply substitute an effective Poisson's ration of $\bar{v} = v/(1+v)$.

```fortran
FUNCTION UK(dxr,r,E,ny,Cdim)
!----------------------------------------------------
!    FUNDAMENTAL SOLUTION FOR DISPLACEMENTS
!    isotropic material (Kelvin solution)
!----------------------------------------------------
IMPLICIT NONE
REAL,INTENT(IN)    :: dxr(:)           !    rx/r etc.
REAL,INTENT(IN)    :: r                !    r
REAL,INTENT(IN)    :: E                !    Young's modulus
REAL,INTENT(IN)    :: ny               !    Poisson's ratio
INTEGER,INTENT(IN):: Cdim              !    Cartesian dimension
REAL:: UK(Cdim,Cdim)                   !    Function returns array
REAL:: G,c,c1,onr,clog,conr            !    Temps
G= E/(2.0*(1+ny))
c1= 3.0 - 4.0*ny
SELECT CASE (Cdim)
CASE (2)         !       Plane strain solution
   c= 1.0/(8.0*Pi*G*(1.0 - ny))
   clog= -c1*LOG(r)
   UK(1,1)= c*(clog + dxr(1)*dxr(1))
   UK(2,2)= c*(clog + dxr(2)*dxr(2))
   UK(1,2)= c*dxr(1)*dxr(2)
   UK(2,1)= UK(1,2)
CASE(3)          !       Three-dimensional solution
   c= 1.0/(16.0*Pi*G*(1.0 - ny))
   conr=c/r
   UK(1,1)= conr*(c1 + dxr(1)*dxr(1))
   UK(2,2)= conr*(c1 + dxr(2)*dxr(2))
   UK(3,3)= conr*(c1 + dxr(3)*dxr(3))
   UK(1,2)= conr*dxr(1)*dxr(2)
   UK(1,3)= conr*dxr(1)*dxr(3)
   UK(2,1)= UK(1,2)
   UK(2,3)= conr*dxr(2)*dxr(3)
   UK(3,1)= UK(1,3)
   UK(3,2)= UK(2,3)
CASE DEFAULT
END SELECT
RETURN
END FUNCTION UK
```

Function TK requires one more parameter to be specified: the vector normal to the boundary (normal vector).

```fortran
FUNCTION TK(dxr,r,Vnor,ny,Cdim)
!----------------------------------------------
!    FUNDAMENTAL SOLUTION FOR TRACTIONS
!    isotropic material (Kelvin solution)
!----------------------------------------------
IMPLICIT NONE
REAL,INTENT(IN)      :: dxr(:)            ! rx/r etc.
REAL,INTENT(IN)      :: r                 ! r
REAL,INTENT(IN)      :: Vnor(:)           ! normal vector
REAL,INTENT(IN)      :: ny                ! Poisson's ratio
INTEGER,INTENT(IN)   :: Cdim              ! Cartesian dimension
REAL                 :: TK(Cdim,Cdim)     ! Function returns array
REAL                 :: c2,c3,costh,Conr  ! Temps
c3= 1.0 - 2.0*ny
Costh= DOT_PRODUCT (Vnor,dxr)
SELECT CASE (Cdim)
CASE (2)            !     plane strain
   c2= 1.0/(4.0*Pi*(1.0 - ny))
   Conr= c2/r
   TK(1,1)= -(Conr*(C3 + 2.0*dxr(1)*dxr(1))*Costh)
   TK(2,2)= -(Conr*(C3 + 2.0*dxr(2)*dxr(2))*Costh)
   DO i=1,2
    DO j=1,3
     IF(i /= j) THEN
       TK(i,j)= -(Conr*(2.0*dxr(i)*dxr(j)*Costh &
          - c3*(Vnor(j)*dxr(i) - Vnor(i)*dxr(j))))
     END IF
     END DO
     END DO
 CASE(3)            !     Three-dimensional
    c2= 1.0/(8.0*Pi*(1.0 - ny))
    Conr= c2/r**2
    TK(1,1)= -Conr*(C3 + 3.0*dxr(1)*dxr(1))*Costh
    TK(2,2)= -Conr*(C3 + 3.0*dxr(2)*dxr(2))*Costh
    TK(3,3)= -Conr*(C3 + 3.0*dxr(3)*dxr(3))*Costh
    DO i=1,3
     DO j=1,3
      IF(i /= j) THEN
       TK(i,j)= -Conr*(3.0*dxr(i)*dxr(j)*Costh &
          - c3*(Vnor(j)*dxr(i) - Vnor(i)*dxr(j)))
      END IF
      END DO
      END DO
 CASE DEFAULT
 END SELECT
 END FUNCTION TK
```

Efficient fundamental solutions for anisotropic material have been developed recently by Tonon[2].

4.4   CONCLUSIONS

In this chapter we have dealt with the description of the material response in a mathematical way and have derived solutions of the equations governing the problem for simple loading. The solutions are for point sources or loads in an infinite domain. It has been shown that the implementation of these fundamental solutions into a FORTRAN 90 function is fairly straightforward. A particular advantage of the new facilities in F90 is that two and three-dimensional solutions can be implemented in one FUNCTION with the parameter Cdim determining the dimensionality of the result.

The Kelvin fundamental solution is not the only one which may be used for a boundary element analysis. Indeed, any solution may be used, including ones which satisfy some boundary conditions explicitly. For example, we may include the zero boundary traction conditions that exist at the ground surface above an excavation. Green's functions for a point load in a semi-infinite domain have been worked out, for example, by Melan in two dimensions[3] and Mindlin in three dimensions[4]. Also, Bonnet[5] presents a solution for bonded half-spaces where two different materials may be considered implicitly in the solution.

The fundamental solutions just derived will form the basis for the methods discussed in the next chapter.

4.5   REFERENCES

1. Thomson, W. (Lord Kelvin) (1848) A note on the integration of the equations of equilibrium of an elastic solid, *Cambridge and Dublin Mathematical Journal*, Cambridge University Press.
2. Tonon, F., Pan, E. and Amadei, B. (2000) Green's functions and BEM formulation for 3D anisotropic media. *Computers and Structures*, Vol.79, No.5, 469-482.
3. Melan, E. (1932) Der Spannungszustand der durch eine Einzelkraft im Inneren beanspruchten Halbscheibe. *Z. Angew. Math. Mech.* **12**, 343-346.
4. Mindlin, R.D. (1936) Force at a point in the interior of a semi-infinite solid. *Physics* **7**, 195-202.
5. Bonnet, M. (1995) *Boundary Integral Equation Methods for Solids and Fluids*. Wiley, Chichester.

5

Boundary Integral Equations

There is nothing more practical
than a good theory

I.Kant

5.1 INTRODUCTION

As explained previously, the basic idea of boundary element methods comes from Trefftz[1], who suggested that in contrast to the method of Ritz, only functions satisfying the differential equations exactly should be used to approximate the solution inside the domain. If we use these functions it means, of course, that we only need to approximate the actual boundary conditions. This approach, therefore, has some considerable advantages:

- The solutions that we obtain inside the domain satisfy the differential equations exactly, approximations (or errors) only occur due to the fact that boundary conditions are only satisfied approximately.
- Since functions are defined globally, there is no need to subdivide the domain into elements.
- The solutions also satisfy conditions at infinity, therefore, there is no problem dealing with infinite domains, where the FEM has to use mesh truncation.

The disadvantage is that we need solutions of differential equations to be as simple as possible if we want to reduce computation time. The most suitable solutions are ones involving concentrated sources or loads in infinite domains. As we know from the previous chapter, these solutions also have some rather nasty properties, such as singularities, which pose some problems with integration later.

The original method proposed by Trefftz is not suitable for writing general purpose programs as accuracy is not satisfactory and, as will be seen later, convergence of the method cannot be assured. However, because of the inherent simplicity of the method, it serves well to explain some of the basic principles involved. Therefore, we will first introduce this method on a simple example in heat flow.

We will actually develop our programs using the direct method, which gets its name from the fact that no *fictitious* source or forces need to be computed, as in the Trefftz method, but that unknowns at the boundary are obtained directly. In the development of the integral equations we will use the theorem of Betti; which should be well known to engineers.

5.2   TREFFTZ METHOD

To introduce the Trefftz method, let us look at a simple two-dimensional example in heat flow. Consider an infinite homogeneous domain having conductivity k where heat (q_0) flows only in the vertical (y) direction (Figure 5.1).

According to the Fourier law introduced in Chapter 4, heat flow in the y- direction is related to temperature by the following equation:

$$-k\frac{\partial u}{\partial y} = q_0 \tag{5.1}$$

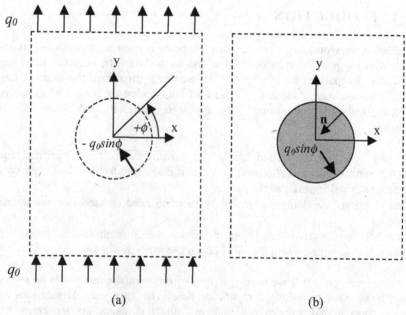

Figure 5.1   Heat flow in an infinite domain (a) without (b) with isolator

Solving the differential equation, the temperature at point $Q(x,y)$ is obtained as

$$u^{(a)}(Q) = -\frac{q_0}{k}y + C \qquad (5.2)$$

If we assume the temperature at the centre of the circle to be zero, then $C = 0$.

We now place a cylindrical isolator in the flow and compute how the flow pattern and temperature change. The property of the isolator is such that no flow can occur in a direction perpendicular to its boundary, that is

$$\frac{\partial u}{\partial \mathbf{n}} = n_x \frac{\partial u}{\partial x} + n_y \frac{\partial u}{\partial y} = 0 \qquad (5.3)$$

where $\mathbf{n}$ $\{n_x, n_y\}$ is the vector normal to the boundary of the isolator. Note that the positive direction of this vector is pointing from the infinite domain into the isolator. For the solution in equation (5.2), we find that this condition is not satisfied because the flow in the direction normal to the isolator boundary (dotted line in Figure 5.1(a)) is computed as):

$$t^{(a)} = -k\frac{\partial u^{(a)}}{\partial \mathbf{n}} = k\, n_y \frac{q_0}{k} = -q_0 \sin\phi \qquad (5.4)$$

If we want to find out the change that the isolator makes to the flow/temperature distribution, then we can think of the problem as being divided into two parts: the first being the trivial one, whose solution we have just obtained, the second one being the solution of the problem where the following boundary condition is applied:

$$t^{(b)} = k\frac{\partial u^{(b)}}{\partial \mathbf{n}} = -q_0 n_y = q_0 \sin\phi = -n_y q_o \qquad (5.5)$$

One can easily verify that if the two solutions are added, then the temperature distribution for the problem of flow past an isolator is obtained. The final solution is therefore

$$u(Q) = u^{(a)}(Q) + u^{(b)}(Q) \qquad (5.6)$$

We now solve the boundary value problem (b) by the Trefftz method.

To apply the Trefftz method, we quite arbitrarily select N points on the boundary of the isolator where we satisfy the boundary conditions, equation (5.5) and another set of points where we apply *fictitious* sources. The reason these are called *fictitious* is that they are not actually present, but can be thought of as parameters of the global approximation functions. We have to be careful with the location of these points and this will be the major drawback of the method. The source points must be placed in such a way that they do not interfere with the result points. In our case, the best place is inside the isolator. Also, we must not place points P too close to the boundary points Q because, as we know, when P approaches Q, the fundamental solutions approach infinity.

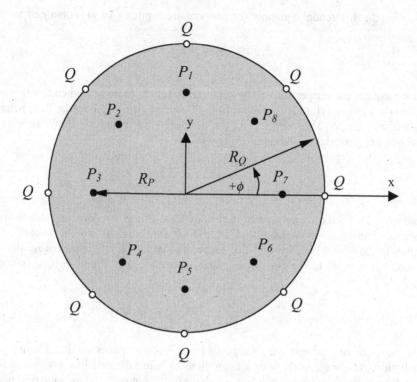

Figure 5.2   Points (P) for fictitious sources and (Q) where boundary conditions are to be satisfied

In Figure 5.2 we show an example of the choice of locations for load points P_i and boundary points Q_i. We place points Q at quarter points on the boundary of the isolator with radius R_Q and points P at a circle, with radius R_P inside the isolator.

In the Trefftz method, we attempt to satisfy the given boundary conditions by adjusting the magnitude of the fictitious sources F_i applied at P_i. Noting that the fundamental solutions for the flow in direction $\mathbf{n}$, which we derived in the last chapter, is $T(P,Q)$, the boundary condition at point Q_1 can be satisfied by

$$t^{(b)}(Q_1) = \sum_1^8 T(P_i, Q_1) F_i \tag{5.7}$$

Here $T(P_i, Q_1)$ is the flow in direction $\mathbf{n}(Q_1)$ at point Q_1 due to a source at P_i. This is also sometimes referred to as an *influence coefficient*. We can now write a similar equation for each boundary point Q_i, a total, therefore, of eight equations:

$$t^{(b)}(Q_1) = q_0 \sin 90^0 = \sum_{i=1}^8 T(P_i, Q_1) F_i$$

$$t^{(b)}(Q_2) = q_0 \sin 135^0 = \sum_{i=1}^8 T(P_i, Q_2) F_i \quad \text{etc.} \tag{5.8}$$

We obtain a system of simultaneous equations which we can solve for unknown *fictitious* sources F_i. Obviously, the number of fictitious sources we can obtain depends on the number of equations we can write and hence, on the number of boundary points Q_j. It is necessary, therefore, to have at least the same number of source points as we have field points.

Once we have solved the system of simultaneous equations and calculated the *fictitious* sources F_i, then the temperature at any point Q on the boundary of the isolator and in the domain (but outside the isolator) can be computed by

$$u(Q) = u^{(a)}(Q) + u^{(b)}(Q)$$

$$\text{where} \quad u^{(b)}(Q) = \sum_1^8 U(P_i, Q) F_i \tag{5.9}$$

The flow at a point Q in the x- and y- directions may be obtained by

$$q_x = q_x^{(b)}; \quad q_y = q_0 + q_y^{(b)}$$

$$\text{where} \quad q_x^{(b)} = -k \frac{\partial u^{(b)}(Q)}{\partial x} = -k \sum_1^8 \frac{\partial U(P_i, Q)}{\partial x} F_i \tag{5.10}$$

$$q_y^{(b)} = -k \frac{\partial u^{(b)}(Q)}{\partial y} = -k \sum_1^8 \frac{\partial U(P_i, Q)}{\partial y} F_i$$

5.3 PROGRAM 5.1: FLOW AROUND CYLINDER, TREFFTZ METHOD

The program shown here allows us to numerically solve the flow around a cylinder problem with a variable number of source points, and thus get a better understanding of the Trefftz method and its limitations. We use the Laplace_lib, which contains the fundamental solutions of the Laplace equations governing our problem, and the Utility_lib containing the subroutine for solving equations by the **USE** statement. Next we read some information about the problem such as heat inflow, conductivity, number of source/boundary points and radius of the cylinder. We finally, quite arbitrarily, specify that the source points are located on a circle with radius Rp which has to be smaller than the radius of the cylinder. Later we can do numerical experiments on the effect of distance between source and boundary points on accuracy of results. Since the size of the arrays for storing the equation system is dependent on the number of source points specified, we allocate them at run-time. Next, we loop over all boundary points (DO loop Field_points) and all source points (DO loop Source_points) to generate the matrix of *influence coefficients* and the right-hand side. The points Q and P are assumed to be equally distributed over the circle.

```fortran
PROGRAM Trefftz
!-----------------------------------
!
!    Program to compute the heat flow past a cylindrical isolator
!    in a 2-D infinite domain using the Trefftz method
!
!-----------------------------------
USE Laplace_lib ; USE Utility_lib
IMPLICIT NONE                            ! declare all variables
REAL                  :: q               ! inflow/outflow
REAL                  :: k               ! Thermal conductivity
INTEGER               :: npnts           ! Number of points P,Q
REAL                  :: rq              ! radius of isolator
REAL                  :: rp              ! radius of source points
REAL(KIND=8),ALLOCATABLE ::  Lhs(:,:)    ! left hand side
REAL(KIND=8),ALLOCATABLE ::  Rhs(:)      ! right hand side
REAL(KIND=8),ALLOCATABLE ::  F(:)        ! fictitious sources
REAL                  :: dxr(2)          ! rx/r ,ry/r
REAL                  :: vnorm(2)        ! normal vector
REAL                  :: Delth,Thetq,Thetp,xq,yq,xp,yp,xi,yi,r,uq
INTEGER               :: npq,npp,ninpts,nin
OPEN(UNIT=10,FILE='INPUT.DAT',STATUS='OLD',ACTION='READ')
OPEN(UNIT=11,FILE='OUTPUT.DAT',STATUS='UNKNOWN',ACTION='WRITE')
READ(10,*)  q,k,npnts,rq,rp
WRITE(11,*) ' Program 2: heat flow past cylinder Trefftz method'
WRITE(11,*) '   Heat inflow/outflow=  ',q
WRITE(11,*) '    Thermal conductivity= ',k
WRITE(11,*) '    Number of Points P,Q= ',npnts
WRITE(11,*) '    Radius of Isolator=   ',rq
WRITE(11,*) '    Radius of Sources =   ',rp
ALLOCATE  (Lhs(npnts,npnts),Rhs(npnts),F(npnts)) !
Delth= 2*Pi/npnts ! increment in angle theta between points (rad)
Thetq= Pi/2.0     ! angle theta to first field point Q1
Field_points: &
DO npq= 1,npnts
   Rhs(npq)= q * SIN(Thetq)         ! right hand side
   xq= rq*COS(Thetq)                ! x-coordinate of field point
   yq= rq*SIN(Thetq)                ! y-coordinate of field point
   vnorm(1)= -COS(Thetq)            ! normal vector to Q
   vnorm(2)= -SIN(Thetq)
   Thetq= Thetq + Delth             ! angle to next field point Q
   Thetp= Pi/2.0                    ! angle to first source point P1
   Source_points:  &
   DO npp= 1,npnts
      xp= rp*COS(Thetp)             ! x-coordinate of source point
      yp= rp*SIN(Thetp)             ! y-coordinate of source point
      dxr(1)= xp-xq
      dxr(2)= yp-yq
      r= SQRT(dxr(1)**2 + dxr(2)**2)   ! dist. field/source pnt
      dxr= dxr/r                       ! normalise vector dxr
      Lhs(npq,npp)= T(r,dxr,vnorm,2)   !
      Thetp= Thetp + Delth             ! angle to next point P
```

```
      END DO   &
      Source_points
   END DO &
   Field_points
   Lhs= - Lhs    ! to eliminate negative diagonal coeff.
   Rhs= - Rhs    ! (Note: done to avoid error message in SOLVE)
   !  Solve system of equations: calculate F out of Lhs and Rhs
   CALL Solve(Lhs,Rhs,F)
   !    Postprocessing - Boundary values of temperature
   WRITE(11,*)   ''
   WRITE(11,*)   'Temperatures at Boundary points:'
   Thetq= Pi/2.0  ! angle to first field point Q1
   Field_points1: &
   DO npq= 1,npnts
      uq= 0.0
      xq= rq*COS(Thetq)          ! x-coordinate of field point
      yq= rq*SIN(Thetq)          ! y-coordinate of field point
      Thetq= Thetq + Delth       ! angle to next field point Q
      Thetp= Pi/2.0              ! angle to first source point P1
      Source_points1: &
      DO npp= 1,npnts
         xp= rp*COS(Thetp)       ! x-coordinate of source point
         yp= rp*SIN(Thetp)       ! y-coordinate of source point
         dxr(1)= xp-xq
         dxr(2)= yp-yq
         r= SQRT(dxr(1)**2 + dxr(2)**2)
         uq= uq + U(r,k,2)*F(npp)
         Thetp= Thetp + Delth    ! angle to next source point P
      END DO  &
      Source_points1
      uq=uq-q/k*yq
      WRITE(11,*) 'Temperature at field point',npq,' =',uq
   END DO  &
   Field_points1
   !    Postprocessing - Interior points
   WRITE(11,*)   ''
   WRITE(11,*)   'Temperatures at interior points:'
   READ(10,*) ninpts             ! read number of interior points
   Int_points:  &
   DO nin= 1,ninpts
      READ(10,*) xi,yi           ! coordinates of interior points
      uq= 0.0
      Thetp= Pi/2.0             ! angle to first source point P1
      Source_points2:  &
      DO npp= 1,npnts
         xp= rp*COS(Thetp)       ! x-coordinate of source point
         yp= rp*SIN(Thetp)       ! y-coordinate of source point
         dxr(1)= xp-xi
         dxr(2)= yp-yi
         r= SQRT(dxr(1)**2 + dxr(2)**2)
         uq= uq + U(r,k,2)*F(npp)
```

```
      Thetp= Thetp + Delth      ! angle to next source point P
   END DO &
   Source_points2
   uq=uq-q/k*yi
   WRITE(11,*) 'Temperature at x=',xi,', y=',yi,' =',uq
END DO &
Int_points
STOP
END PROGRAM Trefftz
```

The system of equations is solved next with the utility program SOLVE. The values of temperature are computed at boundary and interior points, the coordinates of which are specified by the input. Both involve a summation of influences (i.e. fundamental solutions multiplied with the fictitious source intensities).

INPUT DATA for program Trefftz

1.0   Problem specification
       q,k, npnts, rq,rp

                          q ... Heat inflow
                          k ... Thermal conductivity
                          npnts ... Number of points P,Q
                          rq ... Radius of isolator
                          rp ... Radius of sources

2.0   Interior point specification
       Npoints   Number of interior points

3.0   Interior point coordinates (Npoints cards)
       x,y   x,y coordinates of interior points

5.3.1 Sample input and output

Here we show an example of the input for an isolator of radius 1.0 with 32 points P and Q where the source points P are situated along a circle with a radius 0.7.

File INPUT.DAT

```
1.0 1.0 32 1.0 0.7
18
0. -5.
0. -4.5
0. -4.
0. -3.5
0. -3.
```

```
0. -2.5
0. -2.
0. -1.5
0. -1.
0. 1.
0. 1.5
0. 2.
0. 2.5
0. 3.
0. 3.5
0. 4.
0. 4.5
0. 5.
```

File OUTPUT.DAT

```
Program 2 : heat flow past a cylinder with Trefftz method
Heat inflow/outflow=        1.00000
Thermal conductivity=       1.00000
Number of Points P,Q=            32
Radius of Isolator=         1.00000
Radius of Sources =      0.700000
Temperatures at Boundary points:
Temperature at field point          1 =    -1.99996
Temperature at field point          2 =    -1.96060
Temperature at field point          3 =    -1.84570
Temperature at field point          4 =    -1.65967
Temperature at field point          5 =    -1.40969
Temperature at field point          6 =    -1.10540
Temperature at field point          7 =   -0.758505
Temperature at field point          8 =   -0.382387
Temperature at field point          9 =    0.846584E-02
Temperature at field point         10 =    0.398992
Temperature at field point         11 =    0.774144
Temperature at field point         12 =    1.11947
Temperature at field point         13 =    1.42165
Temperature at field point         14 =    1.66906
Temperature at field point         15 =    1.85215
Temperature at field point         16 =    1.96388
Temperature at field point         17 =    1.99993
Temperature at field point         18 =    1.95891
Temperature at field point         19 =    1.84242
Temperature at field point         20 =    1.65493
Temperature at field point         21 =    1.40367
Temperature at field point         22 =    1.09833
Temperature at field point         23 =    0.750663
Temperature at field point         24 =    0.374072
Temperature at field point         25 =   -0.169342E-01
Temperature at field point         26 =   -0.407287
```

```
Temperature at field point              27 =   -0.781947
Temperature at field point              28 =   -1.12648
Temperature at field point              29 =   -1.42760
Temperature at field point              30 =   -1.67371
Temperature at field point              31 =   -1.85533
Temperature at field point              32 =   -1.96546

Temperatures at interior points:
Temperature at x=    0.000000  , y=   -5.00000    =    5.19999
Temperature at x=    0.000000  , y=   -4.50000    =    4.72221
Temperature at x=    0.000000  , y=   -4.00000    =    4.24999
Temperature at x=    0.000000  , y=   -3.50000    =    3.78570
Temperature at x=    0.000000  , y=   -3.00000    =    3.33332
Temperature at x=    0.000000  , y=   -2.50000    =    2.89999
Temperature at x=    0.000000  , y=   -2.00000    =    2.49998
Temperature at x=    0.000000  , y=   -1.50000    =    2.16664
Temperature at x=    0.000000  , y=   -1.00000    =    1.99997
Temperature at x=    0.000000  , y=    1.00000    =   -1.99997
Temperature at x=    0.000000  , y=    1.50000    =   -2.16664
Temperature at x=    0.000000  , y=    2.00000    =   -2.49998
Temperature at x=    0.000000  , y=    2.50000    =   -2.89999
Temperature at x=    0.000000  , y=    3.00000    =   -3.33332
Temperature at x=    0.000000  , y=    3.50000    =   -3.78570
Temperature at x=    0.000000  , y=    4.00000    =   -4.24999
Temperature at x=    0.000000  , y=    4.50000    =   -4.72221
Temperature at x=    0.000000  , y=    5.00000    =   -5.19999
```

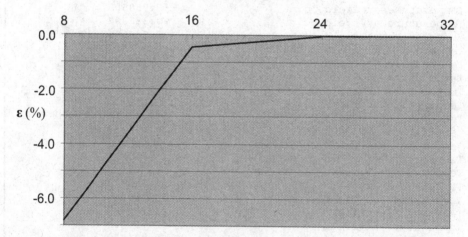

Figure 5.3   Plot of error (ε) in % made in computing the temperature at the top of the isolator plotted against the number of points P (Trefftz method)

The error in the computation of the temperature at the top of the circular isolator (point Q_1) is plotted in Figure 5.3. It can be seen that with 24 elements very accurate results can be obtained.

5.4   DIRECT METHOD

As we have seen from the simple example, the Trefftz method is not suitable for general purpose programming. Not only is the convergence not assured, but it is also not very user-friendly because, in addition to specifying points where boundary conditions are to be satisfied, we have to specify a second set of points where the fictitious forces are to be applied. This is certainly not acceptable, especially if we want to analyse three-dimensional problems.

5.4.1   Theorem of Betti and integral equations

An alternative to the Trefftz method is the direct method. Here we use the well known Betti theorem rather elegantly to get rid of the need to compute *fictitious* sources or forces. We also abolish the need for an additional set of points by placing the source points P to coincide with the field points Q.

This means that the method will become more complicated than Trefftz's, because we will now have to solve a set of integral equations and cope with integrals, which are singular. The direct method, however, is much more user-friendly than Trefftz's method and has the advantage that convergence can be guaranteed. We explain the direct method with an example in elasticity, as engineers associate the Betti theorem with that type of problem. However, we will see that the integral equations derived are equally valid for potential problems.

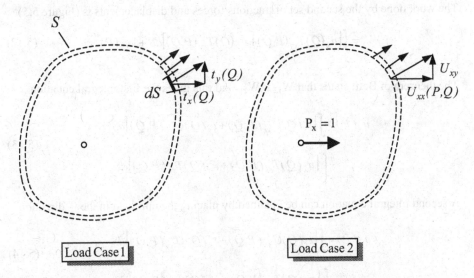

Figure 5.4   Application of Betti's theorem, tractions and displacements for computing W_{12}

Consider an infinite domain with two types of 'loading': *load case* number 1 we assume to be the case we want to solve; and *load case* number 2, a case where only a unit load in the x-direction is specified at a point P (see Figure 5.4). Along a dotted line we show for load case 1 the stresses defined as forces per unit length of the line (dS). These are the *tractions*, which we introduced in the previous chapter, at point Q, with components $t_x(Q)$ and $t_y(Q)$. For load case 2, we show the displacements at point Q on S, which are the fundamental solutions $U_{xx}(P,Q)$ and $U_{xy}(P,Q)$.

A few points must be clarified at this stage: as already mentioned in Chapter 4, we must cut through the continuum to show stresses. Here we cut along a dotted line, which must form a closed contour, but which has been chosen quite arbitrarily. By this cut the continuum is divided into two parts: the interior and exterior domains.

Note that for the following derivation it does not matter which domain is considered and, therefore, the integral equations are valid for infinite as well as finite domains. The only difference will be in the definition of the *outward normal* vector. For the exterior problem (infinite domain), the outward normal will point into the void created by removing the interior region. For the interior domain, the vector will point the opposite way. We will see that the contour chosen here can subsequently be taken to be the boundary of the problem to be solved.

The theorem of Betti states that the work done by the load of case 1 along the displacements of case 2 must equal the work done by the loads of case 2 along the displacements of case 1.

If we assume that there are no body forces acting in the domain (these will be introduced later), the work done by the first set of tractions and displacements is (Figure 5.4)

$$W_{12} = \int_S \left[t_x(Q)U_{xx}(P,Q) + t_y(Q)U_{xy}(P,Q) \right] dS \tag{5.11}$$

The work done by the second set of tractions/forces and displacements is (Figure 5.5)

$$W_{21} = \int_S \left[u_x(Q)T_{xx}(P,Q) + u_y(Q)T_{xy}(P,Q) \right] dS + 1u_x(P) \tag{5.12}$$

The theorem of Betti states that $W_{12} = W_{21}$, and this gives the first integral equation

$$u_x(P) = \int_S \left[t_x(Q)U_{xx}(P,Q) + t_y(Q)U_{xy}(P,Q) \right] dS$$
$$- \int_S \left[u_x(Q)T_{xx}(P,Q) + u_y(Q)T_{xy}(P,Q) \right] dS \tag{5.13}$$

A second integral equation can be obtained by placing the unit load in the y- direction

$$u_y(P) = \int_S \left[t_x(Q)U_{yx}(P,Q) + t_y(Q)U_{yy}(P,Q) \right] dS$$
$$- \int_S \left[u_x(Q)T_{yx}(P,Q) + u_y(Q)T_{yy}(P,Q) \right] dS \tag{5.14}$$

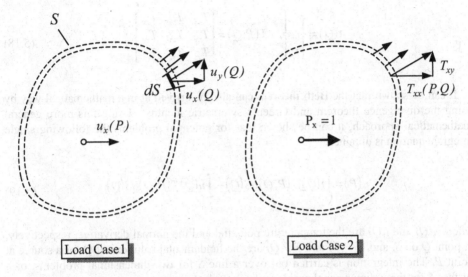

Figure 5.5 Application of Betti's theorem, tractions and displacements for computing W_{21}

using matrix algebra we can combine equations (5.13) and (5.14)

$$\mathbf{u}(P) = \int_S \mathbf{t}^T(Q)\mathbf{U}(P,Q)dS - \int_S \mathbf{u}^T(Q)\mathbf{T}(P,Q)dS \qquad (5.15)$$

where

$$\mathbf{u}(Q) = \begin{Bmatrix} u_x \\ u_y \end{Bmatrix}, \quad \mathbf{U}(P,Q) = \begin{bmatrix} U_{xx} & U_{yx} \\ U_{xy} & U_{yy} \end{bmatrix}$$

$$\mathbf{t}(Q) = \begin{Bmatrix} t_x \\ t_y \end{Bmatrix}, \quad \mathbf{T}(P,Q) = \begin{bmatrix} T_{xx} & T_{yx} \\ T_{xy} & T_{yy} \end{bmatrix} \qquad (5.16)$$

Equations (5.15) represent a system of two (for the two-dimensional problem discussed here) integral equations which directly relate tractions $\mathbf{t}$ and displacements $\mathbf{u}$, thereby removing the need to compute *fictitious* forces.

It is easy to show that for three-dimensional problems, three integral equations (5.15) can be obtained where S is a surface and

$$\mathbf{u}(Q) = \begin{Bmatrix} u_x \\ u_y \\ u_z \end{Bmatrix}, \quad \mathbf{U}(P,Q) = \begin{bmatrix} U_{xx} & U_{yx} & U_{zx} \\ U_{xy} & U_{yy} & U_{zy} \\ U_{xz} & U_{yz} & U_{zz} \end{bmatrix} \qquad (5.17)$$

and

$$\mathbf{t}(Q) = \begin{Bmatrix} t_x \\ t_y \\ t_z \end{Bmatrix}, \quad \mathbf{T}(P,Q) = \begin{bmatrix} T_{xx} & T_{yx} & T_{zx} \\ T_{xy} & T_{yy} & T_{zy} \\ T_{xz} & T_{yz} & T_{zz} \end{bmatrix}$$

(5.18)

It can be shown that the Betti theorem can also be arrived at in a mathematical way by using the divergence theorem and Green's symmetric identity[2]. Using this more general mathematical approach, it can be shown that for potential problems the following single integral equation is obtained:

$$u(P) = \int_S t(Q) U(P,Q) dS(Q) - \int_S u(Q) T(P,Q) dS(Q)$$

(5.19)

where $u(Q)$ and $t(Q)$ are the temperature/potential and the normal derivative, respectively, at point Q on S, and $U(P,Q)$ and $T(P,Q)$ are the fundamental solutions at Q for a source at point P. The integration is carried out over a line S for two-dimensional problems, or a surface S for three-dimensional problems.

5.4.2 Limiting values of integrals as P coincides with Q

We have now succeeded in avoiding computing the fictitious forces, but have not succeeded yet in making the method more user-friendly since we still need two sets of points: points P where the unit sources/loads are applied and points Q where we have to satisfy boundary conditions. Ideally, we would like to have only one set of points on the line where the points Q are specified. The problem is that some integrals in equation s (5.15) or (5.19) only exist in the sense of a limiting value as P approaches Q.

This is explained in Figure 5.6 for two-dimensional potential problems. Here, we examine what happens when points P and Q coincide. We define a region of exclusion around point P, with radius ε and integrate around it. The integrals in equation (5.18) can

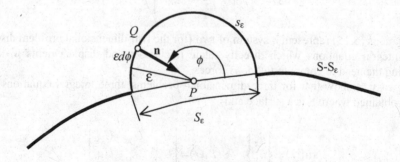

Figure 5.6 Diagram explaining the limiting value of integrals for two-dimensional potential problems

now be split up into integrals over $S-S_\varepsilon$, that is, the part of the curve without the exclusion zone and into integrals over s_ε, that is, the circular boundary. As ε is taken to zero it does not matter if we integrate over s_ε or S_ε, as shown in Figure 5.5.

The right-hand side of equation (5.19) is written as:

$$\int_S t(Q)U(P,Q)dS(Q) - \int_S u(Q)T(P,Q)dS(Q)$$

$$= \int_{S-S_\varepsilon} t(Q)U(P,Q)dS(Q) - \int_{S-S_\varepsilon} u(Q)T(P,Q)dS(Q) \qquad (5.20)$$

$$+ \int_{s_\varepsilon} t(Q)U(P,Q)dS(Q) - \int_{s_\varepsilon} u(Q)T(P,Q)dS(Q)$$

We examine the integrals over s_ε further. Using polar coordinates, as shown, we change the integration limits of the first integral for a smooth surface at P to 0 and π and substitute for the fundamental solution U. Furthermore, as in the limit P will be coincident with Q, we can assume $t(Q)=t(P)$ and $u(Q)=u(P)$. Then we have

$$\int_{s_\varepsilon} t(Q)U(P,Q)dS(Q)=t(P)\int_0^\pi \frac{1}{2\pi k}\ln\left(\frac{1}{\varepsilon}\right)\varepsilon \, d\phi = t(P)\pi\frac{1}{k}\ln\left(\frac{1}{\varepsilon}\right)\varepsilon \qquad (5.21)$$

The integral approaches zero as ε approaches zero. Therefore,

$$\lim_{\varepsilon \to 0}\int_{s_\varepsilon} t(Q)U(P,Q)dS(Q) = 0 \qquad (5.22)$$

The last integral in (5.20) can be written as

$$\int_{s_\varepsilon} u(Q)T(P,Q)dS(Q)=u(P)\int_0^\pi \frac{\cos\theta}{2\pi\varepsilon}\varepsilon \, d\phi = u(P)\int_0^\pi \frac{-1}{2\pi}d\phi = -\frac{1}{2}u(P) \qquad (5.23)$$

As ε cancels out, we do not have to take the limiting value of this integral. The integral equation that has to be used for the case where the source points are located on the continuous line S, is given by

$$\frac{1}{2}u(P)=\lim_{\varepsilon \to 0}\left[\int_{S-S_\varepsilon} t(Q)U(P,Q)dS(Q) - \int_{S-S_\varepsilon} u(Q)T(P,Q)dS(Q)\right] \qquad (5.24)$$

For a three-dimensional problem, we take the zone of exclusion to be a sphere, as shown in Figure 5.6.

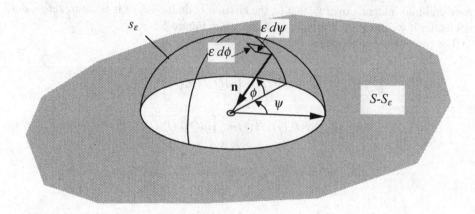

Figure 5.7   Computation of integrals for the case that $P=Q$, three-dimensional case

One can easily verify that the first integral tends to zero as ε approaches zero. The second integral can be computed as

$$\int_{S_\varepsilon} u(Q)T(P,Q)dS(Q) = u(P)\int_0^{2\pi}\int_0^\pi \frac{\cos\theta}{4\pi\varepsilon^2}\,\varepsilon\,d\phi\,\varepsilon\,d\psi = -\frac{1}{2}u(P) \qquad (5.25)$$

which gives the same result as previously.

Obviously, the same limiting procedure can be made for elasticity problems. It is left to the reader to prove that if $P = Q$ the integral equation (5.15) can be rewritten as

$$\frac{1}{2}\mathbf{u}(P) = \lim_{\varepsilon\to 0}\left[\int_{S-S_\varepsilon}\mathbf{t}^T(Q)\mathbf{U}(P,Q)dS(Q) - \int_{S-S_\varepsilon}\mathbf{u}^T(Q)\mathbf{T}(P,Q)dS(Q)\right] \qquad (5.26)$$

If the boundary is not smooth but has a corner, as shown in Figure 5.7, then equation (5.23) has to be modified.

The integration limits are changed and now depend on the angle γ:

$$\int_{S_\varepsilon} u(Q)T(P,Q)dS(Q) = u(P)\int_0^\gamma \frac{\cos\theta}{2\pi\varepsilon}\varepsilon d\phi = u(P)\int_0^\gamma \frac{-1}{2\pi}d\phi = -\frac{\gamma}{2\pi}u(P) \qquad (5.27)$$

A more general integral equation, which also caters for non-smooth boundaries, can be written for potential problems as:

$$cu(P) = \lim_{\varepsilon \to 0} \left[\oint_{S-S_\varepsilon} t(Q)U(P,Q)dS(Q) - \oint_{S-S_\varepsilon} u(Q)T(P,Q)\,dS(Q) \right] \tag{5.28}$$

The reader can easily verify that for 2-D problems

$$c = 1 - \frac{\gamma}{2\pi} \tag{5.29}$$

where γ is defined as the angle subtended at P by s_ε in the limit as ε tends to zero. For 3-D problems the same reasoning can be applied but for general corners expressions are more complicated.

For two- and three-dimensional elasticity problems, we may write a more general form of equation (5.25)

$$\mathbf{cu}(P) = \lim_{\varepsilon \to 0} \left[\int_{S-S\varepsilon} \mathbf{t}^T(Q)\mathbf{U}(P,Q)dS(Q) - \int_{S-S\varepsilon} \mathbf{u}^T(Q)\mathbf{T}(P,Q)dS(Q) \right] \tag{5.30}$$

where $\mathbf{c}$ is a 2 x 2 or 3 x 3 matrix often referred to as "free term".

5.4.3 Solution of integral equations

Using the direct method we have produced a set of integral equations which relates the temperature/potential to the normal gradient or the displacement to the traction at any point Q on the boundary.

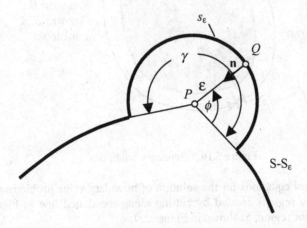

Figure 5.8   Limiting value of integral when P is located on a corner

Since we are now able to place the source points coincidental with the points where the boundary conditions are to be satisfied, we no longer need to be concerned about these points. Indeed, in the direct method, the fictitious sources no longer play any role.

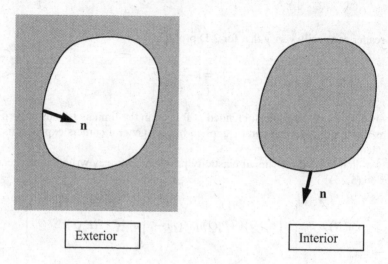

Figure 5.9   Exterior and interior regions obtained by separating domain along dotted line

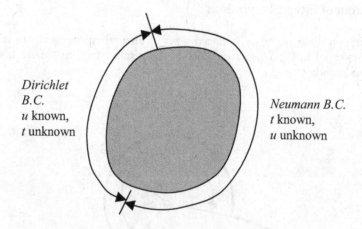

Figure 5.10   Boundary conditions

To use the integral equations for the solution of boundary value problems we consider only one of the two regions created by cutting along the dotted line in Figure 5.4: the interior or the exterior region, as shown in Figure 5.9.

With respect to the integral equations, the only difference between them is the direction of the outward normal **n**, which is assumed to point away from the solid. The interior region is a *finite* region, the exterior an *infinite* region.

For potential problems, we obtain one integral equation per source point P. For elasticity problems we get two or three integral equations per source point, depending on the dimensionality of the problem. Theoretically, if we want to satisfy the boundary conditions exactly at all points on the boundary, we would need an infinite number of points P (and Q). In practice, we will solve the integral equations numerically and attempt to either satisfy the boundary conditions at a limited number of points Q, or specify that some norm of the error in satisfaction of the boundary conditions is a minimum.

For a boundary value problem, either u or t is specified and the other is the unknown to be determined by solving the integral equations. The boundary condition where potential u or displacement **u** is specified is also known as the *Dirichlet* boundary condition, whereas the specification of flow t or traction **t** is often referred to as a *Neumann* boundary condition (see Figure 5.10).

Before we deal with the numerical solution of the integral equations, we must discuss the integrals a little further. As indicated, limiting values of the integrals have to be taken, as the region of exclusion around point P is reduced to zero.

The fundamental solutions or *kernels* of T and U have different types of singularities, which affect this limiting process. The kernel U varies according to $\ln r$ in two dimensions and with $1/r$ in three dimensions and is known as *weakly singular*. As we see later the integration of this function poses no great problems. Kernel T has a $1/r$ singularity in two dimensions and a $1/r^2$ singularity in three dimensions.

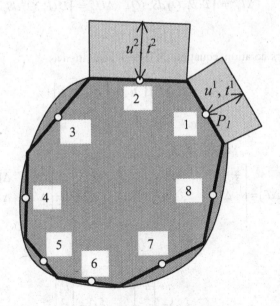

Figure 5.11   Solution of integral equations by linear segments

This is also known by mathematicians as *strongly singular*. The integral of this function only exists in the sense of a *Cauchy principal value*. In engineering terms this means that the integral can be evaluated on a contour around point P only. We will discuss this further in the chapter on numerical implementation.

In the simplest case we may solve the integral equations by dividing the boundary for two-dimensional problems into straight line segments over which the values of u and t are assumed to be constant We assume point P to be located at the centre of each segment. In the example shown in Figure 5.11 we assume the solution of a two-dimensional potential problem with eight segments where either u or t are specified.

We see that this very simple *discretisation* into constant elements violates the continuity conditions between elements. However, we will see by numerical experiments, that the method converges, that is, exact results are obtained as the number of elements tends to infinity. The integrals can now be evaluated over each element separately and the contributions added, that is, equation (5.28) can be rewritten as eight equations (one for each point P)

$$\frac{1}{2}u^e + \sum_{e=1}^{8}\Delta T_i^e u^e = \sum_{e=1}^{8}\Delta U_i^e t^e \quad for \quad i=1,2\ldots8 \tag{5.31}$$

where u^e and t^e is the temperature and flow at the centre of element e. Note that as there is a smooth surface at the centres of the elements (at points P_i) c is assigned 1/2. The integrals over the segments are defined as

$$\Delta T_i^e = \int_{S_e} T(P_i,Q)\,dS_e(Q), \quad \Delta U_i^e = \int_{S_e} U(P_i,Q)\,dS_e(Q) \tag{5.32}$$

Using matrix notation, equation (5.31) can be written as

$$[\Delta T]\{u\} = [\Delta U]\{t\} \tag{5.33}$$

where

$$element \quad nodes \quad \rightarrow$$

$$[\Delta T] = \begin{bmatrix} \frac{1}{2}+\Delta T_1^1 & \Delta T_1^2 & \ldots \\ \Delta T_2^1 & \frac{1}{2}+\Delta T_2^2 & \ldots \\ \ldots & \ldots & \ldots \end{bmatrix}, \quad [\Delta U] = \begin{bmatrix} \Delta U_1^1 & \Delta U_1^2 & \ldots \\ \Delta U_2^1 & \Delta U_2^2 & \ldots \\ \ldots & \ldots & \ldots \end{bmatrix} \tag{5.34}$$

and

$$\{u\} = \begin{Bmatrix} u^1 \\ u^2 \\ . \end{Bmatrix}, \quad \{t\} = \begin{Bmatrix} t^1 \\ t^2 \\ . \end{Bmatrix} \tag{5.35}$$

If we consider the solution of the heat flow problem which we solved by the Trefftz method, then we have a problem where flow $\{t\}_0$ is specified at the boundary and temperatures are unknown. A discretisation of this problem into linear elements is shown in Figure 5.12.

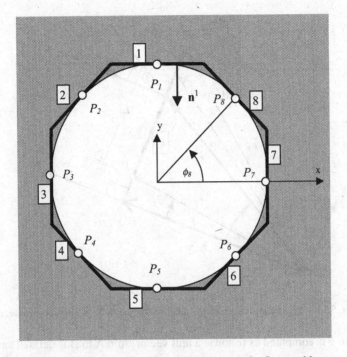

Figure 5.12   Discretisation into linear elements for flow problem

The system of equations can be written as

$$[\Delta T]\{u\} = \{F\} \quad \text{with} \quad \{F\} = [\Delta U]\{t\}_0 \tag{5.36}$$

where the vector $\{t\}_0$ is given by

$$\{t\}_0 = q_0 \begin{Bmatrix} -n_y^1 \\ -n_y^2 \\ \cdot \end{Bmatrix} = q_0 \begin{Bmatrix} sin\phi_1 \\ sin\phi_2 \\ \cdot \end{Bmatrix} \tag{5.37}$$

The integrals which have to be evaluated analytically are

$$\Delta T_i^e = \int\limits_{S_e} T(P_i, Q)\, dS_e(Q) = \int\limits_{S_e} \frac{cos\,\theta}{2\pi\, r}\, dS_e$$

$$\Delta U_i^e = \int\limits_{S_e} U(P_i, Q)\, dS_e(Q) = \int\limits_{S_e} \frac{1}{2\pi} ln\frac{1}{r}\, dS_e \tag{5.38}$$

The integrals can be evaluated using a local coordinate system x', y' through point P and polar coordinates, as shown[3] in Figure 5.13, where θ is defined anticlockwise from a line perpendicular to the element e with start node A and end node B.

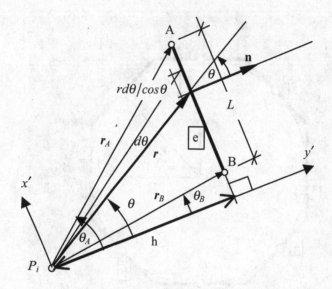

Figure 5.13   Polar coordinate system used for the analytic evaluation of integral ΔT_i^e

The angle θ is computed as follows: a unit vector from A to B is defined as:

$$\mathbf{v}_{AB} = \frac{1}{L} \begin{Bmatrix} x_A - x_B \\ y_A - y_B \end{Bmatrix}$$
(5.39)

The vector normal to element $\mathbf{n}$ is computed by taking the vector x-product of $\mathbf{v}_{AB}$ with the z-axis. This gives

$$\mathbf{n} = \frac{1}{L} \begin{Bmatrix} y_A - y_B \\ -(x_A - x_B) \end{Bmatrix}$$
(5.40)

The cosine and sine of θ are then computed by

$$cos\,\theta = \mathbf{n} \bullet \mathbf{r} \frac{1}{r}$$
(5.41)

and

$$sin\,\theta = \mathbf{v}_{AB} \bullet \mathbf{r} \frac{1}{r}$$
(5.42)

θ can be computed by

$$\theta = \cos_{.}^{-1}\left(\mathbf{n} \bullet \mathbf{r} \frac{1}{r}\right) SIGN\,(\sin\theta) \tag{5.43}$$

The first integral is evaluated as:

$$\Delta T_i^e = \int_{\theta_A}^{\theta_B} \frac{\cos\theta}{2\pi\,r}\,\frac{r d\theta}{\cos\theta} = \int_{\theta_A}^{\theta_B} \frac{1}{2\pi}\,d\theta = \frac{\theta}{2\pi}\Big|_{\theta_A}^{\theta_B} = \frac{1}{2\pi}(\theta_B - \theta_A) \tag{5.44}$$

If P_i is at the centre of element e (i.e. $i = e$) then we have to take the Cauchy principal value of the integral. As shown in Figure 5.14, the integration is carried out over the region of exclusion.

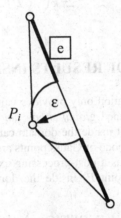

Figure 5.14 Cauchy principal value computation as P_i approaches the centre of element e

The reader may verify that, because of the antisymmetry of T shown in Figure 4.4, we obtain

$$\Delta T_i^i = 0 \tag{5.45}$$

The second integral is computed as

$$\Delta U_i^e = \int_{\theta_A}^{\theta_B} \frac{1}{2\pi\,k}\ln\frac{1}{r}\,\frac{r d\theta}{\cos\theta} = -\int_{\theta_A}^{\theta_B} \frac{h}{2\pi\,k}\ln\left(\frac{h}{\cos\theta}\right)\frac{h d\theta}{\cos^2\theta} =$$

$$-\frac{h}{2\pi}\left[\tan\theta\left(\ln\left(\frac{h}{\cos\theta}\right)-1\right)+\theta\right]_{\theta_A}^{\theta_B} \tag{5.46}$$

where $r = h/\cos\theta$ has been substituted.

For programming purposes, it may be convenient to write this expression in terms of r and θ:

$$\Delta U_i^e = -\frac{1}{2\pi\,k}\left[\; r\sin\theta\,(\ln r -1)+\theta\,r\cos\theta \;\right]_{\theta_A,r_A}^{\theta_B,r_B}$$

(5.47)

If P_i is at the centre of element e of length L, then we have

$$h \rightarrow 0, \quad \theta_B = -\theta_A \rightarrow {}^{\pi}\!/_2, \quad r_B = r_A = {}^{L}\!/_2$$

(5.48)

and the diagonal coefficient is computed as

$$\Delta U_2^2 = -\frac{L}{2\pi\,k}\left(ln\left(\frac{L}{2}\right)-1\right)$$

(5.49)

5.5   COMPUTATION OF RESULTS INSIDE THE DOMAIN

The solution of the integral equation only provides values of u and t on the boundary of the domain. Since we have defined *global shape functions* in the form of fundamental solutions, the results at any point inside the domain can be readily computed. In contrast to the FEM, where results at all node or Gauss points are computed as part of the solution, we compute the interior results as a post-processing exercise. To compute, for example, the temperature/potential at a point P_a inside the domain, we simply rewrite equation (5.19)

$$u(P_a)=\int_S t(Q)U(P_a,Q)dS(Q)-\int_S u(Q)T(P_a,Q)dS(Q)$$

(5.50)

or in discretised form using the subdivisioninto line segments of Figure 5.12

$$u(P_a)= \sum_{e=1}^{8} \Delta T^e(P_a)u^e - \sum_{e=1}^{8} \Delta U^e(P_a)t^e$$

(5.51)

where

$$\Delta T^e = \int_{S_e} T(P_a,Q)\,dS_e(Q) \quad , \quad \Delta U^e = \int_{S_e} U(P_a,Q)\,dS_e(Q)$$

(5.52)

The flows at P_a in the x- and y- directions are computed by taking the derivative of (5.50)

$$q_x(P_a)=-k\frac{\partial u}{\partial x}(P_a)=-k\left(\int_S t(Q)\frac{\partial U}{\partial x}(P_a,Q)dS(Q)-\int_S u(Q)\frac{\partial T}{\partial x}(P_a,Q)dS(Q)\right)$$

$$q_y(P_a)=-k\frac{\partial u}{\partial y}(P_a)=-k\left(\int_S t(Q)\frac{\partial U}{\partial y}(P_a,Q)dS(Q)-\int_S u(Q)\frac{\partial T}{\partial y}(P_a,Q)dS(Q)\right)$$

(5.53)

where the derivatives of U have been presented previously and the derivatives of T are given for two-dimensional problems as

$$\frac{\partial T}{\partial x} = \frac{\partial}{\partial x}\left[n_x \frac{\partial U}{\partial x} + n_y \frac{\partial U}{\partial y}\right] = \frac{\cos 2\theta}{\pi r^2}$$

(5.54)

$$\frac{\partial T}{\partial y} = \frac{\partial}{\partial y}\left[n_x \frac{\partial U}{\partial x} + n_y \frac{\partial U}{\partial y}\right] = \frac{\sin 2\theta}{\pi r^2}$$

For constant boundary elements, equation (5.53) and the mesh in Figure 5.12 this can be replaced by

$$q_x(P_a) = -k\left(\sum_{e=1}^{8} \Delta S_{xa}^e t^e - \sum_{e=1}^{8} \Delta R_{xa}^e u^e\right)$$

(5.55)

$$q_y(P_a) = -k\left(\sum_{e=1}^{8} \Delta S_{ya}^e t^e - \sum_{e=1}^{8} \Delta R_{ya}^e u^e\right)$$

where the integrals

$$\Delta S_{xa}^e = \int_{S_e} \frac{\partial U}{\partial x}(P_a, Q)dS(Q), \quad \Delta S_{ya}^e = \int_{S_e} \frac{\partial U}{\partial y}(P_a, Q)dS(Q)$$

$$\Delta R_{xa}^e = \int_{S_e} \frac{\partial T}{\partial x}(P_a, Q)dS(Q), \quad \Delta R_{ya}^e = \int_{S_e} \frac{\partial T}{\partial y}(P_a, Q)dS(Q)$$

(5.56)

can be evaluated analytically over element e.

Using the notation in Figure 5.13 with node P_i replaced by P_a we can evaluate the integrals analytically in terms of the local coordinates x' and y'.

$$\Delta S_{x'a}^e = \frac{1}{2\pi k}(\theta_B - \theta_A)$$

$$\Delta S_{y'a}^e = \frac{1}{2\pi k}\ln(\cos\theta_B / \cos\theta_A)$$

(5.57)

$$\Delta R_{x'a}^e = \frac{1}{2\pi h}(\cos\theta_B \sin\theta_B - \cos\theta_A \sin\theta_A)$$

$$\Delta R_{y'a}^e = \frac{1}{2\pi h}(\cos^2\theta_B - \cos^2\theta_A)$$

The contribution of element e to the flux in the x'- and y'- direction is given as:

$$q_{x'}^e(P_a) = -k\left(\Delta S_{x'a}^e\, t^e - \Delta R_{x'a}^e\, u^e\right)$$
$$q_{y'}^e(P_a) = -k\left(\Delta S_{y'a}^e\, t^e - \Delta R_{y'a}^e\, u^e\right) \tag{5.58}$$

This has to be transformed into global directions x, y by

$$q_x^e(P_a) = q_{x'}^e n_x - q_{y'}^e n_y$$
$$q_y^e(P_a) = q_{x'}^e n_y + q_{y'}^e n_x \tag{5.59}$$

where n_x, n_y are the components of the vector normal to element e (**n**).
The final fluxes are computed by summing all element contributions

$$q_x(P_a) = \sum_{e=1}^{E} q_x^e; \quad q_y(P_a) = \sum_{e=1}^{E} q_y^e \tag{5.60}$$

5.6   PROGRAM 5.2: FLOW AROUND CYLINDER, DIRECT METHOD

We can now write a computer program for the solution of the flow around a cylinder problem which was previously solved with the Trefftz method. The input section of the program is very similar to Program 5.2, except that no source points have to be specified. The circle is divided into *nseg* straight line segments. At the centre of each segment the boundary condition t_0 is specified. The coefficient matrices, equation (5.34), are set up with the results of the analytical integration, as computed in Section 5.4.3. In setting up the coefficient matrices, we distinguish between diagonal and off-diagonal coefficients. The diagonal coefficients are computed for the case where points P_i are the element centre over which the integartion is carried out.

```
PROGRAM Direct_Method
!---------------------------------------
!   Program to compute the heat flow  past a cylindrical isolator
!   in an 2-D infinite domain using the direct BE method
!   with constant line segments
!---------------------------------------
USE Utility_lib                 !   subroutine to solve equations
REAL            :: q            !   inflow/outflow
REAL            :: k            !   Thermal conductivity
INTEGER         :: nseg         !   Number of segments
REAL            :: rq           !   radius of isolator (inner)
REAL            :: rqo          !   radius of isolator (outer)
REAL(KIND=8),ALLOCATABLE :: Lhs(:,:) ! [DT]
REAL(KIND=8),ALLOCATABLE :: F(:)  ! {F}
REAL(KIND=8),ALLOCATABLE :: u(:)  ! Temp at segment centers
REAL,ALLOCATABLE :: Rhs(:,:)      ! [DU]
```

```fortran
REAL,ALLOCATABLE :: t0(:)                    !      Applied flows
REAL,ALLOCATABLE :: xA(:,:),xB(:,:) !     Start/end coords of seg
REAL,ALLOCATABLE :: xS(:,:)          !      Coords of points Pi
REAL,ALLOCATABLE :: Ve(:,:),Vn(:,:)  !  Vectors A-B and n
REAL :: vrA(2),vrB(2)   ! Vectors to point A and B of seg
REAL :: lens            !  Length of segment
C= 0.5/Pi
!    Read in job information
OPEN(UNIT=10,FILE='INPUT.DAT',STATUS='OLD',ACTION='READ')
OPEN(UNIT=11,FILE='OUTPUT.DAT',STATUS='UNKNOWN',ACTION='WRITE')
READ(10,*) q,k,nseg,rq
WRITE(11,*) 'Heat flow past a cylinder (direct BE method)'
WRITE(11,*) 'Input values:'
WRITE(11,*) ' Heat inflow/outflow= ',q
WRITE(11,*) ' Thermal conductivity=',k
WRITE(11,*) ' Radius of Isolator=  ',rq
WRITE(11,*) ' Number of segments=  ',nseg
!   allocate arrays
ALLOCATE (Lhs(nseg,nseg),Rhs(nseg,nseg),F(nseg))
ALLOCATE (xA(2,nseg),xB(2,nseg),t0(nseg),u(nseg))
ALLOCATE (xS(2,nseg),ve(2,nseg),vn(2,nseg))
C1=0.5/(Pi*k)
Delth= 2.0*Pi/nseg         !      increment in angle theta
rqo=rq/COS(Delth/2.0)  !      outer radius of isolator
Thet= (Pi-Delth)/2.0
!   Compute start/end  coordinates of segments
xA(1,1)= rqo*COS(Thet)
xA(2,1)= rqo*SIN(Thet)
Segments: &
DO ns= 1,nseg-1
 Thet= Thet + Delth
 xB(1,ns)= rqo*COS(Thet)
 xB(2,ns)= rqo*SIN(Thet)
 xA(1,ns+1)= xB(1,ns)
 xA(2,ns+1)= xB(2,ns)
END DO &
Segments
xB(1,nseg)= xA(1,1)
xB(2,nseg)= xA(2,1)
!   Compute centres of segments
Segments1: &
DO ns= 1,nseg
 xS(1,ns)= (xB(1,ns) + xA(1,ns))/2.0
 xS(2,ns)= (xB(2,ns) + xA(2,ns))/2.0
END DO &
Segments1
!   Compute applied tractions at centers of elements
Thet= Pi/2.0
Segments2: &
DO ns= 1,nseg
 t0(ns)= q*SIN(Thet)
 Thet= Thet + Delth
END DO  &
```

```
Segments2
!    Assemble matrices DT and DU
Segments3: &
DO ns=1,nseg
   lens= dist(xA(:,ns),xB(:,ns),2)
   !    Vector parallel and normal to segment A-B
   dx= xA(1,ns) - xB(1,ns)
   dy= xA(2,ns) - xB(2,ns)
   ve(1,ns)= dx/lens
   ve(2,ns)= dy/lens
   vn(1,ns)= ve(2,ns)
   vn(2,ns)=-ve(1,ns)
Points_Pi: &
DO np=1,nseg
  rA= Dist(xA(:,ns),xS(:,np),2)
  rB= Dist(xB(:,ns),xS(:,np),2)
  vrA(1)= xA(1,ns)- xS(1,np)
  vrA(2)= xA(2,ns)- xS(2,np)
  vrB(1)= xB(1,ns)- xS(1,np)
  vrB(2)= xB(2,ns)- xS(2,np)
  COSThA= DOT_PRODUCT(vn(:,ns),vrA)/rA
  COSThB= DOT_PRODUCT(vn(:,ns),vrB)/rB
  SINThA= DOT_PRODUCT(ve(:,ns),vrA)/rA
  SINThB= DOT_PRODUCT(ve(:,ns),vrB)/rB
  ThetA= ACOS(COSThA)*SIGN(1.0,SinThA)
  ThetB= ACOS(COSThB)*SIGN(1.0,SinThB)
    IF(np == ns) THEN             ! Diagonal coefficients
      Lhs(np,np)=  0.5
      Rhs(np,np)= lens*C1*(LOG(lens/2.0)-1.0)
    ELSE                          ! off-diagonal coeff.
      Lhs(np,ns)= C*(ThetB-ThetA)
      Rhs(np,ns)= C1*(rB*SINThB*(LOG(rB)-1)+ThetB*rB*COSThB &
                  - rA*SINThA*(LOG(rA)-1)-ThetA*rA*COSThA)
    END IF
  END DO  &
  Points_Pi
END DO &
Segments3
F= MATMUL(Rhs,t0)     !    compute right hand side vector
CALL Solve(Lhs,F,u)   !    solve system of equations
!    output computed temperatures
WRITE(11,*) 'Temperatures at segment centers:'
Segments4: &
DO ns= 1,nseg
  WRITE(11,'(A,I5,A,F10.3)') &
' Segment',ns,'   T=',u(ns)-q/k*xS(2,ns)
END DO &
Segments4
DEALLOCATE (xS)
!   Compute Temperatures and flows at interior points
READ(10,*,IOSTAT=IOS) Npoints
IF(NPoints == 0 .OR. IOS /= 0) THEN
```

```
  PAUSE 'program Finished'
  STOP
END IF
ALLOCATE (xS(2,NPoints)) !    re-use array Xs
WRITE(11,*) &
'Temperatures(T) and flow (q-x,q-y) at interior points:'
DO n=1,NPoints
  READ(10,*) xS(1,n),xS(2,n)
END DO
Interior_points: &
DO np=1,Npoints
   up= 0.0
   qx= 0.0
   qy= 0.0
   Segments5 : &
   DO ns=1,nseg
     rA= Dist(xA(:,ns),xS(:,np),2)
     rB= Dist(xB(:,ns),xS(:,np),2)
     vrA(1)= xA(1,ns)- xS(1,np)
     vrA(2)= xA(2,ns)- xS(2,np)
     vrB(1)= xB(1,ns)- xS(1,np)
     vrB(2)= xB(2,ns)- xS(2,np)
     COSThA= -DOT_PRODUCT(vn(:,ns),vrA)/rA
     COSThB= -DOT_PRODUCT(vn(:,ns),vrB)/rB
     SINThA= -DOT_PRODUCT(ve(:,ns),vrA)/rA
     SINThB= -DOT_PRODUCT(ve(:,ns),vrB)/rB
     H= RA*CosThA
     ThetA= ACOS(COSThA)*SIGN(1.0,SinThA)
     ThetB= ACOS(COSThB)*SIGN(1.0,SinThB)
     IF(ThetB-ThetA > Pi) ThetA= 2.0*Pi + ThetA ! θB-θA < 180°
     dT= C*(ThetB-ThetA)
     dU= C1*(rB*SINThB*(LOG(rB)-1)+ThetB*rB*COSThB &
     - rA*SINThA*(LOG(rA)-1)-ThetA*rA*COSThA)
     dSx= C/k*(ThetB-ThetA)
     Fact= CosthB/CosthA
     IF(Fact > 0.0) THEN
        dSy= -C/k*LOG(Fact)
     ELSE
        dSy= 0.
     END IF
     dRx= -C/H*(costhB*SINThB - cosThA*sinThA)
     dRy= C/H*(costhB**2 - cosThA**2)
     up= up + dU*t0(ns) - dT*u(ns)
     qxp= -k*(dSx*t0(ns)-dRx*u(ns))               !    q-x'
     qyp= -k*(dSy*t0(ns)-dRy*u(ns))               !    q-y'
     qx= qx + qxp*vn(1,ns) - qyp*vn(2,ns)
     qy= qy + qxp*vn(2,ns) + qyp*vn(1,ns)
   END DO &
   Segments5
   Up= Up - q/k*xS(2,np) !    superimpose solutions
   qy= qy + q
```

```
WRITE(11,'(5(A,F10.3))') &
'x=',xS(1,np),', y=',xS(2,np),', T=',up,',  q-x=',qx,',  q
y=',qy
END DO  &
Interior_points
STOP
END PROGRAM Direct_Method
```

INPUT DATA for program Direct_method

1.0   Problem specification
   q,k, nseg, rq

          q ... Heat inflow
          k ... Thermal conductivity
          nseg ... Number of segments
          rq ... Radius of isolator

2.0   Interior point specification
   Npoints

          Number of interior points

3.0   Interior point coordinates (Npoints cards)
   x,y

          x,y coordinates of interior points

5.6.1 Sample input and output

Here we show the input file for the calculation of the problem in Figure 5.12 with 16 segments and interior points along a horizontal and vertical line.

File INPUT.DAT :

```
1.0 1.0 16 1.0
28
0. 1.
0. 1.5
0. 2.
0. 2.5
0. 3.
0. 3.5
4.
0. 4.5
0. 5.
1. 0.
1.5 0.
2. 0.
2.5 0.
0.
3.5 .0
4. 0.
```

```
4.5 0.
5. 0.
5.5 0.
6. 0.
6.5 0.
7. 0.
7.5 0.
8. 0.
8.5 0.
9. 0.
9.5 0.
10.
```

File OUTPUT.DAT:

```
Heat flow past a cylinder (direct BE method)
Input values:
Heat inflow/outflow=      1.00000
Thermal conductivity=     1.00000
Radius of Isolator=       1.00000
Number of segments=            16
Temperatures at segment centers:
 Segment    1   T=     -2.026
 Segment    2   T=     -1.872
 Segment    3   T=     -1.433
 Segment    4   T=     -0.775
 Segment    5   T=      0.000
 Segment    6   T=      0.775
 Segment    7   T=      1.433
 Segment    8   T=      1.872
 Segment    9   T=      2.026
 Segment   10   T=      1.872
 Segment   11   T=      1.433
 Segment   12   T=      0.775
 Segment   13   T=      0.000
 Segment   14   T=     -0.775
 Segment   15   T=     -1.433
 Segment   16   T=     -1.872
Temperatures(T) and flow (q-x,q-y) at interior points:
x=0.000, y=1.000, T=    -2.026, q-x=     0.000, q-y=      0.032
x=0.000, y=1.500, T=    -2.184, q-x=     0.000, q-y=      0.543
x=0.000, y=2.000, T=    -2.513, q-x=     0.000, q-y=      0.743
x=0.000, y=2.500, T=    -2.910, q-x=     0.000, q-y=      0.836
x=0.000, y=3.000, T=    -3.342, q-x=     0.000, q-y=      0.886
x=0.000, y=3.500, T=    -3.793, q-x=     0.000, q-y=      0.916
x=0.000, y=4.000, T=    -4.257, q-x=     0.000, q-y=      0.936
x=0.000, y=4.500, T=    -4.728, q-x=     0.000, q-y=      0.949
x=0.000, y=5.000, T=    -5.205, q-x=     0.000, q-y=      0.959
```

```
x=1.000,  y=0.000,  T=      0.000,   q-x=      0.000,   q-y=      1.391
x=1.500,  y=0.000,  T=      0.000,   q-x=      0.000,   q-y=      1.455
x=2.000,  y=0.000,  T=      0.000,   q-x=      0.000,   q-y=      1.257
x=2.500,  y=0.000,  T=      0.000,   q-x=      0.000,   q-y=      1.164
x=3.000,  y=0.000,  T=      0.000,   q-x=      0.000,   q-y=      1.114
x=3.500,  y=0.000,  T=      0.000,   q-x=      0.000,   q-y=      1.084
x=4.000,  y=0.000,  T=      0.000,   q-x=      0.000,   q-y=      1.064
x=4.500,  y=0.000,  T=      0.000,   q-x=      0.000,   q-y=      1.051
x=5.000,  y=0.000,  T=      0.000,   q-x=      0.000,   q-y=      1.041
x=5.500,  y=0.000,  T=      0.000,   q-x=      0.000,   q-y=      1.034
x=6.000,  y=0.000,  T=      0.000,   q-x=      0.000,   q-y=      1.029
x=6.500,  y=0.000,  T=      0.000,   q-x=      0.000,   q-y=      1.024
x=7.000,  y=0.000,  T=      0.000,   q-x=      0.000,   q-y=      1.021
x=7.500,  y=0.000,  T=      0.000,   q-x=      0.000,   q-y=      1.018
x=8.000,  y=0.000,  T=      0.000,   q-x=      0.000,   q-y=      1.016
x=8.500,  y=0.000,  T=      0.000,   q-x=      0.000,   q-y=      1.014
x=9.000,  y=0.000,  T=      0.000,   q-x=      0.000,   q-y=      1.013
x=9.500,  y=0.000,  T=      0.000,   q-x=      0.000,   q-y=      1.011
x=10.000  y=0.000,  T=      0.000,   q-x=      0.000,   q-y=      1.010
```

The error in the temperature at segment 1 versus the number of segments is plotted in Figure 5.14. It can be seen that the error falls below 1% for 24 segments. A plot of the flow in vertical direction along a horizontal line versus the number of segments is shown in Figure 5.16. The theoretical value of q_y should approach the value of 2.0 exactly on the boundary. It can be seen that as the boundary is approached, the values are significantly in error, and that this error has to do with the element size adjacent to the interior point.

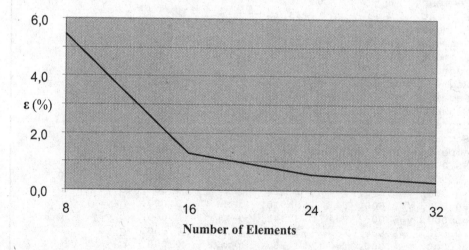

Figure 5.15   Error ε in % for the temperature at segment 1 (Direct method)

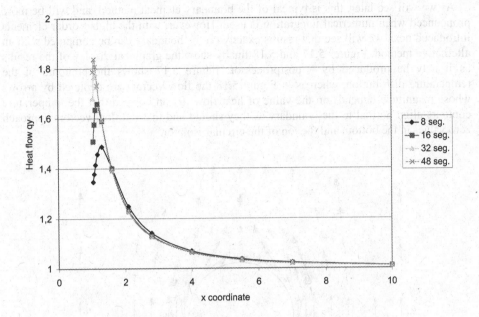

Figure 5.16   Variation of y-component of heat flow vector along horizontal line

Figure 5.17   Flow past a cylindrical isolator: contour lines of temperature

As we will see later, this is typical of the boundary element method, and will be more pronounced when numerical integration is used. However, with the higher order elements introduced next, we will see that results exactly on the boundary can be computed with an alternative method. Figures 5.17 and 5.18 finally show the graphical display of the results as it may be produced by a postprocessor. Figure 5.17 shows the contours of the temperature distribution, whereas in Figure 5.18 the flow vectors are depicted by arrows whose magnitude depends on the value of heat flow. It can be seen that the temperature contours align normal to the boundary as they should and that the flow vectors approach zero values at the bottom and the top of the circular isolator.

Figure 5.18   Flow past a cylindrical isolator: flow vectors

5.7   CONCLUSIONS

In this chapter we have introduced the Trefftz and boundary integral equation methods. Although we found that the Trefftz method is not suitable for general purpose programming, it can, because of its simplicity, be used to demonstrate the basic principles involved. A short program can be written and used for numerical experiments.

As the original idea by Trefftz, conceived in the days before computers, was not found to be suitable, improvements to the methods were sought. This lead initially to the so-called indirect method, in which the sources are distributed over a line or area instead of being concentrated at a point. This allowed us, using similar limiting procedures as shown in this chapter, to place the sources at the boundary, therefore alleviating the need for two sets of points. We have not discussed this method here because it has been largely superseded by the direct method, which avoids the computation of *fictitious* sources/forces altogether.

Using the well known theorem of Betti, we developed boundary integral equations relating tractions to displacements, or temperatures/potentials to normal gradients. We found that using a limiting procedure, the source points can be placed on the boundary to be coincidental with the points where we satisfy given boundary conditions, thereby rendering the method usable for general purpose programming. However, we find that evaluating some boundary integrals will cause difficulties, since the integrands tend to infinity at certain points. Some of the integrals exist only in the sense of a principal value. Indeed, the difficult mathematics involved, which may have prevented matching the success of the FEM in the early days, stems from the difficulty in evaluating these integrals. If simple elements, that is, line segments, such as those used here for solving the 2-D heat flow problem or triangular elements, such as they may be used for three-dimensional problems are defined, where the known boundary condition and the unknown are assumed to be constant, then the integration can be carried out analytically, although for 3-D elasticity the integrals become quite involved. However, even for the simple heat flow example, we find that these constant elements are not very accurate, and many elements are needed to model a smooth surface.

To the author's best knowledge, it was Lachat and Watson[4] who first thought of introducing isoparametric boundary elements, which were of the same type as the ones already in use at that time in the FEM and are thought to be due to Ergatoudis[5], although the basic concept can be found in old mathematics books. The method, previously known as the *Boundary Integral Equation* method, now became the *Boundary Element Method (BEM)*. Now, of course, analytical integration is no longer a feasible way of computing the coefficients of the system of equations and we have to revert to numerical integration. For engineers, who usually find no pleasure in writing pages of analytical evaluation of integrals, this of course was a Godsend. Using the Gauss integration method introduced in Chapter 3, the evaluation of the integrals can now be reduced to evaluating sums. However, because of the nature of the integrals we must be very careful that the accuracy is adequate. In contrast to the FEM, where less may be better (i.e. the application of reduced integration for the evaluation of element stiffness) we will find that the BEM is much less forgiving when it comes to the accuracy of the integrals.

The boundary element method using higher order isoparametric elements is the method used almost exclusively in modern general purpose computer programs. We will therefore deal in some depth with the numerical implementation of the method in the next chapter.

5.8   EXERCISES

Exercise 5.1
Using Program 5.1 (Trefftz method), find out the influence of the following on the accuracy of results:

(a) when the distance between source points P and field points Q is reduced to ½ and ¼ of the value used in Section 5.3.

(b) when the number of points P,Q is increased to twice and three times the value used in Section 1.

Exercise 5.2

Expand Program 5.1 (Trefftz Method), so that flow vectors **q** may be computed at interior points.

Exercise 5.3

Use program 5.2 (Direct_method) to compute the heat flow problem solved by the Trefftz method. Investigate the influence of the number of segments on results by using 8, 16 and 32 segments. Plot the norm of the error to show convergence.

Exercise 5.4

Modify Program 5.1, so that potential problems for general boundary shapes can be analysed by allowing points P and Q to be specified as input instead of being generated automatically. Test the program by analysing the flow past an elliptical isolator.

Exercise 5.5

Modify Program 5.2, so that potential problems for general boundary shapes can be analysed by allowing boundary segments and boundary conditions to be specified as input. Test the program by analysing the flow past an elliptical isolator.

5.9   REFERENCES

1. Trefftz, E. (1926) *Ein Gegenstück zum Ritzschen Verfahren*. Proc. 2nd Int. Congress in Applied Mechanics, Zürich, p.131.
2. Beer, G. and Watson, J.O. (1995) *Introduction to Finite and Boundary Element Methods for Engineers*. Wiley, Chichester.
3. Banerjee, P.K. (1994) *Boundary Element Methods in Engineering Science*. 2nd ed. McGraw Hill, New York.
4. Lachat, J.C. and Watson, J.O. (1976) Effective numerical treatment of boundary integral equations. *Int. J. Num. Meth. Eng.* **10**, 991-1005.
5. Ergadoudis, J.G., Irons, B.M. and Zienkiewicz, O.C. (1968) Curved, isoparametric quadrilateral elements for finite element analysis. *Int. J. Solids & Struct.* **4**, 31-42.

6

Boundary Element Methods – Numerical Implementation

There is nothing more powerful than an idea whose time has come

V. Hugo

6.1 INTRODUCTION

In the previous chapter, we derived boundary integral equations that relate the known boundary conditions to the unknowns. These integral equations can only be solved numerically. The simplest numerical implementation is using line elements, where the knowns and unknowns are assumed to be constant inside the element. This enables the integrals over the elements to be evaluated analytically. In the previous chapter, we presented this type of element for the solution of two-dimensional potential problems. The analytical evaluation, however, would become quite cumbersome for two- and three-dimensional elasticity problems. Constant elements were used in the early days of the development where the method was known as the *Boundary Integral Equation* (BIE) Method[1]. This parallels the development of the FEM where triangular and tetraheder elements, with exact integration, were used. In 1968, Ergatoudis and Irons[2] suggested that isoparametric finite elements and numerical integration could be used to obtain better results, with fewer elements. The concept of higher order elements and numerical integration is very appealing to engineers because it alleviates the need for tedious analytical integration and, more importantly, it allows the writing of general purpose software with a choice of element types. Indeed, this concept will allow us to develop one single program which can solve two and three-dimensional problems in elasticity and potential flow, or any other problem for which we can supply a fundamental solution.

The concept of using isoparametric boundary elements, first introduced by Lachat and Watson[3], prompted a change of name to *Boundary Element Methods*. Therefore, this chapter is about the numerical implementation of this method, using the concepts already discussed in detail in Chapter 3.

6.2   DISCRETISATION AND ISOPARAMETRIC ELEMENTS

Here we introduce the concept of solving integral equations by using higher order elements, where the known and the unknown boundary values are assumed to vary between the nodes of an element, either as linear or quadratic functions. Recall from Chapter 3 that we have, for a one-dimensional isoparametric element and for potential problems, the following interpolations:

$$\mathbf{x}(\xi)=\sum N_n(\xi)\mathbf{x}_n^e \quad \textit{Geometry}$$

$$u(\xi)=\sum N_n(\xi)u_n^e \quad \textit{Temperature / Potential} \tag{6.1}$$

$$t(\xi)=\sum N_n(\xi)t_n^e \quad \textit{Flux}$$

Consider the example in Figure 6.1, where the boundary of a two-dimensional potential problem is divided into linear isoparametric elements.

In equation (6.1) we define a local numbering of the element node n as explained in Chapter 3 for each element. In order to enforce continuity conditions, we define a global numbering of the nodes. That is, we define a global vector containing the temperatures at all nodes:

$$\mathbf{u} = \begin{Bmatrix} u_1 \\ u_2 \\ \vdots \end{Bmatrix} \tag{6.2}$$

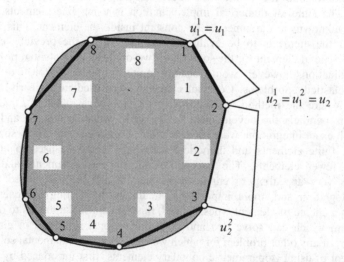

Figure 6.1   Discretisation of two-dimensional problem into linear boundary elements

The local and global numbering are related to each other by the element *connectivity* or *incidences*. For example, element 1 has connectivity vector {1,2}, which means that the values of u for the two nodes of the element appear at the first and second position in the global vector $\mathbf{u}$. Although we usually wish to enforce continuity of u, this is not necessary for t, the boundary flux, which may be discontinuous.

We now consider the numerical treatment of the integral equation:

$$cu(P)=\lim_{\varepsilon\to0}\left[\int_{S-S_\varepsilon}t(Q)U(P,Q)dS(Q)-\int_{S-S_\varepsilon}u(Q)T(P,Q)dS(Q)\right] \qquad (6.3)$$

Substituting equations (6.1) for $t(Q)$ and $u(Q)$, and splitting the integrals into a sum of integrals over elements gives (leaving out the limiting value process, which we now implicitly assume)

$$cu(P)=\sum_{e=1}^{E}\int_{S_e}\left(\sum_{n=1}^{N}N_n t_n^e\right)U(P,\xi)dS(\xi)-\sum_{e=1}^{E}\int_{S_e}\left(\sum_{n=1}^{N}N_n u_n^e\right)T(P,\xi)dS(\xi) \qquad (6.4)$$

where E is the total number of elements and N is the number of nodes per element. The process is generally known as *discretisation*. Since t_n^e and u_n^e, being nodal values, are constant with respect to the integration, they can be taken out of the integral, and equation (6.4) can be rewritten as:

$$cu(P)=\sum_{e=1}^{E}\sum_{n=1}^{N}t_n^e\int_{S_e}N_n(\xi)U(P,\xi)dS(\xi)-\sum_{e=1}^{E}\sum_{n=1}^{N}u_n^e\int_{S_e}N_n(\xi)T(P,\xi)dS(\xi) \qquad (6.5)$$

The integration has now been changed to a sum of integrations of Kernel shape function products over elements and Q has been replaced by the local coordinate ξ.

Theoretically, Betti's theorem should be valid for any location P and, therefore, we can write equation (6.5) for an infinite number of points P_i. In practice, we select a limited number of points only. Since for potential problems either t or u must be known on the boundary, there will be as many unknowns as there are nodes. In the simplest numerical method, also known as *point collocation*, we therefore obtain the necessary integral equations by placing points P_i at I nodes of the mesh :

$$cu(P_i)=\sum_{e=1}^{E}\sum_{n=1}^{N}t_n^e\int_{S_e}N_n(\xi)U(P_i,\xi)dS(\xi)$$

$$-\sum_{e=1}^{E}\sum_{n=1}^{N}u_n^e\int_{S_e}N_n(\xi)T(P_i,\xi)dS(\xi) \quad i=1,2\ldots I \qquad (6.6)$$

This would mean, however, that the theorem by Betti is only satisfied for certain locations of P and, therefore, we may introduce some error by this assumption. We will discuss this in more detail later. Other more accurate methods exist for example, where we attempt to reduce the error in the satisfaction of the Betti theorem over a region to a minimum. These methods are also known as *weighted residual methods*, of which the *Galerkin* method is a special case, where the shape functions are used as weighting functions. The *Galerkin* method will not be discussed here because it has been found that the additional complexity and computational effort introduced does not result in the expected increase in accuracy[4].

Equation (6.6) can be rewritten as

$$cu(P_i) + \sum_{e=1}^{E}\sum_{n=1}^{N} \Delta T_{ni}^e u_n^e = \sum_{e=1}^{E}\sum_{n=1}^{N} \Delta U_{ni}^e t_n^e \quad i = 1,2....I \tag{6.7}$$

where

$$\Delta U_{ni}^e = \int_{S_e} N_n(\xi) U(P_i,\xi) dS(\xi) \quad , \quad \Delta T_{ni}^e = \int_{S_e} N_n(\xi) T(P_i,\xi) dS(\xi) \tag{6.8}$$

As explained previously, limiting values of the integrals have to be taken.

For elasticity problems, the integral equation which has to be discretised is given as

$$\mathbf{cu}(P) = \lim_{\varepsilon \to 0} \left[\int_{S-S\varepsilon} \mathbf{t}^T(Q)\mathbf{U}(P,Q)dS(Q) - \int_{S-S\varepsilon} \mathbf{u}^T(Q)\mathbf{T}(P,Q)dS(Q) \right] \tag{6.9}$$

In discretised form, this equation is written as

$$\mathbf{cu}(P_i) + \sum_{e=1}^{E}\sum_{n=1}^{N} \Delta \mathbf{T}_{ni}^e \mathbf{u}_n^e = \sum_{e=1}^{E}\sum_{n=1}^{N} \Delta \mathbf{U}_{ni}^e \mathbf{t}_n^e \tag{6.10}$$

where, for two-dimensional problems,

$$\Delta \mathbf{U}_{ni}^e = \int_{S_e} N_n(\xi) \mathbf{U}(P_i,\xi) dS(\xi), \quad \Delta \mathbf{T}_{ni}^e = \int_{S_e} N_n(\xi) \mathbf{T}(P_i,\xi) dS(\xi) \tag{6.11}$$

and for three-dimensional problems

$$\Delta \mathbf{U}_{ni}^e = \int_{S_e} N_n(\xi,\eta) \mathbf{U}(P_i,\xi,\eta) dS(\xi,\eta)$$

$$\Delta \mathbf{T}_{ni}^e = \int_{S_e} N_n(\xi,\eta) \mathbf{T}(P_i,\xi,\eta) dS(\xi,\eta) \tag{6.12}$$

Since there are two or three integral equations per location P_i, we now get 2 M or 3 M equations depending on the Cartesian dimension. As we will see later in the section on assembly, equation (6.10) can be written in matrix form, where coefficients are assembled in a similar way as in the FEM. For this it is convenient to store the coefficients for element into arrays $[\Delta U]^e$ and $[\Delta T]^e$, which for potential problems are defined as

$$\rightarrow elem\,nodes$$

$$[\Delta U]^e = \begin{bmatrix} \Delta U_{11} & \Delta U_{21} & \cdots \\ \Delta U_{12} & \Delta U_{22} & \cdots \\ \vdots & \vdots & \cdots \end{bmatrix} \quad and \quad [\Delta T]^e = \begin{bmatrix} \Delta T_{11} & \Delta T_{21} & \cdots \\ \Delta T_{12} & \Delta T_{22} & \cdots \\ \vdots & \vdots & \cdots \end{bmatrix} \downarrow coll.\,pnts \quad (6.13)$$

The arrays are of size N x I for potential problems, where N is the number of element nodes and I is the number of collocation points. For elasticity problems, the arrays are of size $2N$ x $2I$, for two-dimensional problems and $3N$ x $3I$, for three-dimensional problems.

In the following section, we deal with the numerical integration of Kernel shape function products over elements.

6.3   INTEGRATION OF KERNEL SHAPE FUNCTION PRODUCTS

The evaluation of integrals (6.8) or (6.11) over isoparametric elements is probably the most crucial aspect of the numerical implementation of BEM, and this is much more involved than in the FEM. The problem lies in the fact that the functions which have to be integrated exhibit singularities at certain points in the elements.

6.3.1   Singular integrals and rigid body motions

How an integral can be evaluated will depend on the type of singularity. In general, we can say that a *weakly singular* (functions of order $ln r$ for 2-D and $1/r$ for 3-D problems) can be evaluated using numerical integration, that is the Gauss Quadrature discussed in Chapter 3. However, care has to be taken that an appropriate accuracy is obtained, by adapting the number of integration points depending on the closeness of the collocation point to the region of integration. Theoretically, the integrals of functions which are *strongly singular* (functions of order $1/r$ for 2-D problems and $1/r^2$ for 3-D problems) are improper integrals, and only exist as *Cauchy* principal values[5] (These correspond to kernels T).

However, we can show that for integration on a flat surface approaching the collocation point the symmetric part of the kernel T is zero, and the anti-symmetric part approaches $+\infty$ on one side and $-\infty$ on the other side of the point. If we assume a symmetric integration region at point P_i, therefore, the integral of the anti-symmetric part also becomes zero. To explain this consider a problem in 2-D elasticity with a flat surface at point P_i as shown in Figure 6.2. For this problem the angle between vector **r** and **n** is 90^0 and therefore $cos\theta$ is zero.

According to equation (4.53),

$$T_{xx} = \frac{C_2}{r}\left[C_3 + 2\left(\frac{r_x}{r}\right)^2\right]\cos\theta = 0$$

$$T_{xy} = \frac{C_2}{r}\left[2\frac{r_x}{r}\frac{r_y}{r}\cos\theta + C_3\left(n_x\frac{r_y}{r} - n_y\frac{r_x}{r}\right)\right] = \frac{C}{r}\left(n_x\frac{r_y}{r} - n_y\frac{r_x}{r}\right)$$

(6.14)

From the distribution of the anti-symmetric part of T_{xy} shown in Figure 6.2, we can see that the integral of T_{xy} over a flat integration region which is symmetric with respect to point P_i will give zero value. As a consequence, the diagonal coefficients only contain the 'free term' $\mathbf{c}$ as computed in equation (5.30). One can easily devise a scheme whereby we assume a flat boundary very near to the collocation point and use normal Gauss integration over parts which exclude this flat region, so that we do not have to worry about computing the *Cauchy* principal value of the integral. However, we still have to compute the 'free term' which for general corners in a 3-D analysis tends to be complicated.

A simpler alternative exists and is commonly used in the boundary element literature. The concept is that we do not need to actually compute the integrals because matrix $\mathbf{A}$ may be determined from the fact that for a pure rigid body translation of an elastic domain, there must be no change in shape of the body, and therefore applied tractions must be zero. To generate a rigid body translation for a two-dimensional finite domain, we substitute $u_x = 1$ and $u_y = 0$ (translation in the x- direction) and $u_x = 0$ and $u_y = 1$ (translation in the y- direction) for all nodes and set all tractions to zero. Equation (6.10) can then be written as

$$\mathbf{A}(P_i) = \mathbf{c}(P_i) + \sum_{e=1}^{E}\sum_{\substack{n=1 \\ g(n)=i}}^{N}\Delta\mathbf{T}_{ni}^{e} = -\sum_{e=1}^{E}\sum_{\substack{n=1 \\ g(n)\neq i}}^{N}\Delta\mathbf{T}_{ni}^{e}$$

(6.15)

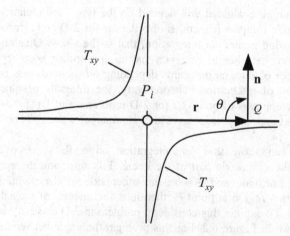

Figure 6.2   Variation of T_{xy} over a flat boundary

In equation (61.5) $g(n)$ means the global node no of the nth node of element e and $\mathbf{A}$ is a matrix of diagonal coefficients. For an infinite domain we cannot apply a rigid body translation. However, if we consider a two-dimensional domain to be bounded by a circle of radius R (see Figure 6.3), where R is approaching infinity, then we may apply a rigid body translation, but we must consider in addition to the integrals which extend over the surface of problem S also that over the boundary S_R of the auxiliary surface, that is

$$\mathbf{A}(P_i) = -\sum_{e=1}^{E} \sum_{\substack{n=1 \\ g(n) \neq i}}^{N} \Delta \mathbf{T}_{ni}^{e} - \int_{S_R} \mathbf{T}(P_i, Q)\, dS \qquad (6.16)$$

The second integral is often referred to as the *azimuthal* integral[4]. For two-dimensional elasticity problems, a typical integral is given by (see Figure. 6.3)

$$\int_{S_R} T_{xx}(P,Q)\, dS = \int_{0}^{2\pi} C_2 \left(C_3 + 2\cos\phi^2 \right)(-1)\, d\phi = -1 \qquad (6.17)$$

and

$$\int_{S_R} T_{xy}(P,Q)\, dS = \int_{0}^{2\pi} \frac{C_2}{R} [2\cos\phi\, \sin\phi\,(-1) - C_3(\sin\phi\,\cos\phi - \cos\phi\,\sin\phi)] R\, d\phi = 0 \qquad (6.18)$$

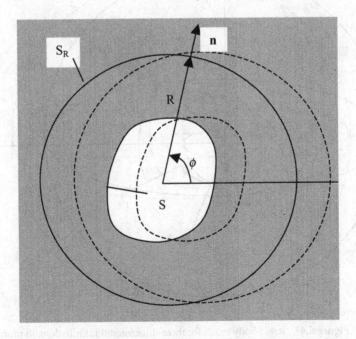

Figure 6.3 Rigid body translation in x-direction of an infinite domain

The *azimuthal* integral of matrix **T** can be written as

$$\int_{S_R} \mathbf{T}\,(P,Q)dS = -\mathbf{I} \tag{6.19}$$

where **I** is a 2 x 2 unit matrix.

We can see that since R cancels out, the integral is valid for any radius of the sphere, including a radius of infinity, so the method of computing **A** by rigid body translation is also valid for infinite domains, as long as the *azimuthal* integral is added.

For three-dimensional elasticity problems, the infinite domain is assumed to be a sphere of radius R and typical values of the *azimuthal* integral are computed by (see Figure 6.4):

$$\int_{S_R} T_{xx}(P,Q)dS = \int_0^{2\pi}\int_0^{2\pi} \frac{C_2}{R^2}\left(C_3 + 3\cos^2\phi\right)(-1)Rd\psi Rd\phi = -1 \tag{6.20}$$

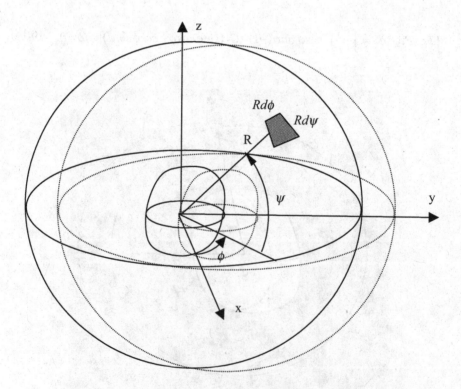

Figure 6.4   Rigid body mode for three-dimensional infinite domain problem

and

$$\int_{S_R} T_{xy}(P,Q)dS =$$

(6.21)

$$\int_0^{2\pi}\int_0^{2\pi} \frac{C_2}{R^2}\left[3\cos\phi\sin\phi(-1)-C_3(\cos\phi\sin\phi-\sin\phi\cos\phi)\right]Rd\psi Rd\phi = 0$$

so equation (6.18) is equally valid for three-dimensional problems, except that **I** is a 3 x 3 unit matrix.

For the case where the domain is semi-infinite, then the integration limits of the integral have to be changed to 0 to π and it can be easily verified that (Figure 6.5)

$$\int_{S_R} \mathbf{T}(P,Q)dS = -\tfrac{1}{2}\mathbf{I}$$

(6.22)

For potential problems, we may consider a concept similar to the rigid body motion, by assuming that for uniform temperature at all nodes of the boundary of the body and no internal heat generation there can be no heat flow. Rewriting equation (6.7) to separate diagonal and off-diagonal terms, we have

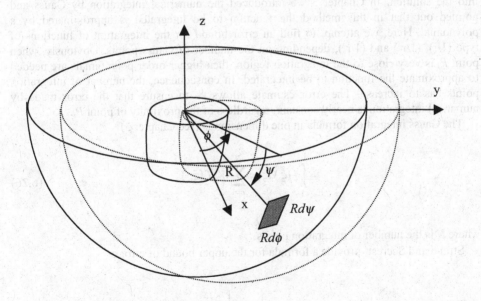

Figure 6.5   Rigid body mode for semi-infinite domain problem

$$A(P_i)u(P_i) + \sum_{\substack{e=1}}^{E} \sum_{\substack{n=1 \\ g(n)\neq i}}^{N} \Delta T_{ni}^{e}\, u_{n}^{e} = \sum_{e=1}^{E} \sum_{n=1}^{N} \Delta U_{ni}^{e}\, t_{n}^{e} \tag{6.23}$$

where

$$A(P_i) = c + \sum_{\substack{e=1}}^{E} \sum_{\substack{n=1 \\ g(n)=i}}^{N} \Delta T_{ni}^{e} \tag{6.24}$$

For a finite region, the 'uniform temperature' mode gives us the following equation for the diagonal coefficient $A(P_i)$:

$$A(P_i) = -\sum_{\substack{e=1}}^{E} \sum_{\substack{n=1 \\ g(n)\neq i}}^{N} \Delta T_{ni}^{e} \tag{6.25}$$

The reader can easily verify that for infinite and semi-infinite regions the *azimuthal* integral, which has to be subtracted from A, is -1 and -½, respectively.

6.3.2 Numerical integration

It has already been mentioned that it is very important to maintain an adequate accuracy of the numerical integration. If this is not done, then significant errors may be introduced into the solution. In Chapter 3 we introduced the numerical integration by Gauss and pointed out that in this method the function to be integrated is approximated by a polynomial. Here, we attempt to find an error bound for the integration of functions of type $(1/r)$, $(1/r^2)$ and $(1/r^3)$, depending on the number of Gauss points. Obviously, when point P_i is very close to the integration region, then higher order polynomials are needed to approximate the function to be integrated. In consequence, the number of integration points has to increase. The error estimate allows us to ensure that the error made by numerical integration is nearly constant, regardless of the proximity of point P_i.

The Gauss integration formula in one dimension is (see Chapter 3)

$$\int_{-1}^{1} f(\xi)d\xi = \sum_{n=1}^{N} W_n f(\xi_n) \tag{6.26}$$

where N is the number of integration points.

Stroud and Secrest[7] provide a formula for the upper bound of error ε

$$\varepsilon \leq 2\frac{4}{(2)^{2N}(2N)!}\left|\frac{\partial^{2N}}{\partial \xi^{2N}} f(\xi)\right| \tag{6.27}$$

Considering the integration over an element of length L with point P_i located at a distance R on the side (Figure 6.6), and taking $f(\xi)=1/r$, we obtain

$$\left| \frac{\partial^{2N}}{\partial \xi^{2N}} f(\xi) \right| = 2 \frac{(2N)! \, L^{2N}}{(2)^{2N} r^{2N+1}} \tag{6.28}$$

and for the integration error

$$\varepsilon \le \frac{4}{(4r/L)^{2N}} \tag{6.29}$$

Two extreme values may be substituted for r:

- $r = R$ give maximum error
- $r = R+L$ gives minimum error

It can be shown that, if we substitute $r/L = R/L + 0.35$, then there is a reasonable agreement between the number of integration points obtained by solving formula (6.29) and the actual number of integration points needed to achieve a certain precision ε. As an example, in Figure 6.7 we show a comparison between the predicted number of Gauss points and the actual number needed to integrate function $1/r$ to precision ε (normalised difference between exact and approximate result). Although this method is not mathematically exact, but based on empiricism, we will use this for working out the number of Gauss points needed for maintaining a constant error, regardless of the proximity of P_i. In Table 6.1 we tabulate the recommended number of Gauss points N required for integrating functions $(1/r)$, $(1r^2)$ and $(1/r^3)$. Although the actual functions to be integrated are much more complicated, because they involve products of the fundamental solution with shape functions and the Jacobian, experience has shown that values used in the table lead to an acceptable accuracy of results and this will be demonstrated later. Other more elaborate integration schemes can be found in References 4 and 6.

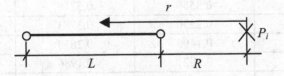

Figure 6.6   Integration over one-dimensional element

The table indicates the number of Gauss points (up to 8) needed for integrating the functions to an accuracy of 10^{-3}. We see that the required number of Gauss points increases quite rapidly, as the ratio of distance of P_i to edge of element and length of integration (R/L) decreases. According to the table, there are limiting values of R/L of

0.0698, 0.1512 and 0.2249 for *(1/r)*, *(1/r²)* and *(1/r³)* type singularity. This means that if point P_i is closer than that to the element, then we either have to use a higher integration order or we have to subdivide the element into subregions of integration.

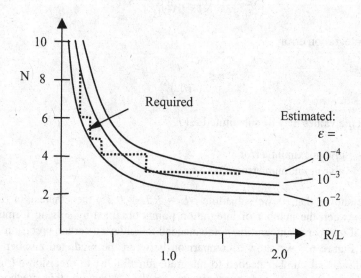

Figure 6.7   Estimated number of integration points needed to integrate function $1/r$ to a specified precision ε (dotted line is the required number of points needed for precision 10^{-3})

Table 6.1   Number of Gauss points needed to integrate function f to accuracy of 10^{-3}

N	R/L		
	$f = 1/r$	$f = 1/r^2$	$f = 1/r^3$
2	1.6382	2.6230	3.5627
3	0.6461	1.0276	1.3857
4	0.3550	0.5779	0.7846
5	0.2230	0.3783	0.5212
6	0.1490	0.2679	0.3767
7	0.1021	0.1986	0.2864
8	0.0698	0.1512	0.2249

Cases where the point is very close to the element occur when there is a drastic change in element size or the boundary surfaces are very close to each other, for example, in the case of the thin beam discussed later. Although the implementation of integration with subdivision is fairly straightforward, care has to be taken not to go to extremes with the value of R/L, because we must avoid cases where points P_i are too unevenly distributed, since the coefficient matrix may become nearly singular. On the example of a cantilever beam, it will be demonstrated that for very slender beams, the results deteriorate despite an accurate evaluation of the integrals using subdivision.

In the subsequent development of the code, the subdivision of elements has not been considered to avoid increasing the complexity. Long term experience has shown that for evaluation of the coefficient matrices this is acceptable, as extreme values of R/L can and should be avoided, for example, by preventing extreme mesh gradation or providing a finer mesh when surfaces are very close to each other. The restriction to a maximum of eight integration points and no subdivision will mean that the closest a point can be to an element of a characteristic length L is about 0.07L, for two-dimensional problems and about 0.15L, for three-dimensional ones, although practical experience shows that acceptable results can be obtained even below these values. As will be seen later, element subdivision will be needed if internal results are computed, because of the higher order of Kernels.

We convert Table 6.1 into a **FUNCTION** Ngaus which returns the number of Gauss points according to the value of R/L.

```fortran
INTEGER FUNCTION Ngaus(RonL,ne)
!-----------------------------------------------------------
!     Function returns number of Gauss points needed
!     to integrate a function 1/r^ne
!-----------------------------------------------------------
REAL , INTENT(IN)        :: RonL !  R/L
INTEGER , INTENT(IN)     :: ne   !  exponent (1,2,3)
REAL            :: Rlim(7)  !  array to store values of table
SELECT CASE(ne)
CASE(1)
 Rlim= (/1.6382, 0.6461, 0.3550, 0.2230, 0.1490, 0.1021, 0.0698/)
CASE(2)
 Rlim= (/2.6230, 1.0276, 0.5779, 0.3783, 0.2679, 0.1986, 0.1512/)
CASE(3)
 Rlim= (/3.5627, 1.3857, 0.7846, 0.5212, 0.3767, 0.2864, 0.2249/)
CASE DEFAULT
END SELECT
!   Determine minimum no of Gauss points needed
DO   N=1,7
   IF(RonL >= Rlim(N)) THEN
       Ngaus= N+1
          EXIT
   END IF
END DO
IF(Ngaus == 0) THEN        ! Point is too close to the surface
   Ngaus=8
END IF
RETURN
END FUNCTION Ngaus
```

6.3.3 Numerical integration over one-dimensional elements

Here we discuss the integration over one-dimensional finite boundary elements. The integrals which have to be evaluated over the isoparametric element, shown in Figure 6.8, are for potential problems

$$\Delta U_{ni}^e = \int_{-1}^{1} N_n(\xi) U(P_i,\xi) J(\xi) d\xi, \quad \Delta T_{ni}^e = \int_{-1}^{1} N_n(\xi) T(P_i,\xi) J(\xi) d\xi \tag{6.30}$$

where $U(P_i,\xi)$ and $T(P_i,\xi)$ are the fundamental solutions at $Q(\xi)$ for a source at point P_i, $J(\xi)$ the Jacobian is given by equation (3.38), and $N_n(\xi)$ are linear or quadratic shape functions.

When point P_i is not one of the element nodes, both integrals can be evaluated by Gauss Quadrature and the integrals in equation (6.30) can be replaced by two sums

$$\Delta T_{ni}^e \approx \sum_{m=1}^{M} N_n(\xi_m) T(P_i,\xi_m) J(\xi_m) W_m$$

$$\Delta U_{ni}^e \approx \sum_{m=1}^{M} N_n(\xi_m) U(P_i,\xi_m) J(\xi_m) W_m \tag{6.31}$$

where the number of integration points M are determined from the proximity of P_i.

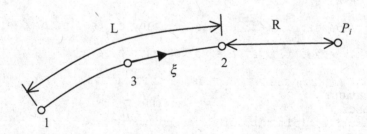

Figure 6.8   One dimensional element, integration where P_i is not one of the element nodes

When P_i is one of the element nodes, functions U and T become singular. Consider the two cases in Figure 6.9:

(a) P_i is located at point 1 and n in the above equations is 2. This means that although kernels T and U tend to infinity as point 1 is approached, the shape function tends to zero, so the integral of product $N_n(\xi)U(P_i,\xi)$ and $N_n(\xi)T(P_i,\xi)$ tend to a finite value.

Thus, for the case where P_i is not at node n of the element, the integral can be evaluated with the above formulae without any problems.

(b) P_i is located at point 2 and n in the above equations is 2. In this case, kernels T and U tend to infinity, the shape function to unity and products $N_n(\xi)U(P_i,\xi)$ and $N_n(\xi)T(P_i,\xi)$ also tend to infinity. Since kernel U has a singularity of order $ln(1/r)$, the first product cannot be integrated using Gauss Quadrature. The integral of the second product only exists as a Cauchy principal value, but the diagonal terms of the coefficient matrix can be evaluated using "rigid body motion', i.e. by applying equations (6.15) or (6.25).

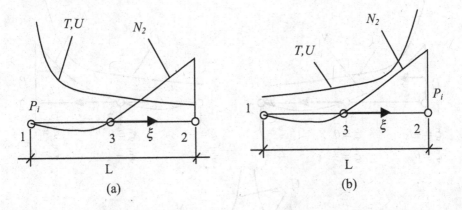

Figure 6.9 Integration when P_i is one of the element nodes

For the integration of the product with $ln(1/r)$, we can use a modified Gauss Quadrature called the *Gauss-Laguerre*[7] integration

$$\int_0^1 f(\bar{\xi})\,ln(\tfrac{1}{\xi})\,d\xi \approx \sum_{m=1}^M W_m f(\bar{\xi}_m) \qquad (6.32)$$

where M is the number of integration points.

The weights and coordinates are given by the Subroutine Gauss_Laguerre_coor, which is listed on page 135 before the subroutines for integration. Note that for this integration scheme $\bar{\xi} = 0$ at the singular point and the limits are from 0 to 1, so a change in coordinates has to be made before equation (6.32) can be applied.

This change in coordinate is given by (see Figure 6.10):

$$\xi = 2\bar{\xi} - 1 \quad when \quad P_i \ is \ at \ node \ \ 1$$
$$\xi = 1 - 2\bar{\xi} \quad when \quad P_i \ is \ at \ node \ \ 2$$
$$(6.33)$$

For the case where we integrate over a quadratic element, the integrand is discontinuous if P_i is located at the midside node. The integration has to be split into two regions, one over $-1<\xi<0$, the other over $0<\xi<1$. For the computation of product $N_n(\xi)U(P_i,\xi)$, the intrinsic coordinates for the two sub-regions are computed by (see Figure 6.10):

$$\begin{aligned} \xi &= -\bar{\xi} \quad \text{for subregion 1}\\ \xi &= \bar{\xi} \quad \text{for subregion 2} \end{aligned} \tag{6.34}$$

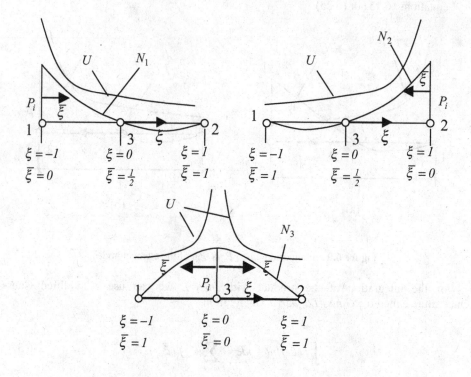

Figure 6.10 Integration when P_i and n coincide

To evaluate the first integral in equation (6.31) we must substitute for r as a function of ξ. For a linear element we may simply write $r = J\xi$, where J is the Jacobian:

$$\Delta U_{ni}^e = \int_0^1 N_n(\xi)\frac{1}{2\pi k}\ln\left(\frac{1}{J\xi}\right)J\frac{d\xi}{d\bar{\xi}}\,d\bar{\xi} =$$

$$\int_0^1 N_n(\xi)\frac{1}{2\pi k}\ln\left(\frac{1}{\xi}\right)J\frac{d\xi}{d\bar{\xi}}\,d\bar{\xi} + \int_0^1 N_n(\xi)\frac{1}{2\pi k}\ln\left(\frac{1}{J}\right)J\frac{d\xi}{d\bar{\xi}}\,d\bar{\xi} \tag{6.35}$$

The first integral may be evaluated with Gauss-Laguerre:

$$\int_0^1 N_n(\xi)\frac{1}{2\pi k}ln\left(\frac{1}{\xi}\right)J(\xi)\frac{d\xi}{d\overline{\xi}}\,d\overline{\xi}$$

$$\approx \sum_{m=1}^{M} N_n(\xi_m)\frac{1}{2\pi k}\,J(\xi_m)\,W_m\frac{d\xi}{d\overline{\xi}}$$

(6.36)

whereas the second part is integrated with normal Gauss Quadrature. The Jacobian $\partial\xi/\partial\overline{\xi}$ can be easily obtained by differentiation of equations (6.33) and (6.34).

The second integral in equation (6.36) can be evaluated using normal Gauss Quadrature. For quadratic elements, the substitution for r in terms of ξ is more complicated. One may approximately substitute $r=a\xi$, where a is the length of a straight line between the two nodes of the element. This should give a small error for elements which are nearly straight. A more accurate computation r as a function of ξ is discussed in Reference [6].

A SUBROUTINE which provides the coordinates and weights for a GausLaguerre integration is given below.

```
SUBROUTINE Gauss_Laguerre_coor(Cor,Wi,Intord)
!-------------------------------------
! Returns Gauss_Laguerre coordinates and Weights
!   for up to 8 Gauss points
!-------------------------------------
IMPLICIT NONE
REAL, INTENT(OUT)    :: Cor(8)  ! Gauss point coordinate
REAL, INTENT(OUT)    :: Wi(8)   ! weights
INTEGER,INTENT(IN)   :: Intord  ! integration order
SELECT CASE (Intord)
CASE (1)
    Cor(1)= 0.5
    Wi(1) = 1.0
CASE(2)
    Cor(1)= .112008806 ; Cor(2)=.602276908
    Wi(1) = .718539319 ; Wi(2) =.281460680
CASE(3)
    Cor(1)= .063890793 ; Cor(2)= .368997063 ; Cor(3)= .766880303
    Wi(1) = .513404552 ; Wi(2) = .391980041 ; Wi(3) =.0946154065
CASE(4)
    Cor(1)= .0414484801 ; Cor(2)=.245274914 ; Cor(3)=.556165453
    Cor(4)= .848982394
    Wi(1) = .383464068 ; Wi(2) =.386875317 ; Wi(3) =.190435126
    Wi(4) = .0392254871
CASE(5)
    Cor(1)= .0291344721 ; Cor(2)= .173977213
```

```
   Cor(3)= .411702520 ; Cor(4)=.677314174
   Cor(5)= .894771361
   Wi(1) = .297893471 ; Wi(2) = .349776226
   Wi(3) = .234488290 ; Wi(4) = .0989304595
   Wi(5) = .0189115521
CASE(6)
   Cor(1)= .021634005 ; Cor(2)= .129583391 ; Cor(3)=.314020449
   Cor(4)= .538657217 ; Cor(5)= .756915337 ; Cor(6)=.922668851
   Wi(1) = .238763662 ; Wi(2) = .308286573 ; Wi(3) =.245317426
   Wi(4) = .142008756 ; Wi(5) = .055454622 ; Wi(6)=.0101689586
CASE(7)
   Cor(1)= .016719355 ; Cor(2)= .100185677 ; Cor(3)=.246294246
   Cor(4)= .433463493 ;
   Cor(5)= .632350988 ; Cor(6)=.811118626 ; Cor(7)= .940848166
   Wi(1) = .196169389 ; Wi(2) = .270302644 ; Wi(3) = .239681873
   Wi(4) = .165775774
   Wi(5) = .088943227 ; Wi(6) =.0331943043 ; Wi(7)= .0059327870
CASE(8)
   Cor(1)= .0133202441 ; Cor(2)=.0797504290 ; Cor(3)=.197871029
   Cor(4)= .354153994
   Cor(5)= .529458575 ; Cor(6)= .701814529 ; Cor(7)= .849379320
   Cor(8)= .953326450
   Wi(1) = .164416604 ; Wi(2) = .237525610 ; Wi(3) = .226841984
   Wi(4) = .175754079
   Wi(5) = .112924030 ; Wi(6) =.0578722107 ; Wi(7) =.0209790737
   Wi(8) =.00368640710
CASE DEFAULT
   CALL Error_Message('Gauss points not in range 1-8')
END SELECT
END SUBROUTINE Gauss_Laguerre_coor
```

A SUBROUTINE Integ2P is shown below which integrates the kernel/shape function products over one-dimensional isoparametric elements for potential problems.

```
SUBROUTINE Integ2P (Elcor,Inci,Nodel,Ncol,xP,k,dUe,dTe)
!-------------------------------------------------
!  Computes  Element contributions[dT]e and [dU]e
!   for 2-D potential problems
!  by numerical integration
!
!-------------------------------------------------
IMPLICIT NONE
REAL, INTENT(IN):: Elcor(:,:)   ! Element coordinates
INTEGER, INTENT(IN) :: Inci(:)  ! Element Incidences
INTEGER, INTENT(IN) :: Nodel    ! No. of Element Nodes
INTEGER , INTENT(IN):: Ncol     ! Number of points Pi
REAL, INTENT(IN)   :: xP(:,:)   ! Array with coll. point coords.
REAL , INTENT(IN) :: k          ! Permeability/Conductivity
REAL , INTENT(OUT):: dUe(:,:),dTe(:,:)
```

```
REAL :: epsi= 1.0E-4 !    Small value for comparing cords
REAL ::Eleng,Rmin,RonL,Glcor(8),Wi(8),Ni(Nodel),Vnorm(2),GCcor(2)
REAL :: UP,Jac,dxr(2),TP,r,pi,c1,c2,xsi,eta,dxdxb
INTEGER :: i,m,n,Mi,nr,ldim,cdim,nreg
 pi=3.14159265
ldim= 1
 cdim=ldim+1
  CALL Elength(Eleng,Elcor,Nodel,ldim) ! Element Length
!------------------------------------------------------------------
!   Integration off-diagonal coeff.  -> normal Gauss Quadrature
!------------------------------------------------------------------
dUe= 0.0 ; dTe= 0.0           ! Clear arrays for summation
Colloc_points: &
DO i=1,Ncol
 Rmin= Min_dist(Elcor,xP(:,i),Nodel,ldim,inci)! Distance to Pi
 RonL= Rmin/Eleng   ! R/L Mi= Ngaus(RonL,1) ! Number of Gauss
 points for (1/r) sing.
 Call Gauss_coor(Glcor,Wi,Mi)        ! Assign coords/Weights
 Gauss_points: &
 DO m=1,Mi
  xsi= Glcor(m)
    CALL Serendip_func(Ni,xsi,eta,ldim,Nodel,Inci)
    Call Normal_Jac(Vnorm,Jac,xsi,eta,ldim,Nodel,Inci,elcor)
    CALL Cartesian(GCcor,Ni,ldim,elcor) ! Coords of Gauss pt
    r= Dist(GCcor,xP(:,i),cdim)            ! Dist. P,Q
    dxr= (GCcor-xP(:,i))/r                 ! rx/r , ry/r
    UP= U(r,k,cdim) ; TP= T(r,dxr,Vnorm,cdim)    ! Kernels
    Node_points: &
    DO n=1,Nodel
       IF(Dist(Elcor(:,n),xP(:,i),cdim) < epsi) EXIT ! Pi is n
       dUe(i,n)= dUe(i,n) + Ni(n)*UP*Jac*Wi(m)
       dTe(i,n)= dTe(i,n) + Ni(n)*TP*Jac*Wi(m)
    END DO &
    Node_points
 END DO &
 Gauss_points
END DO &
Colloc_points
!---------------------------------
!
!    Diagonal terms of dUe
!
!---------------------------------
c1= 1/(2.0*pi*k)
Colloc_points1: &
DO i=1,Ncol
 Node_points1: &
 DO n=1,Nodel
    IF(Dist(Elcor(:,n),xP(:,i),cdim) > Epsi) CYCLE ! Pi not n
    Nreg=1
    IF(n == 3) nreg= 2
```

```fortran
!-----------------------------------------------
!      Integration of logarithmic term
!-----------------------------------------------
Subregions: &
DO nr=1,Nreg
   Mi= 4
   Call Gauss_Laguerre_coor(Glcor,Wi,Mi)
   Gauss_points1:&
   DO m=1,Mi
     SELECT CASE (n)
       CASE (1)
          xsi= 2.0*Glcor(m)-1.0
          dxdxb= 2.0
       CASE (2)
         xsi= 1.0 -2.0*Glcor(m)
         dxdxb= 2.0
       CASE (3)
          dxdxb= 1.0
          IF(nr == 1) THEN
            xsi= -Glcor(m)
          ELSE
            xsi= Glcor(m)
          END IF
       CASE DEFAULT
       END SELECT
      CALL Serendip_func(Ni,xsi,eta,1,Nodel,Inci)
      Call Normal_Jac(Vnorm,Jac,xsi,eta,1,Nodel,Inci,elcor)
      dUe(i,n)= dUe(i,n) + Ni(n)*c1*Jac*dxdxb*Wi(m)
    END DO &
   Gauss_points1
  END DO &
  Subregions
!-----------------------------------------------
!      Integration of non logarithmic term
!-----------------------------------------------
Mi= 2
Call Gauss_coor(Glcor,Wi,Mi)   ! Assign coords/Weights
Gauss_points2: &
DO m=1,Mi
  SELECT CASE (n)
      CASE (1:2)
        c2=-LOG(Eleng)*c1
      CASE (3)
        c2=LOG(2/Eleng)*c1
      CASE DEFAULT
  END SELECT
  xsi= Glcor(m)
  CALL Serendip_func(Ni,xsi,eta,ldim,Nodel,Inci)
  Call Normal_Jac(Vnorm,Jac,xsi,eta,ldim,nodel,Inci,elcor)
  dUe(i,n)= dUe(i,n) + Ni(n)*c2*Jac*Wi(m)
END DO &
```

```
   Gauss_points2
  END DO &
  Node_points1
 END DO &
 Colloc_points1
 RETURN
 END SUBROUTINE Integ2P
```

The integration scheme just discussed is equally applicable to elasticity problems, except that when integrating the functions with kernel **U** when P_i is one of the nodes of the element we have to consider that only U_{xx} and U_{yy} have a logarithmic and non-logarithmic part. The logarithmic part is integrated with Gauss-Laguerre, for example:

$$\Delta U_{xx\,ni}^{\,e} = \int_0^1 N_n(\xi) \frac{(1-v)(3-4v)}{4\pi\,E\,(1-v)} ln\left(\frac{1}{\xi}\right)\!\left(P_i,\xi\right) J(\xi) \frac{d\xi}{d\overline{\xi}}\,d\overline{\xi}$$

$$\approx \sum_{m=1}^{M} N_n(\xi_m) \frac{(1-v)(3-4v)}{4\pi\,E\,(1-v)} J(\xi_m) W_m \frac{d\xi}{d\overline{\xi}}$$

(6.37)

whereas the non-logarithmic part is integrated using Gauss Quadrature.

A **SUBROUTINE** for integrating over one-dimensional elements for elasticity can be written. The main differences to the previous subroutine are that the Kernels **U** and **T** are now 2 x 2 matrices and we have to add two more **DO** loops for the direction of the load at P_i and the direction of the displacement/traction at $Q(\xi)$. A structure chart of SUBROUTINE Integ2E is shown in Figure 6.11.

For the implementation of symmetry as discussed in Chapter 7, two additional parameters have been used: ISYM and NDEST. The first parameter contains the symmetry code, as will be explained in the next chapter; the second is an array that is used to eliminate variables which have zero value because they arelocated on a symmetry plane.

Note that the storage of coefficients is by degree of freedom number rather than node number. There are two columns per node and two rows per collocation point. The storage of the element coefficients $[\Delta U]^e$ is as follows:

$$,\!\rightarrow\ element\ \ nodes$$

$$[\Delta U]^e = \begin{bmatrix} \Delta U_{xx11} & \Delta U_{xy11} & \Delta U_{xx21} & \Delta U_{xy21} & \cdots \\ \Delta U_{yx11} & \Delta U_{yy11} & \Delta U_{yx21} & \Delta U_{yy21} & \cdots \\ \Delta U_{xx12} & \Delta U_{xy12} & \Delta U_{xx22} & \Delta U_{xy22} & \cdots \\ \vdots & \vdots & \vdots & \vdots & \ddots \end{bmatrix} \downarrow coll.\ pnts$$

(6.38)

where the first two subscripts denote the direction of the unit load, and the displacement and subsequent subscripts denote the local (element) node number and the collocation point number.

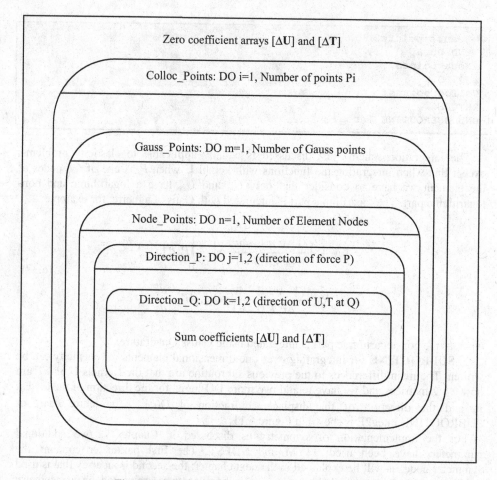

Figure 6.11 Structure chart for SUBROUTINE Integ2E

```
SUBROUTINE
&Integ2E(Elcor,Inci,Nodel,Ncol,xP,E,ny,dUe,dTe,Ndest,Isym)
!-----------------------------------------------------------
!    Computes  [dT]e and [dU]e for 2-D elasticity problems
!    by numerical integration
!-----------------------------------------------------------
IMPLICIT NONE
REAL, INTENT(IN)      :: Elcor(:,:) !  Element coordinates
INTEGER, INTENT(IN) :: Inci(:)      !  Element Incidences
INTEGER, INTENT(IN) :: Nodel        !  No. of Element Nodes
INTEGER , INTENT(IN):: Ncol         !  Number of points Pi
REAL, INTENT(IN)  :: E,ny           !  Elastic constants
REAL, INTENT(IN)      :: xP(:,:)    !  Array with coords. Of Pi
REAL(KIND=8), INTENT(OUT) :: dUe(:,:),dTe(:,:) ! element coeff.
INTEGER , INTENT(IN):: Isym         !  Symmetry code
```

```fortran
INTEGER , INTENT(IN):: Ndest(:,:)  !   Node destination vector
REAL :: epsi= 1.0E-4  ! Small value for comparing cords
REAL :: Eleng,Rmin,RonL,Glcor(8),Wi(8),Ni(Nodel)
REAL :: Vnorm(2),GCcor(2)
REAL :: Jac,dxr(2),UP(2,2),TP(2,2), xsi, eta, r, dxdxb,Pi,C,C1
INTEGER :: i,j,k,m,n,Mi,nr,ldim,cdim,iD,nD,Nreg
Pi=3.14159265359
C=(1.0+ny)/(4*Pi*E*(1.0-ny))
ldim= 1                              ! Local dimension
cdim=ldim+1
CALL Elength(Eleng,Elcor,Nodel,ldim)  ! Element Length
dUe= 0.0 ; dTe= 0.0            ! Clear arrays for summation
!-----------------------------------------------------------------
!   Integration off-diagonal coeff.  -> normal Gauss Quadrature
!-----------------------------------------------------------------
 Colloc_points:&
 DO i=1,Ncol
! Distance coll. point and element
 Rmin= Min_dist(Elcor,xP(:,i),Nodel,ldim,inci)
 RonL= Rmin/Eleng                 ! R/L
 Mi= Ngaus(RonL,1)  !  Number of Gauss points for o(1/r)
 Call Gauss_coor(Glcor,Wi,Mi)  ! Assign coords/Weights
 Gauss_points:&
 DO m=1,Mi
  xsi= Glcor(m)
  CALL Serendip_func(Ni,xsi,eta,ldim,Nodel,Inci)
  CALL Normal_Jac(Vnorm,Jac,xsi,eta,ldim,Nodel,Inci,elcor)
  CALL Cartesian(GCcor,Ni,ldim,elcor)
  r= Dist(GCcor,xP(:,i),cdim)    ! Dist. P,Q
  dxr= (GCcor-xP(:,i))/r           ! rx/r , ry/r
  UP= UK(dxr,r,E,ny,Cdim) ; TP= TK(dxr,r,Vnorm,ny,Cdim)
  Node_points:&
   DO n=1,Nodel
     Direction_P:&
     DO j=1,2
       iD= 2*(i-1) + j ! line number in array
       Direction_Q:&
       DO k= 1,2
       nD= 2*(n-1) + k ! column number in array
       IF(Dist(Elcor(:,n),xP(:,i),cdim) > epsi) THEN  ! n∈P_i
           dUe(iD,nD)= dUe(iD,nD) + Ni(n)*UP(j,k)*Jac*Wi(m)
           dTe(iD,nD)= dTe(iD,nD) + Ni(n)*TP(j,k)*Jac*Wi(m)
       ELSE IF (j /= k) THEN
         dUe(iD,nD)= dUe(iD,nD) + Ni(n)*UP(j,k)*Jac*Wi(m)
       ELSE
         ! non-log part of U
         dUe(iD,nD)= dUe(iD,nD) + Ni(n)*C*dxr(j)*dxr(k)*Jac*Wi(m)
       END IF
       END DO &
       Direction_Q
       END DO &
```

```fortran
            Direction_P
      END DO &
      Node_points
   END DO &
  Gauss_points
END DO&
Colloc_points
!
!-------------------------------------------------------
!      Computation of diagonal coefficients
!-------------------------------------------------------
C= C*(3.0-4.0*ny)
Colloc_points1:&
DO i=1,Ncol
 Node_points1:&
 DO n=1,Nodel
   IF(Dist(Elcor(:,n),xP(:,i),cdim) > Epsi) CYCLE ! n ∉ P_i
   Nreg=1
   IF (n == 3) nreg= 2
   Subregions:&
   DO nr=1,Nreg
      Mi= 4
      Call Gauss_Laguerre_coor(Glcor,Wi,Mi)
      Gauss_points1:&
      DO m=1,Mi
       SELECT CASE (n)
       CASE (1)
         xsi= 2.0*Glcor(m)-1.0
         dxdxb= 2.0
       CASE (2)
         xsi= 1.0 -2.0*Glcor(m)
         dxdxb= 2.0
       CASE (3)
         dxdxb= 1.0
       IF(nr == 1) THEN
         xsi= -Glcor(m)
       ELSE
         xsi= Glcor(m)
       END IF
      CASE DEFAULT
      END SELECT
      CALL Serendip_func(Ni,xsi,eta,ldim,Nodel,Inci)
      Call Normal_Jac(Vnorm,Jac,xsi,eta,ldim,Nodel,Inci,elcor)
      Direction1: &
      DO j=1,2
       iD= 2*(i-1) + j  !  line number in array
       nD= 2*(n-1) + j  !  column number in array
       dUe(iD,nD)= dUe(iD,nD) + Ni(n)*C*Jac*dxdxb*Wi(m)
      END DO&
      Direction1
```

```
   END DO &
  Gauss_points1
END DO &
Subregions
Mi= 2
Call Gauss_coor(Glcor,Wi,Mi)! Assign coords/Weights
Gauss_points2:&
DO m=1,Mi
SELECT CASE (n)
CASE (1:2)
  C1=-LOG(Eleng)*C
CASE (3)
  C1=LOG(2/Eleng)*C
CASE DEFAULT
END SELECT
xsi= Glcor(m)
CALL Serendip_func(Ni,xsi,eta,ldim,Nodel,Inci)
CALL Normal_Jac(Vnorm,Jac,xsi,eta,ldim,Nodel,Inci,elcor)
Direction2:&
DO j=1,2
   iD= 2*(i-1) + j !  line number in array
   nD= 2*(n-1) + j !  column number in array
   dUe(iD,nD)= dUe(iD,nD) + Ni(n)*C1*Jac*Wi(m)
END DO &
Direction2
END DO &
Gauss_points2
END DO &
Node_points1
END DO &
Colloc_points1
RETURN
END SUBROUTINE Integ2E
```

6.3.4 Numerical integration for two-dimensional elements

Here we discuss numerical integration over two-dimensional isoparametric finite boundary elements. We find that the basic principles are very similar to integration over one-dimensional elements in that we consider the case where P_i is not one of the nodes of an element and where it is. Starting with potential problems, the integrals which have to be evaluated (see Figure 6.12) are:

$$\Delta U_{ni}^e = \int\limits_{-1}^{1}\int\limits_{-1}^{1} N_n(\xi,\eta)\, U(P_i,\xi,\eta)\, J(\xi,\eta)\, d\xi\, d\eta$$

$$\Delta T_{ni}^e = \int\limits_{-1}^{1}\int\limits_{-1}^{1} N_n(\xi,\eta)\, T(P_i,\xi,\eta)\, J(\xi,\eta)\, d\xi\, d\eta$$

(6.39)

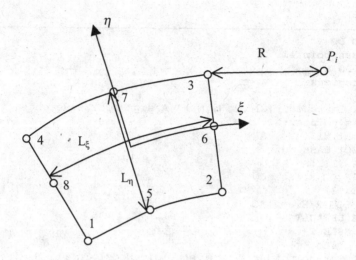

Figure 6.12 Two-dimensional isoparametric element

When P_i is not one of the element nodes, then the integrals can be evaluated using numerical integration formulae which we get by applying Gauss Quadrature in the ξ and η direction. This gives

$$\Delta U_{ni}^e \approx \sum_{m=1}^{M}\sum_{k=1}^{K} N_n(\xi_m,\eta_k)\,U(P_i,\xi_m,\eta_k)\,J(\xi_m,\eta_k)\,W_m W_k$$

$$\Delta T_{ni}^e \approx \sum_{m=1}^{M}\sum_{k=1}^{K} N_n(\xi_m,\eta_k)\,T(P_i,\xi_m,\eta_k)\,J(\xi_m,\eta_k)\,W_m W_k \tag{6.40}$$

The number of integration points in the ξ direction M is determined from Table 6.1, where L is taken as the size of the element in the ξ direction, L_ξ, and the number of points in η direction K is determined by substituting for L the size of the element in η direction (L_η) in Figure 6.12.

When P_i is a node of the element but not node n, then kernel U approaches infinity as $(1/r)$ but the shape function approaches zero, so product $N_n U$ may be determined using Gauss Quadrature. Kernel T approaches infinity as $(1/r^2)$ and cannot be integrated using the above scheme. When P_i is node n of the element, then product $N_n U$ cannot be evaluated with Gauss Quadrature. The integral of the product $N_n T$ only exists as a *Cauchy* principal value, but since this corresponds to a diagonal coefficient it can be evaluated using rigid body motion. For the evaluation of the second integral in equation (6.39), when P_i is a node of the element but not node n, and for evaluating the first integral, when P_i is node n, we propose to split up the element into triangular subelements, as shown in Figures 6.13 and 6.14. For each subelement we introduce a local coordinate system that is chosen in such a way that the transformation approaches zero at node P_i.

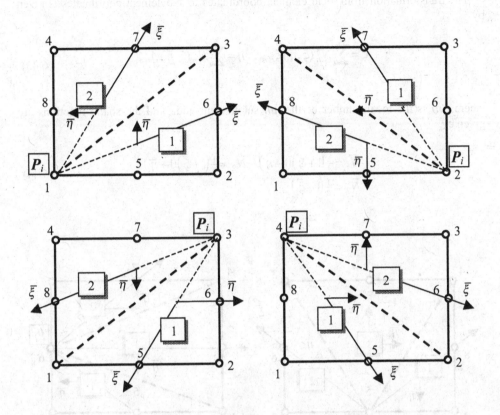

Figure 6.13 Subelements for numerical integration when P_i is a corner node of element

Numerical integration formulae have to be applied over two or three sub elements depending if P_i is a corner or mid-side node.

Using this scheme, the first integral in equation (6.39) is rewritten as

$$\Delta U_{ni}^e = \sum_{\substack{g(n)=P_i}}^{2(3)} \int_{-1}^{1}\int_{-1}^{1} N_n(\xi,\eta)\, U(P_i,\xi,\eta)\, J(\xi,\eta)\, \frac{\partial(\xi,\eta)}{\partial(\overline{\xi},\overline{\eta})}\, d\overline{\xi}\, d\overline{\eta} \qquad (6.41)$$

The equation for numerical evaluation of the integral using Gauss Quadrature is given by

$$\Delta U_{ni}^e \approx \sum_{\substack{g(n)=P_i}}^{2(3)} \sum_{m=1}^{M} \sum_{k=1}^{K} N_n(\overline{\xi}_m,\overline{\eta}_k)\, U(P_i,\overline{\xi}_m,\overline{\eta}_k)\, J(\overline{\xi}_m,\overline{\eta}_k)\, \overline{J}(\overline{\xi}_m,\overline{\eta}_k)\, W_m W_k \qquad (6.42)$$

where $\overline{J}(\overline{\xi},\overline{\eta})$ is the Jacobian of the transformation from $\overline{\xi},\overline{\eta}$ to ξ,η.

The transformation from local element coordinates to subelement coordinates is given by

$$\xi = \sum_{n=1}^{3} \overline{N}_n(\overline{\xi}, \overline{\eta}) \xi_{l(n)}, \quad \eta = \sum_{n=1}^{3} \overline{N}_n(\overline{\xi}, \overline{\eta}) \eta_{l(n)} \tag{6.43}$$

where $l(n)$ is the local number of the nth subelement node and the shape functions are given by

$$\overline{N}_1 = \tfrac{1}{4}(1+\overline{\xi})(1-\overline{\eta}), \quad \overline{N}_2 = \tfrac{1}{4}(1+\overline{\xi})(1+\overline{\eta})$$
$$\overline{N}_3 = \tfrac{1}{2}(1-\overline{\xi}) \tag{6.44}$$

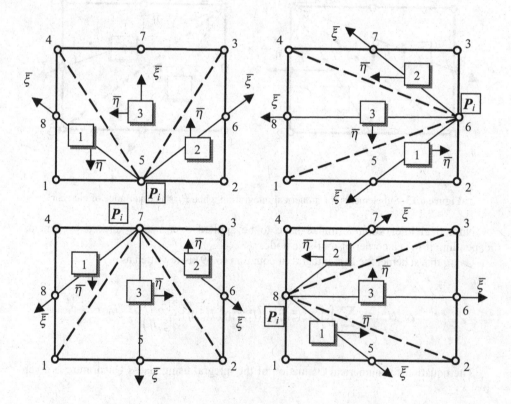

Figure 6.14 Subelements for numerical integration when P_i is a midside node of element

Tables 6.2 and 6.3 give the local node numbers $l(n)$ in equation (6.43), depending on the number of the subelement and the position of P_i.

Table 6.2 Local node numbers of subelement nodes P_i at corner nodes

P_i at node	Subelement 1			Subelement 2		
	$n=1$	$n=2$	$n=3$	$n=1$	$n=2$	$n=3$
1	2	3	1	3	4	1
2	3	4	2	4	1	2
3	1	2	3	4	1	3
4	1	2	4	2	3	4

The Jacobian matrix of the transformation (6.43) is given by

$$\overline{\mathbf{J}} = \begin{bmatrix} \dfrac{\partial \xi}{\partial \overline{\xi}} & \dfrac{\partial \eta}{\partial \overline{\xi}} \\ \dfrac{\partial \xi}{\partial \overline{\eta}} & \dfrac{\partial \eta}{\partial \overline{\eta}} \end{bmatrix} \tag{6.45}$$

where

$$\frac{\partial \xi}{\partial \overline{\xi}} = \sum_{n=1}^{3} \frac{\partial \overline{N}_n}{\partial \overline{\xi}}(\overline{\xi},\overline{\eta})\,\xi_n, \quad \frac{\partial \xi}{\partial \overline{\eta}} = \sum_{n=1}^{3} \frac{\partial \overline{N}_n}{\partial \overline{\eta}}(\overline{\xi},\overline{\eta})\,\eta_n$$

$$\frac{\partial \eta}{\partial \overline{\xi}} = \sum_{n=1}^{3} \frac{\partial \overline{N}_n}{\partial \overline{\xi}}(\overline{\xi},\overline{\eta})\,\eta_n, \quad \frac{\partial \eta}{\partial \overline{\eta}} = \sum_{n=1}^{3} \frac{\partial \overline{N}_n}{\partial \overline{\eta}}(\overline{\xi},\overline{\eta})\,\eta_n \tag{6.46}$$

Table 6.3 Local node numbers of subelement nodes P_i at midside nodes

P_i at node	Subelement 1			Subelement 2			Subelement 3		
	$n=1$	$n=2$	$n=3$	$n=1$	$n=2$	$n=3$	$n=1$	$n=2$	$n=3$
5	4	1	5	2	3	5	3	4	5
6	1	2	6	3	4	6	4	1	6
7	4	1	7	2	3	7	1	2	7
8	1	2	8	3	4	8	2	3	8

The Jacobian is given by

$$\overline{J} = det\,|\overline{\mathbf{J}}| = \frac{\partial \xi}{\partial \overline{\xi}}\frac{\partial \eta}{\partial \overline{\eta}} - \frac{\partial \eta}{\partial \overline{\xi}}\frac{\partial \xi}{\partial \overline{\eta}} \tag{6.47}$$

The reader may verify that for $\overline{\xi} = -1$, this Jacobian is zero.

The integration scheme is applicable without modification to elasticity problems. In equation (6.41) we simply replace the scalars U and T with matrices $\mathbf{U}$ and $\mathbf{T}$.

A subprogram, which calculates the element coefficient arrays $[\Delta U]^e$ and $[\Delta T]^e$ for potential problems or $[\Delta U]^e$ and $[\Delta T]^e$ for elasticity problems, can be written based on the theory discussed. The diagonal coefficients of $[\Delta T]^e$ cannot be computed by integration over elements of kernel shape function products. As has already been discussed in Section 6.3.1, these can be computed using the additional conditions obtained from the consideration of rigid body modes. The implementation of this procedure will be discussed in the next chapter. In Subroutine Integ3 we distinguish between elasticity and potential problems by the input variable Ndof (number of degrees of freedom per node), which is set to 1 for potential problems and to 3 for elasticity problems.

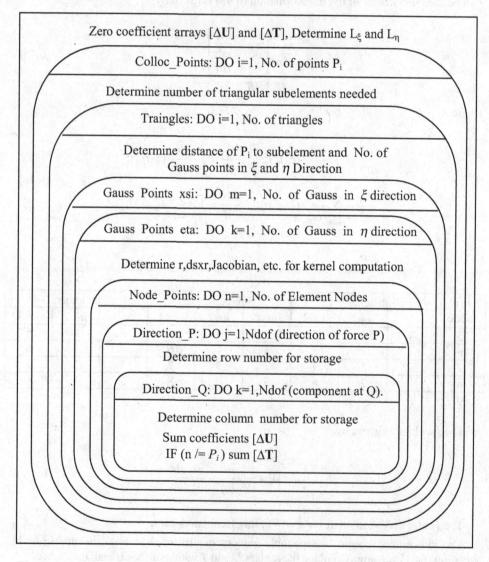

Figure 6.15 Structure chart for computation of [ΔT] and if [ΔU] if P_i is one of the element nodes

Subroutine INTEG3 is divided into two parts. The first part deals with integration when P_i is not one of the nodes of the element over which the integration is made. Gauss integration in two directions is used here. The integration of ΔT_{ni}^e and ΔU_{ni}^e is carried out concurrently but should actually be treated separately because the functions to be integrated have different degrees of singularity and, therefore, require a different number of Gauss points. For simplicity both are integrated using the number of Gauss points required for the higher order singularity. Indeed, the subroutine presented has not been programmed very efficiently, but for the purpose of this book simplicity was the paramount factor. Additional improvements in efficiency can, for example, be made by carefully examining if the operations in the DO loops actually depend on the DO loop variable. If they do not, then that operation should be taken outside of the corresponding DO loop. Substantial savings can be made here for a program that involves up to seven implied DO loops and which has to be executed for all boundary elements.

The second part of the SUBROUTINE, deals with the case where P_i is one of the nodes of the element which we integrate over. To deal with the singularity of the integrand, the element has to be subdivided into 2 or 3 triangles, as explained previously. Since there are a lot of implied DO loops involved, we show a structure chart of this part of the program in Figure 6.15.

```
SUBROUTINE
  & Integ3(Elcor,Inci,Nodel,Ncol,xPi,Ndof,E,ny,dUe,dTe,Ndest,Isym)
!-----------------------------------------------------------------
!     Computes  [dT]e and [dU]e for 3-D problems
!     by numerical integration
!-----------------------------------------------------------------
IMPLICIT NONE
REAL, INTENT(IN)     :: Elcor(:,:)     !   Element coordinates
INTEGER, INTENT(IN)  :: Inci(:)        !   Element Incidences
INTEGER, INTENT(IN)  :: Nodel          !   No. of Element Nodes
INTEGER , INTENT(IN) :: Ncol           !   Number of points Pi
REAL , INTENT(IN)    :: xPi(:,:)       !   coords. Of Pi
INTEGER , INTENT(IN) :: Ndof           !   D.o.F. /node (1 or 3)
REAL , INTENT(IN)    :: E,ny           !   Elastic constants
INTEGER, INTENT(IN)  :: Ndest(:,:)     !   for Symmetry
INTEGER, INTENT(IN)  :: Isym           !   Symmetry code
REAL(KIND=8) , INTENT(OUT) :: dUe(:,:),dTe(:,:) ! coefficients
REAL :: Elengx,Elenge,Rmin,RLx,RLe,Glcorx(8),Wix(8)
REAL :: Glcore(8),Wie(8),Weit,r
REAL :: Ni(Nodel),Vnorm(3),GCcor(3),dxr(3)
REAL :: Jac,Jacb,xsi,eta,xsib,etab
REAL :: UP(Ndof,Ndof),TP(Ndof,Ndof),ko  !   Arrays for Kernels
INTEGER :: i,m,n,k,ii,jj,ntr,Mi,Ki,id,nd,lnod,Ntri
INTEGER :: ldim= 2        !  Element dimension
INTEGER :: Cdim= 3        !  Cartesian dimension
ELengx= Dist((Elcor(:,3)+Elcor(:,2))/2.,&
             (Elcor(:,4)+Elcor(:,1))/2.,Cdim)  ! Lxsi
ELenge= Dist((Elcor(:,2)+Elcor(:,1))/2.,&
             (Elcor(:,3)+Elcor(:,4))/2.,Cdim)  ! Leta
```

```
dUe= 0.0 ; dTe= 0.0        !   Clear arrays for summation!
!-------------------------------------------------------------------
!     Part 1 : Pi is not one of the element nodes
!-------------------------------------------------------------------
Colloc_points:&
DO i=1,Ncol
  IF(.NOT. ALL(Inci /= i)) CYCLE        ! Incidence array contains I
  Rmin= Min_dist(Elcor,xPi(:,i),Nodel,ldim,Inci)
  Mi= Ngaus(Rmin/Elengx,2)  ! No. of G.P. in ξ dir. o(1/r²)
  Call Gauss_coor(Glcorx,Wix,Mi)  ! G.P. coords/Weights ξ-dir

  Ki= Ngaus(Rmin/Elenge,2)  ! No. of G.P. in ξ dir. o(1/r²)
  Call Gauss_coor(Glcore,Wie,Ki)  ! G.P. coords/Weights η-dir
  Gauss_points_xsi:&
  DO m=1,Mi
   xsi= Glcorx(m)
   Gauss_points_eta:&
   DO k=1,Ki
     eta= Glcore(k)
     Weit= Wix(m)*Wie(k)
     CALL Serendip_func(Ni,xsi,eta,ldim,Nodel,Inci)
     CALL Normal_Jac(Vnorm,Jac,xsi,eta,ldim,Nodel,Inci,elcor)
     CALL Cartesian(GCcor,Ni,ldim,elcor)
     r= Dist(GCcor,xPi(:,i),Cdim)        ! Dist. P,Q
     dxr= (GCcor-xPi(:,i))/r             ! rx/r , ry/r
     IF(Ndof .EQ. 1) THEN
       UP= U(r,ko,Cdim) ; TP= T(r,dxr,Vnorm,Cdim) ! Potential
     ELSE
       UP= UK(dxr,r,E,ny,Cdim) ; TP= TK(dxr,r,Vnorm,ny,Cdim)! Elast
     END IF
     Direction_P:&
     DO ii=1,Ndof
       IF(Isym == 0) THEN
         iD= Ndof*(i-1) + ii  !  line number in dU,dT
       ELSE
         ID= Ndest(i,ii)
       END IF
       Direction_Q:&
       DO jj=1,Ndof
         Node_points: &
         DO n=1,Nodel
           nD= Ndof*(n-1) + jj  !  column number
           dUe(iD,nD)= dUe(iD,nD) + Ni(n)*UP(ii,jj)*Jac*Weit
           dTe(iD,nD)= dTe(iD,nD) + Ni(n)*TP(ii,jj)*Jac*Weit
         END DO &
         Node_points
       END DO &
       Direction_Q
     END DO &
     Direction_P
```

```
   END DO &
   Gauss_points_eta
  END DO &
  Gauss_points_xsi
END DO &
Colloc_points
!-----------------------------------------------------------
!     Part 2 : Pi is one of the element nodes
!-----------------------------------------------------------
Colloc_points1: &
 DO i=1,Ncol
  lnod= 0
  DO n= 1,Nodel  !   Determine which local node is Pi
    IF(Inci(n) == i) THEN
      lnod=n
    END IF
  END DO
  IF(lnod == 0) CYCLE  !  None -> next Pi
  Ntri= 2
  IF(lnod > 4) Ntri=3  !  Number of sub-regions
  Triangles: &
  DO ntr=1,Ntri
    CALL Tri_RL(RLx,RLe,Elengx,Elenge,lnod,ntr)
    Mi= Ngaus(RLx,2)
    Call Gauss_coor(Glcorx,Wix,Mi)
    Ki= Ngaus(RLe,2)
    Call Gauss_coor(Glcore,Wie,Ki)
    Gauss_points_xsi1:&
    DO m=1,Mi
      xsib= Glcorx(m)
      Gauss_points_eta1:&
      DO k=1,Ki
        etab= Glcore(k)
        Weit= Wix(m)*Wie(k)
        ! Coord transf from triang cords to xsi,eta
        CALL Trans_Tri(ntr,lnod,xsib,etab,xsi,eta,Jacb)
        CALL Serendip_func(Ni,xsi,eta,ldim,Nodel,Inci)
        Call Normal_Jac(Vnorm,Jac,xsi,eta,ldim,Nodel,Inci,elcor)
        Jac= Jac*Jacb
        CALL Cartesian(GCcor,Ni,ldim,elcor)
        r= Dist(GCcor,xPi(:,i),Cdim)  !  Dist. P,Q
        dxr= (GCcor-xPi(:,i))/r       !  rx/r , ry/r
        IF(Ndof .EQ. 1) THEN
          UP= U(r,ko,Cdim) ; TP= T(r,dxr,Vnorm,Cdim)
        ELSE
          UP= UK(dxr,r,E,ny,Cdim) ; TP= TK(dxr,r,Vnorm,ny,Cdim)
        END IF
        Direction_P1: &
        DO ii=1,Ndof
          IF(Isym == 0) THEN
            iD= Ndof*(i-1) + ii  ! line number in dU,dT
```

```
         ELSE
           ID= Ndest(i,ii)
         END IFDirection_Q1:     &
         DO jj=1,Ndof
           Node_points1: &
           DO n=1,Nodel
            nD= Ndof*(n-1) + jj
            dUe(iD,nD)= dUe(iD,nD) + Ni(n)*UP(ii,jj)*Jac*Weit
            IF(Inci(n) /= i) THEN   ! only off-diag elements of dTe
              dTe(iD,nD)= dTe(iD,nD) + Ni(n)*TP(ii,jj)*Jac*Weit
            END IF
           END DO Node_points1
         END DO Direction_Q1
       END DO Direction_P1
     END DO Gauss_points_eta1
   END DO Gauss_points_xsi1
  END DO Triangles
 END DO Colloc_points1
 RETURN
 END SUBROUTINE Integ3
```

6.4   CONCLUSIONS

In this chapter we have discussed in some detail numerical methods which can be used to perform the integration of the kernel shape function products over boundary elements. Because of the nature of these functions, special integration schemes had to be devised so that the precision of integration was similar for all locations of P_i relative to the boundary element over which the integration was carried out. If this is not taken into consideration, results obtained from a BEM analysis will be in error and, in extreme cases, meaningless.

The number of integration points which has to be used to obtain a given precision of integration is not easy to determine. Whereas exact error estimates have been worked out by several researchers based on mathematical theory, so far they are only applicable to regular meshes and not to isoparametric elements of arbitrary curved shape. The scheme proposed here for working out the number of integration points has been developed on a semi-empirical basis, but has been found to work well over a period of nearly 20 years. The reader may consider the examples worked out in Chapter 9, to ascertain this accuracy.

We have now developed a library of subroutines which we will need for the writing of a general purpose computer program. All that is left to do is the assembly of coefficient matrices from element contributions, to specify the boundary conditions and to solve the system of equations.

6.5 EXERCISES

Exercise 6.1

Check the integration scheme proposed for one-dimensional boundary elements by performing the integration of equation (6.32) for a straight line element parallel to the x-axis of length 2 for $n = 1$, with two different locations of P_i (Figure 6.16):

(a) along the element
(b) perpendicular to the centre of the element

for values of R/L= 0.5, 0.1, 0.05 by comparing with the analytical solution.

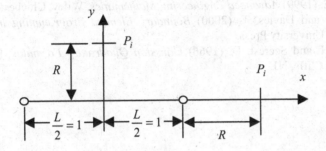

Figure 6.16 Problem for checking accuracy of integration

Exercise 6.2

Check the integration scheme proposed for two-dimensional boundary elements by performing the integration of equation (6.41) for a square element of size 2 x 2 and for $n = 1$ with two different locations of P_i (Figure 6.17)

(c) In the same plane as the element
(d) perpendicular to the centre of the element

for values of R/L= 0.5, 0.1, 0.05 by comparing with the analytical solution.

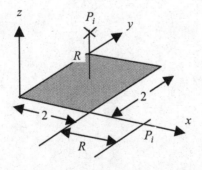

Figure 6.17 Geometry of test problem for checking the accuracy of numerical integration

6.6   REFERENCES

1. Deist, F.H. and Georgiadis, E. (1973) A computer system for three-dimensional stress analysis in elastic media. *Rock Mechanics*, **5**, 189-202.
2. Ergatoudis, J.G., Irons, B.M. and Zienkiewicz, O.C. (1968) Three dimensional analysis of arch dams and their foundations. *Proc. Symp. Arch Dams,* Inst. Civ. Eng.
3. Lachat, J.C. and Watson, J.O. (1976) Effective numerical treatment of boundary integral equations. *Int. J. Num. Meth. Eng.*, **10**, 991-1005.
4. Beer, G. and Watson, J.O. (1991) *Introduction to Finite and Boundary Element Methods for Engineers*. Wiley, Chichester.
5. Kreyszig, E. (1999) *Advanced Engineering Mathematics*, Wiley, Chichester.
6. Gao, X.W. and Davies, T. (2000) *Boundary Element Programming in Mechanics*. Cambridge University Press.
7. Stroud, A.H. and Secrest, D. (1966) *Gaussian Quadrature Formulas*, Prentice-Hall, Englewood Cliffs, NJ.

7

Assembly and Solution

A difficult thing is real pleasure
Seneca

7.1 INTRODUCTION

The previous chapter dealing with the numerical integration of kernel shape function products addressed probably the most important aspect of the boundary element method. We note that this is more involved than the integration used in the FEM for determining the element stiffness matrix. In the current chapter, we will find that subsequent steps in solving the integral equations are fairly straightforward and similar to the FEM, especially with respect to the assembly of element contributions into the global coefficient matrix.

The solution of the system of equations, however, is somewhat different to the FEM as we deal with non-symmetric and fully populated coefficient matrices. However, this will make the task easier, as there is no need to worry about developing special solvers which exploit the sparsity of the system. Indeed, the fact that the equation system is fully populated has been claimed to be one of the main drawbacks of the method. However, because the system of equations obtained is much smaller, it more than compensates for this and, as we will see later, computation times required for the solution are usually much shorter than the FEM. We will also see in Chapter 10 that if we introduce the concept of multiple boundary element regions, sparsity is introduced, which can be exploited by special solvers.

At the end of this chapter we will have all the procedures necessary for a general purpose program which can solve steady state problems in potential flow and elasticity.

The program, however, will only give us values of the unknown at the boundary. As already pointed out, a special feature of the BEM is that results at any point inside the domain can be computed with greater accuracy as a postprocessing exercise. This topic will be dealt with in Chapter 8.

7.2   ASSEMBLY OF SYSTEM OF EQUATIONS

As always, we start with potential problems. In the previous section we discussed the computation of element contributions to equation (6.7), that is

$$cu(P_i) + \sum_{e=1}^{E} \sum_{n=1}^{N} \Delta T_{ni}^e \, u_n^e = \sum_{e=1}^{E} \sum_{n=1}^{N} \Delta U_{ni}^e \, t_n^e \qquad (7.1)$$

We recall the notation used:

$$\textit{Element number}$$
$$\downarrow$$
$$\Delta T_{\underset{\uparrow}{ni\leftarrow}}^{e} \quad \textit{Collocation point} \qquad (7.2)$$
$$\textit{Node Number}$$

Before being able to solve the system of equations, we must replace the double sums by a matrix multiplication of the type

$$[\Delta T]\{u\} = [\Delta U]\{t\} \qquad (7.3)$$

where vectors $\{u\}$, $\{t\}$ contain potential/temperature and fluxes, respectively, in a global numbering system, and $[\Delta T]$, $[\Delta U]$ are global coefficient matrices assembled by *gathering* element contributions. In the global coefficient arrays, rows correspond to collocation points P_i and columns to the global node number. The gathering process is very similar to the assembly process in the FEM, except that whole columns are added. For the gathering process we need the *connectivity* or *incidences* of element e, which indicate the global node numbers of the element.

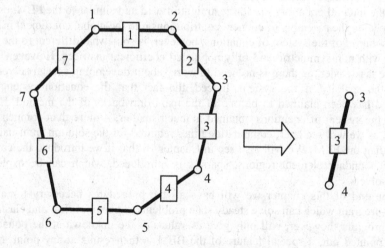

Figure 7.1   2-D BE mesh for explaining assembly (potential problems)

Referring to the simple 2-D mesh with linear elements in Figure 7.1, the incidences of element 3 for example may be given by (/3,4/).

To assemble $[\Delta T]$ columns of the coefficient matrix of all elements are added. For example for element 3, columns of $[\Delta T]^3$, are added to the global matrices as shown

$$\rightarrow \quad Node \quad numbers$$

$$[\Delta T]=\begin{array}{c} 1 \ 2 \quad 3 \quad 4 \quad 5 \ 6 \ 7 \\ \begin{bmatrix} \cdot \ \cdot \ \Delta T_{11}^3 \ \Delta T_{21}^3 \ \cdot \ \cdot \ \cdot \\ \cdot \ \cdot \ \Delta T_{12}^3 \ \Delta T_{22}^3 \ \cdot \ \cdot \ \cdot \\ \cdot \ \cdot \ \Delta T_{13}^3 \ \Delta T_{23}^3 \ \cdot \ \cdot \ \cdot \\ \cdot \ \cdot \ \Delta T_{14}^3 \ \Delta T_{24}^3 \ \cdot \ \cdot \ \cdot \\ \cdot \ \cdot \ \Delta T_{15}^3 \ \Delta T_{25}^3 \ \cdot \ \cdot \ \cdot \\ \cdot \ \cdot \ \Delta T_{16}^3 \ \Delta T_{26}^3 \ \cdot \ \cdot \ \cdot \\ \cdot \ \cdot \ \Delta T_{17}^3 \ \Delta T_{27}^3 \ \cdot \ \cdot \ \cdot \end{bmatrix} \end{array} \quad \downarrow \ P_i \qquad (7.4)$$

Note that numbering of the columns in $[\Delta T]$ is now according to the global node numbering whereas the numbering of $[\Delta T]^e$ is according to the local node numbering.

For elasticity problems there is more than one unknown per node, so columns are numbered according to the degree of freedom, rather than node number. For two-dimensional elasticity problems, each node has two degrees of freedom and the incidences of element 3 in Figure 7.2 for example are expanded to a *destination* vector (/5,6,7,8/), where the numbers 5,6 correspond to the x and y-components of the unknown at node 3.

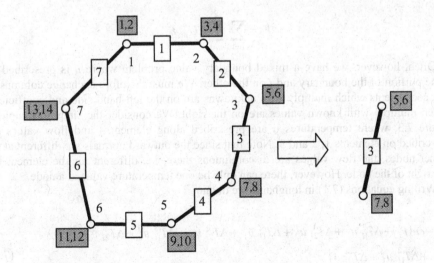

Figure 7.2   2-D mesh for explaining assembly (2-D elasticity problems)

The assembly process for element 3 is then modified to

$$\rightarrow \quad Destination \quad Numbers$$

$$1 \ 2 \ 3 \ 4 \qquad 5 \qquad\qquad 6 \qquad\qquad 7 \qquad\qquad 8 \quad \ldots$$

$$[\Delta \mathbf{T}] = \begin{bmatrix} \cdot & \cdot & \cdot & \cdot & \Delta T^3_{xx11} & \Delta T^3_{yx11} & \Delta T^3_{xx21} & \Delta T^3_{yx21} & \cdot & \cdot & \cdot & \cdot & \cdot \\ \cdot & \cdot & \cdot & \cdot & \Delta T^3_{xy11} & \Delta T^3_{yy11} & \Delta T^3_{xy21} & \Delta T^3_{yy21} & \cdot & \cdot & \cdot & \cdot & \cdot \\ \cdot & \cdot & \cdot & \cdot & \Delta T^3_{xx12} & \Delta T^3_{yx12} & \Delta T^3_{xx22} & \Delta T^3_{yx22} & \cdot & \cdot & \cdot & \cdot & \cdot \\ \cdot & \cdot & \cdot & \cdot & \Delta T^3_{xy12} & \Delta T^3_{yy12} & \Delta T^3_{xy22} & \Delta T^3_{yy22} & \cdot & \cdot & \cdot & \cdot & \cdot \\ \vdots & \vdots & \vdots & \vdots & \vdots & \vdots & \vdots & \vdots & \vdots & \vdots & \vdots & \vdots & \vdots \end{bmatrix} \tag{7.5}$$

Note that destination numbers are now used for numbering the columns. Assuming that, as in the introductory example solved with the Trefftz method, the flux t is known on all boundary nodes and solution u is required, then we assemble the left-hand side, perform the matrix multiplication on the right hand side of equation (7.3) and solve the system of equations. Alternatively, multiplication $[\Delta U]\{t\}$ can be made element by element at the assembly level, without explicitly creating the matrix $[\Delta U]$, therefore saving on storage space. This would also allow us to consider discontinuous distribution of normal gradients or tractions. For the simple example in Figure 7.1, equation (7.1) can be replaced by

$$[\Delta T]\{u\} = \{F\} \tag{7.6}$$

where the coefficients of the right-hand side vector $\{F\}$ are given by

$$F_i = \sum_{e=1}^{7} \sum_{n=1}^{2} \Delta U^e_{ni} \, t^e_n \tag{7.7}$$

Often, however, we have a mixed boundary value problem where u is prescribed on some portion of the boundary and t on the other. We must therefore exchange columns so that coefficients which multiply with unknowns are on the left-hand side and coefficients which multiply with known values are on the right. We consider the simple example in Figure 7.3, where temperatures u are prescribed along element 4 and flow values are prescribed on elements 1, 2 and 3. Note that since the outward normals are different at the corner nodes, the flow values are discontinuous there, i.e. different for the element left and right of the node. However, there can only be one temperature value at a node.

Writing equations (7.1) in longhand, we obtain:

$$cu(P_i) + \Delta T^1_{1i} \, u^1_1 + \Delta T^1_{2i} \, u^1_2 + \Delta T^2_{1i} \, u^2_1 + \Delta T^2_{2i} \, u^2_2 + \Delta T^3_{1i} \, u^3_1 + \Delta T^3_{2i} \, u^3_2$$

$$+ \Delta T^4_{1i} \, u^4_1 + \Delta T^4_{2i} \, u^4_2 \tag{7.8}$$

$$= \Delta U^1_{1i} \, t^1_1 + \Delta U^1_{2i} \, t^1_2 + \Delta U^2_{1i} \, t^2_1 + \Delta U^2_{2i} \, t^2_2 + \Delta U^3_{1i} \, t^3_1 + \Delta U^3_{2i} \, t^3_2 + \Delta U^4_{1i} \, t^4_1 + \Delta U^4_{2i} \, t^4_2$$

The assembly procedure has to be modified, so that we put all known values on the right-hand side and all unknown ones on the left side of the equation.

Known values are:

$$t_1^1; \quad t_2^1; \quad t_1^2; \quad t_2^2; \quad t_1^3; \quad t_2^3$$
$$u_1; \quad u_4 \tag{7.9}$$

Unknown values are:

$$u_2^1 = u_1^2 = u_2; \quad u_2^2 = u_1^3 = u_3$$
$$t_1^4; \quad t_2^4 \tag{7.10}$$

After placing unknown values on the left and known values on the right, equation (7.8) is written as:

$$cu(P_i) + (\Delta T_{2i}^1 + \Delta T_{1i}^2)\, u_2 + (\Delta T_{2i}^2 + \Delta T_{1i}^3)\, u_3$$
$$- \Delta U_{1i}^4\, t_1^4 - \Delta U_{2i}^4\, t_2^4 = \tag{7.11}$$
$$\Delta U_{1i}^1\, t_1^1 + \Delta U_{2i}^1\, t_2^1 + \Delta U_{1i}^2\, t_1^2 + \Delta U_{2i}^2\, t_2^2 + \Delta U_{1i}^3\, t_1^3 + \Delta U_{2i}^3\, t_2^3$$
$$- \Delta T_{1i}^4\, u_1 - \Delta T_{2i}^4\, u_4$$

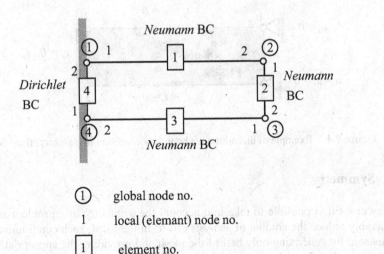

①	global node no.
1	local (elemant) node no.
1	element no.

Figure 7.3 Example of two-dimensional potential problem with mixed boundary conditions

In equation (7.11) a global numbering for the nodes has been implemented for u. Equation (7.11) can now be written for P_i at nodes 1, 2, 3, 4 using matrix algebra as:

$$
\begin{matrix}
\quad\ 1 & \quad\ 2 & \quad\ 3 & \quad\ 4
\end{matrix}
$$

$$
\begin{bmatrix}
-\Delta U_{21}^4 & \Delta T_{21}^1 + \Delta T_{11}^2 & \Delta T_{21}^2 + \Delta T_{11}^3 & -\Delta U_{11}^4 \\
-\Delta U_{22}^4 & \left[\Delta T_{22}^1 + \Delta T_{12}^2\right] & \Delta T_{22}^2 + \Delta T_{12}^3 & -\Delta U_{12}^4 \\
-\Delta U_{23}^4 & \Delta T_{23}^1 + \Delta T_{13}^2 & \left[\Delta T_{23}^2 + \Delta T_{13}^3\right] & -\Delta U_{13}^4 \\
-\Delta U_{24}^4 & \Delta T_{24}^1 + \Delta T_{14}^2 & \Delta T_{24}^2 + \Delta T_{14}^3 & -\Delta U_{14}^4
\end{bmatrix}
\begin{bmatrix}
t_1^4 \\
u_2 \\
u_3 \\
t_2^4
\end{bmatrix} = \{F\}
\qquad (7.12)
$$

The diagonal elements involving ΔT are highlighted by brackets. As explained in 6.3.1., we compute and assemble these diagonal coefficients by considering 'rigid body modes'. The coefficients of the right-hand side vector $\{F\}$ are given by:

$$
F_i = \Delta U_{1i}^1 \, t_1^1 + \Delta U_{2i}^1 t_2^1 + \Delta U_{1i}^2 \, t_1^2 + \Delta U_{2i}^2 t_2^2 + \Delta U_{1i}^3 \, t_1^3 + \Delta U_{2i}^3 \, t_2^3
$$
$$
- \Delta T_{1i}^4 \, u_1 - \Delta T_{2i}^4 \, u_4
\qquad (7.13)
$$

For problems in elasticity, the assembly process for mixed boundary value problems is similar but, since the assembly is by degrees of freedom rather than node numbers, boundary conditions will also depend on the direction. An example of this is given in Figure 7.4.

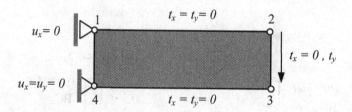

Figure 7.4   Example of discontinuous boundary condition in elasticity: fixed beam

7.2.1 Symmetry

In many cases it is possible to take into account the symmetry of a problem and thereby considerably reduce the amount of analysis effort. In the FEM, such conditions are simply implemented by generating only half of the mesh and providing the appropriate boundary conditions at the plane(s) of symmetry. In the BEM we can take a different approach, alleviating the need to have boundary elements on the symmetry plane. For example, for the problem shown in Figure 7.5, of a circular excavation in an infinite domain, nodes do not exist on the plane of symmetry.

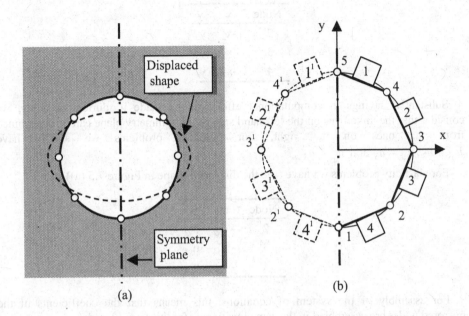

Figure 7.5 Example with one plane of symmetry and mesh used to explain implementation

The approach in dealing with symmetry conditions in the BEM will be explained here. Consider the simple mesh for the analysis of a circular excavation consisting of a total of eight elements, shown in Figure 7.5b. The idea is to input only elements on the right-hand side of the symmetry plane (elements 1 to 4) and to automatically generate the elements on the left (elements 1^1 to 4^1). The incidences of these 'mirrored' elements are:

Element	i	j
1	4	5
2	3	4
3	2	3
4	1	2
1^1	5	4^1
2^1	4^1	3^1
3^1	3^1	2^1
4^1	2^1	1

Note that the sequence of nodes for all mirrored elements is reversed. This is important because it affects the direction of the outward normal vector **n**. The coordinates of nodes at the left of the symmetry plane can be computed from those on the right by:

Node	x	y
2^1	$-x_2$	y_2
3^1	$-x_3$	y_3
4^1	$-x_4$	y_4

Substantial savings in computational effort can be made if, during assembly, we consider that the unknowns on the left-hand side of the symmetry plane can be determined from the ones on the right. For potential problems we simply have $u_{2^1} = u_2$, $u_{3^1} = u_3$ and $u_{4^1} = u_4$.

For elasticity problems we have (see the displaced shape in Figure 7.5 (a))

Node	u_x	u_y
2^1	$-u_{x2}$	u_{y2}
3^1	$-u_{x3}$	u_{y3}
4^1	$-u_{x4}$	u_{y4}

For assembly of the system of equations, this means that the coefficients of the mirrored nodes are assembled in the same location as for the original nodes.

For elasticity problems, the negative signs of the x-component of the displacement have to be considered during assembly. If this assembly procedure is used, then the number of unknowns for the problem is reduced to the nodes on the right-hand side of the symmetry plane and on the plane itself. The only additional computational effort will be the integration of the kernel shape function products over the mirrored elements. If

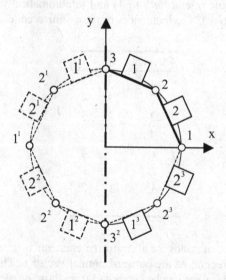

Figure 7.6 Example with two planes of symmetry

conditions of symmetry exist about the x and y axes, then the elements are 'mirrored' twice, as shown in Figure 7.6.

The incidence vectors are now:

Element	i	j
1	2	3
2	1	2
1^1	3	2^1
2^1	2^1	1^1
1^2	2^2	3^2
2^2	1^2	2^2
1^3	3^1	2^3
2^3	2^3	1

Note that for all elements, except 1^2 and 2^2, the incidences are reversed. The coordinates of the 'mirrored' nodes are:

Node	x	y
1^1	$-x_1$	y_1
3^2	x_3	$-y_3$
2^1	$-x_2$	y_2
2^2	$-x_2$	$-y_2$
2^3	x_2	$-y_2$

For potential problems we have $u_{2^1} = u_{2^2} = u_{2^3} = u_2, u_{1^1} = u_1$ and $u_{3^1} = u_3$. For elasticity problems, the displacements at the primed nodes are given by:

Node	u_x	u_y
1^1	$-u_{x1}$	0
3^1	0	$-u_{y3}$
2^1	$-u_{x2}$	u_{y2}
2^2	$-u_{x2}$	$-u_{y2}$
2^3	u_{x2}	$-u_{y2}$

The method can be extended to three-dimensional problems. Up to three planes of symmetry are possible and an element has to be projected seven times. For the mesh in Figure 7.7, we determine the incidences for the mirrored elements keeping a consistent outward normal, as shown (anticlockwise numbering of element 1).

Element	i	j	k	l
1	1	2	3	4
1^1	1	4	3^1	2^1
1^2	1	2^1	3^2	4^1
1^3	1	4^1	3^3	2
1^4	1^1	4^3	3^7	2^3
1^5	1^1	2^2	3^4	4^3
1^6	1^1	4^2	3^5	2^2
1^7	1^1	2^3	3^6	4^2

We note that for mirror image number $n = 1, 3, 4$ and 6, the incidences have to be reversed to maintain a consistent outward normal, as shown in Figure 7.7.

We now discuss the computer implementation of up to three symmetry planes. We specify a symmetry code, where:

Symmetry code m	Symmetry about	No. of mirrored elements
1	y-z plane	1
2	y-z and x-z plane	3
3	All three planes	7

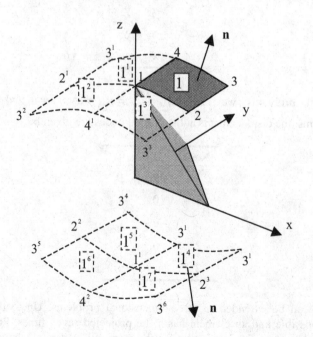

Figure 7.7 Three-dimensional BE mesh with 3 planes of symmetry

For the mirrored nodes we can compute coordinates $\mathbf{x}$, displacements $\mathbf{u}$ and tractions $\mathbf{t}$ from the original nodes by:

$$\mathbf{x}^n = \mathbf{T}^n \mathbf{x}; \quad \mathbf{u}^n = \mathbf{T}^n \mathbf{u}; \quad \mathbf{t}^n = \mathbf{T}^n \mathbf{t} \tag{7.14}$$

where superscript n denotes the mirror image number, as used in Figures 7.5 to 7.7.

Transformation matrices $\mathbf{T}$ are computed as follows. First we define three matrices $\mathbf{T}_m$

$$\mathbf{T}_1 = \begin{bmatrix} -1 & 0 & 0 \\ 0 & 1 & 0 \\ 0 & 0 & 1 \end{bmatrix}; \quad \mathbf{T}_2 = \begin{bmatrix} 1 & 0 & 0 \\ 0 & -1 & 0 \\ 0 & 0 & 1 \end{bmatrix}; \quad \mathbf{T}_3 = \begin{bmatrix} 1 & 0 & 0 \\ 0 & 1 & 0 \\ 0 & 0 & -1 \end{bmatrix} \tag{7.15}$$

In terms of these, the transformation matrices are defined as

$$\mathbf{T}^1 = \mathbf{T}_1; \quad \mathbf{T}^2 = \mathbf{T}_2; \quad \mathbf{T}^3 = \mathbf{T}_1 \mathbf{T}_2;$$
$$\mathbf{T}^4 = \mathbf{T}_3; \quad \mathbf{T}^5 = \mathbf{T}_1 \mathbf{T}_3; \quad \mathbf{T}^6 = \mathbf{T}_2 \mathbf{T}_3; \quad \mathbf{T}^7 = \mathbf{T}_1 \mathbf{T}_2 \mathbf{T}_3 \tag{7.16}$$

For implementation, we provide an additional loop for each element which, depending on the symmetry code, is executed 1, 2, 4 or 8 times. For no symmetry (code 0) we only consider the original element. For code 1 (symmetry about the y-z plane) we consider one mirrored image of the element. For symmetry codes 2 and 3, three and seven mirrored images of the element are considered.

7.2.2 Subroutine MIRROR

Subroutine MIRROR has been written to generate elements across symmetry planes. It returns the incidence, destination and coordinate vector of the mirrored element, as well as multiplication factors for the assembly. In the subroutine we assume that if points are on the symmetry plane, then they have a zero coordinate, and one must ensure that this is actually the case. We note that for subroutines INTEG to work for the mirrored elements, we must change the mirrored node numbers to some arbitrary value. Here we have chosen to add the maximum node number to the incidences. Note that the numbers of the original node must not be changed (for example, node 2 in the example of Figure 7.7). This node can be recognised because it lies on a plane of symmetry:

Node is on	Mirror image (n) which must not be changed
x-z plane	3
y-z plane	1
x-y plane	4
2 planes	1 to 3

Destination vectors of the mirrored elements used for the assembly remain the same as for the original element, except that if incidences are reversed to maintain a consistent outward normal, then the destination vector must also be reversed. Subroutine Reverse is used and it is listed in the Utility Library in Appendix A. Note that in the implementation the transformation matrices in equation (7.16) have already been multiplied and only the diagonal terms are considered.

```fortran
SUBROUTINE Mirror(Isym,nsy,Nodes,Elcor,Fac,Incie,Ldeste &
                  Elres_te,Elres_ue,,Nodel,Ndof,Cdim)
!---------------------------------------------
!      Creates mirror image of element
!---------------------------------------------
INTEGER, INTENT(IN) ::   Isym        ! symmetry indicator
INTEGER, INTENT(IN) ::   nsy         ! symmetry count
INTEGER, INTENT(IN) ::   nodes       ! highest node no
REAL, INTENT(IN OUT)::   Elcor(:,:)  ! Coords (will be modified)
REAL, INTENT(OUT)   ::   Fac(:)      ! Multiplication factors
INTEGER, INTENT(IN OUT):: Incie(:)!  Incidences    (will be
INTEGER, INTENT(IN OUT):: Ldeste(:)! Destinations  modified)
REAL, INTENT(IN OUT)    :: Elres_te(:) ! Element tractions
REAL, INTENT(IN OUT)    :: Elres_te(:) ! Element displacements
INTEGER, INTENT(IN) ::   Nodel       ! Nodes per element
INTEGER, INTENT(IN) ::   Ndof        ! d.o.F. per Node
INTEGER, INTENT(IN) ::   Cdim        ! Cartesian dimension
REAL   :: TD(3) ! Transformation vector (diagonal elements of T)
INTEGER :: n,m,Ison1,Ison2,Ison3,i
Fac(1:nodel*ndof)= 1.0
IF(nsy == 1) RETURN
!      Assign coefficients of TD
SELECT CASE (nsy-1)
   CASE(1)
     TD=(/-1.0,1.0,1.0/)
   CASE(2)
     TD=(/-1.0,-1.0,1.0/)
   CASE(3)
     TD=(/1.0,-1.0,1.0/)
   CASE(4)
     TD=(/1.0,1.0,-1.0/)
   CASE(5)
     TD=(/-1.0,1.0,-1.0/)
   CASE(6)
     TD=(/1.0,-1.0,-1.0/)
   CASE(7)
     TD=(/-1.0,-1.0,-1.0/)
END SELECT
!      generate coordinates and incidences
Nodes0: &
DO n=1,nodel
 Elcor(:,n)= Elcor(:,n)*TD(m)
!   Check if point is on any symmetry plane

 Ison1= 0
 Ison2= 0
 Ison3= 0
```

```
IF(Elcor(1,n)==0.0) Ison1=1
IF(Elcor(2,n)==0.0) Ison2=1
IF(Cdim > 2 .AND. Elcor(3,n)==0.0) Ison3=1
!   only change incidences for unprimed nodes
IF(ison1==1 .AND. nsy-1 ==1) CYCLE
IF(ison2==1 .AND. nsy-1 ==3) CYCLE
IF(ison1+ison2+ison3 > 1 .AND. nsy-1<4) CYCLE
Incie(n)= Incie(n) + Nodes
END DO &
Nodes0
!      generate multiplication factors elast. Problems only
IF(Ndof > 1) THEN
I=0
Nodes1: &
DO n=1,nodel
  Degrees_of_freedom: &
  DO m=1,Ndof
    I=I+1
    Fac(I)= TD(m)
  END DO &
  Degrees_of_freedom
END DO &
Nodes1
END IF
!  Reverse destination vector for selected elem
SELECT CASE (nsy-1)
  CASE (1,3,4,6)
    CALL &
    Reverse(Incie,elcor,ldeste,Elres_te,Elres_ue,Ndof,Cdim,nodel)
  CASE DEFAULT
END SELECT
RETURN
END SUBROUTINE Mirror
```

7.2.3 Subroutine ASSEMBLY

A subprogram for assembling the coefficient matrices using a vector of incidences or destinations, as well as information about the type of boundary and symmetry condition is easily written. The information about the boundary condition is supplied for each node or each degree of freedom of an element, and the code is 0 for *Neumann* and 1 for the *Dirichlet* condition. Care has to be taken where the boundary condition is discontinuous. For example, in Figure 7.3, both temperature and flow values are known at the first node of element 1, but only temperature is known at the second node of element 4 (both nodes equal 1 in global numbering). For the assembly we must therefore specify a global code, in addition to a boundary code for each element. Then, if *Neumann* BC is specified and the global code is *Dirichlet*, both ΔT and ΔU are assembled on the right-hand side. In the parameter list of subroutine vectors, Elres_u and Elres_t are introduced. These will eventually contain all results of an element. At the stage when the SUBROUTINE is called, however, they contain only known (prescribed) values with all other values, being zero.

SUBROUTINE Assembly can be used for the assembly of two or three-dimensional problems in potential flow or elasticity. The incidence vector in potential flow problems and the destination vector in elasticity problems are substituted for LDEST.

```fortran
SUBROUTINE Assembly(Lhs,Rhs,DTe,DUe,Ldest,BCode,Ncode &
           ,Elres_ue,Elres_te,Diag,Ndofe,Ndof,Nodel,Fac)
!-------------------------------------------------
!   Assembles Element contributions DTe , DUe
!   into global matrix Lhs and vector Rhs
!   Also sums off-diagonal coefficients
!   for the computation of diagonal coefficients
!-------------------------------------------------
REAL(KIND=8)                   :: Lhs(:,:),Rhs(:)     ! Global arrays
REAL(KIND=8), INTENT(IN):: DTe(:,:),DUe(:,:)    ! Element arrays
INTEGER , INTENT(IN) :: LDest(:) ! Element destination vector
INTEGER , INTENT(IN) :: BCode(:) ! Boundary code(local)
INTEGER , INTENT(IN) :: NCode(:) ! Boundary code (global)
INTEGER , INTENT(IN) :: Ndofe      ! D.o.F's / Elem
INTEGER , INTENT(IN) :: Ndof       ! D.o.F's / Node
INTEGER , INTENT(IN) :: Nodel       ! Nodes/Element
REAL , INTENT(IN)    :: Elres_ue(:) ! vector u for element
REAL , INTENT(IN)    :: Elres_te(:) ! vector t for element
REAL , INTENT(IN)    :: Fac(:)      ! Mult. factor for symmetry
REAL(KIND=8) :: Diag(:,:) ! Array containing diagonal coeff of DT
INTEGER :: n,Ncol
DoF_per_Element:&
DO m=1,Ndofe
  Ncol=Ldest(m)        !    Column number
  IF(BCode(m) == 0) THEN      !    Neumann BC
    Rhs(:) = Rhs(:) + DUe(:,m)*Elres_te(m)*Fac(m)
  !   The assembly of dTe depends on the global BC
    IF (NCode(Ldest(m)) == 0) THEN
      Lhs(:,Ncol)= Lhs(:,Ncol) + DTe(:,m)*Fac(m)
    END IF
    IF (NCode(Ldest(m)) == 1) THEN
      Rhs(:) = Rhs(:) - DTe(:,m) * Elres_ue(m)*Fac(m)
    END IF
  END IF
  IF(BCode(m) == 1) THEN      !    Dirichlet BC
    Lhs(:,Ncol) = Lhs(:,Ncol) - DUe(:,m)*Fac(m)
    Rhs(:)= Rhs(:) - DTe(:,m) * Elres_ue(m)*Fac(m)
  END IF
END DO &
DoF_per_Element
!
!
```

```
!
!       Sum off-diagonal coefficients
DO n=1,Nodel
   DO k=1,Ndof
      l=(n-1)*Ndof+k
      Diag(:,k)= Diag(:,k) - DTe(:,l)*Fac(m)
   END DO
END DO
RETURN
END SUBROUTINE Assembly
```

Element contributions to the coefficient matrix which have been computed numerically with **SUBROUTINE** Integ have zero values for coefficients ΔT^e_{ni} when the node n coincides with i (i.e. $g(n)= i$), because these coefficients are not computed using numerical integration. In the assembled matrix $[\Delta T]$, these coefficients correspond to diagonal elements. The rigid body motion method (Chapter 6) can be applied to compute these coefficients. For example, for a finite region the assembled diagonal coefficients ΔT_{ii} is given by:

$$\Delta T_{ii} = -\sum_{e=1}^{E} \sum_{\substack{n=1 \\ g(n)\neq i}}^{N} \Delta T^e_{ni} \tag{7.17}$$

The double sum is computed by **SUBROUTINE** Assembly and stored in an array Diag for later use.

7.3   SOLUTION OF SYSTEM OF EQUATIONS

After assembly and the computation of the diagonal coefficients (adding the *azimuthal* integral as required for infinite regions) a system of simultaneous equations is obtained. The difference to the system of equations obtained for the FEM is that it is not symmetric and fully populated. The non-symmetry of the coefficient matrix had engineers, who were used to symmetric stiffness matrices, baffled for a while. The question was why, since we have used the theorem by Betti, which Maxwell used to prove reciprocity and therefore the symmetry of the stiffness matrix, are we not getting a symmetric coefficient matrix? The answer lies in the fact that we are not solving the integral equations exactly, but by numerical approximation. Instead of enforcing the Betti theorem at an infinite number of points, as we should, we select a limited number of points which are nodal points of the mesh. Another aspect is that the kernel **T** for elasticity problems is not symmetric. However if one examines the coefficient matrices carefully one discovers that if boundary elements with linear functions are used, then the non-symmetry is much less pronounced than if elements with quadratic variation are used. The reason for this is not quite clear.

The fact that coefficient matrices are fully populated makes things easier in the sense that we do not need to worry about sparse solvers at this stage. We will see later that when we introduce multiple regions, for example, to cater for non-homogenities or to model cracks or faults, we will also introduce sparseness.

The lack of sparseness of course means that no savings can be made by using special solution schemes, such as band or skyline storage. The number of degrees of freedom, however, should be considerably smaller as compared with the FEM, especially for soil or rock mechanics problems, where the domain has no defined boundaries.

7.3.1 Gauss elimination

The Gauss elimination method is probably the oldest and most used one for solving the system of equations. Consider the following system of equations

$$\mathbf{Au} = \mathbf{F} \tag{7.18}$$

The solution for unknowns $\mathbf{u}$ involves two steps:

STEP 1: Reduction
Here we introduce zeroes below the diagonal elements, so that we end up with an upper triangular coefficient matrix.

For example, consider the nth and ith equation of a system

$$
\begin{aligned}
a_{nn}u_n + \ldots + a_{nj}u_j + \ldots &= F_n \\
a_{in}u_n + \ldots + a_{ij}u_j + \ldots &= F_i
\end{aligned}
\tag{7.19}
$$

To introduce a zero in the nth column of the ith equation we subtract (a_{in}/a_{nn}) times the equation n from equation i:

$$
\begin{aligned}
a_{nn}u_n + \ldots + a_{nj}u_j + \ldots &= F_n \\
0 \quad + \ldots + a_{ij}^{*}u_j + \ldots &= F_i^{*}
\end{aligned}
\tag{7.20}
$$

where

$$a_{ij}^{*} = a_{ij} - \frac{a_{in}}{a_{nn}}a_{nj}$$

$$F_i^{*} = F_i - \frac{a_{in}}{a_{nn}}F_n \tag{7.21}$$

The procedure, which is sometimes referred to as *elimination of variable n,* can be visualised as a repeated modification of the coefficients a to a*, sometimes referred to as *starring* operation. We continue doing the procedure for all the equations until all the coefficients of $\mathbf{A}$ below the diagonal are zero.

STEP 2: Backsubstitution.
The results may now be obtained by computing the unknown from the last equation, which involves one term on the left hand side only.

The formula for computing the nth unknown is given by:

$$u_n = -\frac{1}{a_{nn}} \sum_{i=n+1}^{N} a_{ni} u_i - F_n \qquad (7.22)$$

where N is the number of unknown.

The above procedure is easily converted to a subroutine. Subroutine SOLVE, shown here, assumes that matrices Lhs, Rhs and u have been assigned a size in the main program.

```
SUBROUTINE Solve(Lhs,Rhs,u)
!-------------------------------------------------
!    Solution of system of equations
!    by Gauss Elimination
!-------------------------------------------------
REAL(KIND=8) :: Lhs(:,:)      !   Equation Left hand side
REAL(KIND=8) :: Rhs(:)        !   Equation right hand side
REAL(KIND=8) ::  u(:)         !   Unknown
INTEGER           M           !   Size of system
REAL(KIND=8) ::  FAC
M= UBOUND(u,1) !  Find out what size has been assigned
!  Reduction
Equation_n: &
DO n=1,M-1
  IF(Lhs(n,n) < 1.0E-10 .AND. Lhs(n,n) > -1.0E-10) THEN
    CALL Error_Message('Singular Matrix')
  END IF
  Equation_i: &
  DO I= n + 1,M
      FAC= Lhs(i,n)/Lhs(n,n)
      Lhs(i, n+1 : M)= Lhs(i, n+1 : M) - Lhs(n, n+1 : M)*FAC
      Rhs(i)= Rhs(i) - Rhs(n)*FAC
  END DO  &
  Equation_i
END DO &
Equation_n
!     Backsubstitution
Unknown_n: &
DO n= M,1,-1
  u(n)= -1.0/Lhs(n,n)*(SUM(Lhs(n,n + 1:M)*u(n + 1:M)) - Rhs(n))
END DO &
Unknown_n
RETURN
END SUBROUTINE Solve
```

As already mentioned for the solution of equations involving many subtractions, it is necessary to use REAL (KIND=8) for the arrays, to avoid an accumulation of round-off error. For a 3-D elasticity problem involving 1000 nodes, the space required for storing the coefficient matrix in REAL (KIND=8) is 72 Mbytes. For the solution of large problems on small computers, this space may not be available, and special algorithms must be devised, where part of the matrix is written onto disk. Methods for the partitioned solution of large systems can be found in Reference 1.

For the reduction of the system of equations using the Gauss algorithm we need three implied DO loops. In the implementation the innermost DO loop is written implicitly using the new feature available in FORTRAN 90. The innermost DO loop involves one multiplication and one subtraction and is executed $(M - n)$ times, where M is the number of unknowns and n is the DO loop counter of the outermost DO loop. The DO loop above it involves a division and is also executed $(M - n)$ times. Finally, the outermost DO-loop is executed $M - 1$ times. It can be shown, therefore, that the total number of operations required is $2/3M^3 + \frac{1}{2}M^2 + 1/6M$. For large systems the first term is dominant.

This means that, for example, for a problem in three-dimensional elasticity involving 1000 nodes, approx. 2×10^{10} operations are necessary for the reduction. If we want to analyse these problems in a reasonable time there is clearly a need for more efficient solvers.

Recently, there has been a resurgence of iterative conjugate gradient solvers. The advantage of these solvers is that the number of operations, and hence the solution time, is only proportional to M^2 and that they can be adapted easily to run on parallel computers[2].

$$
\begin{aligned}
&\mathbf{u}^{(0)} \ \ldots \ \text{initial} \quad \text{guess} \quad \text{for} \quad \mathbf{u} \\
&\mathbf{r}^{(0)} = \mathbf{F} - \mathbf{A}\mathbf{u}^{(0)} \\
&\widetilde{\mathbf{r}}^{(0)} = \mathbf{r}^{(0)} \quad ; \quad \mathbf{q}^{(0)} = \mathbf{p}^{(0)} = 0 \quad ; \quad \rho_0 = 1 \\
&j = 1 \\
&\text{DO} \\
&\quad \rho_j = \ \widetilde{\mathbf{r}}^{(0)} \bullet \mathbf{r}^{(j-1)} \quad ; \quad \beta_j = \frac{\rho_j}{\rho_{j-1}} \\
&\quad \mathbf{a}^{(j)} = \mathbf{r}^{(j-1)} + \beta_j \mathbf{q}^{(j-1)} \quad ; \quad \mathbf{p}^{(j)} = \mathbf{a}^{(j)} + \beta_j \left(\mathbf{q}^{(j-1)} + \beta_j \mathbf{p}^{(j-1)} \right) \\
&\quad \mathbf{v}^{(j)} = \mathbf{A}\mathbf{p}^{(j)} \\
&\quad \sigma_j = \widetilde{\mathbf{r}}^{(0)} \bullet \mathbf{v}^{(j)} \quad ; \quad \alpha_j = \frac{\rho_j}{\sigma_j} \\
&\quad \mathbf{q}^{(j)} = \mathbf{a}^{(j)} - \alpha_j \mathbf{v}^{(j)} \quad ; \quad \mathbf{r}^{(j)} = \mathbf{r}^{(j-1)} + \alpha_j \mathbf{A} \left(\mathbf{a}^{(j)} + \mathbf{q}^{(j)} \right) \\
&\quad \mathbf{u}^{(j)} = \mathbf{u}^{(j-1)} + \alpha_j \left(\mathbf{a}^{(j)} + \mathbf{q}^{(j)} \right) \\
&\quad \text{IF} \left(\text{Converged} \right) \text{EXIT} \\
&\quad j = j+1 \\
&\text{END DO}
\end{aligned}
$$

Figure 7.8   Pseudo-code for conjugate gradient method

7.3.2 Conjugate gradient solver

The conjugate gradient method solves the system of equations using successive improvements of the solutions by creating two vector spaces which are mutually orthogonal (conjugate) to each other. For unsymmetrical systems of equations, the algorithm for solving the system of equations (7.18) with a modified conjugate gradient method[3] is shown in Figure 7.8. The method converges faster if a good initial guess for the unknown u can be made. However, a zero entry is possible.

The method involves two vector dot products and two matrix vector multiplied per iteration. Seven auxiliary vectors are used. The criteria for convergence is either that the maximum difference between iterations, or that the norm of the differences is smaller than the tolerance specified. The conversion of this pseudo-code is straightforward, and needs no further explanation.

```fortran
SUBROUTINE CGSolve(Lhs,F,u)
!-------------------------------------------------
! Solution of system of equations
! by conjugate gradient iteration
!-------------------------------------------------
REAL(KIND=8) ::   Lhs(:,:)     ! Equation Left hand side
REAL(KIND=8) ::        F(:)     ! Equation right hand side
REAL(KIND=8) ::        u(:)     ! Unknowns
REAL(KIND=8),ALLOCATABLE :: r(:),rt(:),q(:)
REAL(KIND=8),ALLOCATABLE :: p(:),v(:),a(:),du(:)
REAL(KIND=8 )::   roj,rojm1,betaj,sigmaj,alfaj,Tol
Tol= 1.0E -4   !    Tolerance for convergence
M= UBOUND(F,1)
!   Allocate temporary arrays
ALLOCATE (r(M),rt(M),q(M),p(M),v(M),a(M),du(M))
!   Only perform this for non-zero start vector
IF(.NOT.ALL(u == 0.0)) THEN
   r= F - MATMUL(Lhs,u)
END IF
!   Initialise
rt= r ; q=0.0 ; p=0.0  ; rojm1=1.0
DO
   roj= DOT_PRODUCT(rt,r)  ;  betaj= roj/rojm1 ; rojm1= roj
   a= r + betaj*q ; p= a + betaj*(q + betaj*p)
   v= MATMUL(Lhs,p)
   Sigmaj= DOT_PRODUCT(rt,v) ; alfaj= roj/sigmaj
   q= a - alfaj*v ; r= r + alfaj*MATMUL(Lhs,a+q)
   du=  alfaj*(a+q)
   u= u + du
   IF(MAXVAL(ABS(du/u)) < Tol) EXIT ! stop if max. diff. is < Tol
END DO
!   De-allocate temporary arrays
DEALLOCATE (r,rt,q,p,v,a,du)
RETURN
END SUBROUTINE CGSolve
```

7.3.3 Scaling

When we look at the fundamental solutions for elasticity, we note that the kernel **U** contains the modulus of elasticity whereas **T** does not. Depending of the units used, we expect a large difference in values. As we have seen at the beginning of this chapter, if there is a mixed boundary value problem then there is a mixture of **U** and **T** terms in the assembled coefficient matrix. This may cause problems in the solution of equations, since very small terms would be subtracted from very large ones. Additionally, we note that for 2-D problems kernel U varies with $ln(1/r)$, which approaches $-\infty$ value as $r \rightarrow \infty$.

For the above reasons, scaling of the data is required. Scaling is applied in such a way that all tractions are divided by E and all coordinates by the largest difference between coordinates (this results in a scaled problem size of unity).

7.4   PROGRAM 7.1: GENERAL PURPOSE PROGRAM, DIRECT METHOD, ONE REGION

We now have developed all the necessary tools for writing a general purpose computer program for computing two- and three-dimensional problems in potential flow and elasticity. The first part of the program reads input data. There are three types of data: job specification, geometry and boundary data. They are read in by calling three separate subroutines **Jobin**, **Geomin**, **BCinput**. The job information consists of the Cartesian dimension of the problem (2-D or 3-D), the type of region (finite or infinite), whether it is a potential or elasticity problem, the type of elements used (linear or quadratic), the properties, that is conductivity for potential problems and modulus of elasticity and Poisson's ratio for elasticity problems, and the number of elements/nodes. The geometrical information consists of the coordinates of the nodes and the element incidences. Finally the boundary conditions are input. In the program we assume that all nodes have a *Neumann* boundary condition with zero prescribed value by default. All nodes with *Dirichlet* boundary conditions, and all nodes having a *Neumann* BC, with non-zero prescribed values have to be input. After specification of the BCs, element destination vectors can be set up by a call to Subroutine Destination, listed in Appendix A. As explained previously, the destinations are the addresses of the coefficients in the global arrays. Note that for symmetry it is advantageous to exclude those degrees of freedom which have zero value, and a node destination vector (Ndest) has been included in order to consider this. As explained previously, a global boundary code vector is needed to cater for the case where the boundary code is discontinuous at a node. Scaling as described above is applied by a call to **SUBROUTINE Scal**.

The assembly is made by calling **SUBROUTINE Assemb**. Since the diagonal coefficients are not computed using numerical integration but are determined using the 'rigid body mode' method, all off-diagonal coefficients are summed and, if the region is infinite, the azimuthal integral is added. Off-diagonal coefficients are stored in a vector **Diag**, and the boundary condition codes will determine if these are assembled into the left- or right-hand side. The system of equations is solved next. Using the element destination vector, the results **Elres_u** and **Elres_t** are gathered from the global vector **u1**. As will be seen later, it is convenient for postprocessing to store results element-by-element.

```
PROGRAM General_purpose_BEM
!-----------------------------------------------------------
!      General purpose BEM program
!      for solving elasticity and potential problems
!-----------------------------------------------------------
USE Utility_lib ; USE Elast_lib ; USE Laplace_lib
USE Integration_lib
IMPLICIT NONE
INTEGER, ALLOCATABLE :: Inci(:,:)   ! Element Incidences
INTEGER, ALLOCATABLE :: BCode(:,:), NCode(:) ! Element BC's
INTEGER, ALLOCATABLE :: Ldest(:,:)  ! Element dest. vector
INTEGER, ALLOCATABLE :: Ndest(:,:)  ! Node destination vector
REAL, ALLOCATABLE :: Elres_u(:,:)   ! Results , u
REAL, ALLOCATABLE :: Elres_t(:,:)   ! Results , t
REAL, ALLOCATABLE :: Elcor(:,:)     ! Element coordinates
REAL, ALLOCATABLE :: xP(:,:)        ! Node co-ordinates
REAL(KIND=8), ALLOCATABLE :: dUe(:,:),dTe(:,:),Diag(:,:)
REAL(KIND=8), ALLOCATABLE :: Lhs(:,:),F(:)
REAL(KIND=8), ALLOCATABLE :: u1(:) ! global vector of unknown
CHARACTER (LEN=80) :: Title
INTEGER :: Cdim,Node,m,n,Istat,Nodel,Nel,Ndof,Toa
INTEGER :: Nreg,Ltyp,Nodes,Maxe,Ndofe,Ndofs,ndg,NE_u,NE_t
INTEGER :: nod,nd,i,j,k,l,DoF,Pos,Isym,nsym,nsy
REAL,ALLOCATABLE    :: Fac(:)       ! Factors for symmetry
REAL,ALLOCATABLE    :: Elres_te(:),Elres_ue(:)
INTEGER,ALLOCATABLE :: Incie(:)     ! Incidences 1 element
INTEGER,ALLOCATABLE :: Ldeste(:)    ! Destination vector
REAL :: Con,E,ny,Scat,Scad
!-----------------------------------------------------------
!   Read job information
!-----------------------------------------------------------
OPEN (UNIT=1,FILE='INPUT',FORM='FORMATTED') ! Input
OPEN (UNIT=2,FILE='OUTPUT',FORM='FORMATTED')! Output
Call Jobin(Title,Cdim,Ndof,Toa,Nreg,Ltyp,Con,E,ny, &
           Isym,nodel,nodes,maxe)
Nsym= 2**Isym   !   number of symmetry loops
ALLOCATE(xP(Cdim,Nodes))   ! Array for node coordinates
ALLOCATE(Inci(Maxe,Nodel)) ! Array for incidences
CALL Geomin(Nodes,Maxe,xp,Inci,Nodel,Cdim)
Ndofe= Nodel*Ndof  !     Total degrees of freedom of element
ALLOCATE(BCode(Maxe,Ndofe))  !    Element Boundary codes
ALLOCATE(Elres_u(Maxe,Ndofe),Elres_t(Maxe,Ndofe))
CALL BCinput(Elres_u,Elres_t,Bcode,nodel,ndofe,ndof)
ALLOCATE(Ldest(maxe,Ndofe))  ! Elem. destination vector
ALLOCATE(Ndest(Nodes,Ndof))
!-----------------------------------------------------------
!   Determine Node and Element destination vectors
!-----------------------------------------------------------
CALL Destination(Isym,Ndest,Ldest,xP, &
Inci,Ndofs,nodes,Ndof,Nodel,Maxe)
!-------------------------------------------------
!      Determine global Boundary code vector
!-------------------------------------------------
```

```
ALLOCATE(NCode(Ndofs))
DoF_o_System: &
DOnd=1,Ndofs
 DO Nel=1,Maxe
   DO m=1,Ndofe
     IF (nd == Ldest(Nel,m) .and. NCode(nd) == 0) THEN
       NCode(nd)= NCode(nd)+BCode(Nel,m)
     END IF
   END DO
 END DO
END DO &
DoF_o_System
CALL Scal(E,xP(:,:),Elres_u(:,:),Elres_t(:,:),Cdim,Scad,Scat)
ALLOCATE(dTe(Ndofs,Ndofe),dUe(Ndofs,Ndofe)) ! Elem. coef.
ALLOCATE(Diag(Ndofs,Ndof))                   ! Diagonal coefficients
ALLOCATE(Lhs(Ndofs,Ndofs),F(Ndofs),u1(Ndofs)) ! global arrays
ALLOCATE(Elcor(Cdim,Nodel))                  ! Elem. Coordinates
ALLOCATE(Fac(Ndofe))                         ! Factor symmetry
ALLOCATE(Incie(Nodel))                       ! Element incidences
ALLOCATE(Ldeste(Ndofe))                      ! Element destination
ALLOCATE(Elres_te(Ndofe),Elres_ue(Ndofe))
!------------------------------------------------------------------
!  Compute element coefficient matrices
!------------------------------------------------------------------
Lhs(:,:) = 0.0; F(:) = 0.0; u1(:) = 0.0
Elements_1:&
DO Nel=1,Maxe
 Symmetry_loop:&
 DO nsy= 1,Nsym
   Elcor(:,:)= xP(:,Inci(Nel,:))   ! gather element coordinates
   Incie= Inci(nel,:)              ! incidences
   Ldeste= Ldest(nel,:)            ! and destinations
   Fac(1:nodel*ndof)= 1.0
   Elres_te(:)=Elres_t(Nel,:)
   IF(Isym > 0) THEN
     CALL Mirror(Isym,nsy,Nodes,Elcor,Fac,&
     Incie,Ldeste,Elres_te,Elres_ue &
             ,nodel,ndof,Cdim)
   END IF
   IF(Cdim == 2) THEN
     IF(Ndof == 1) THEN
       CALL Integ2P(Elcor,Incie,Nodel,Nodes&
       ,xP,Con,dUe,dTe,Ndest,Isym)
     ELSE
       CALL Integ2E(Elcor,Incie,Nodel,Nodes&
       ,xP,E,ny,dUe,dTe,Ndest,Isym)
     END IF
   ELSE
     CALL Integ3(Elcor,Incie,Nodel,Nodes,xP,Ndof &
             ,E,ny,Con,dUe,dTe,Ndest,Isym)
   END IF
   CALL Assembly(Lhs,F,DTe,DUe,Ldeste,BCode(Nel,:),Ncode &
```

```
              ,Elres_u(Nel,:),Elres_te,Diag&
,Ndofe,Ndof,Nodel,Fac)
END DO &
Symmetry_loop
END DO &
Elements_1
!-------------------------------------------------------------
!  Add azimuthal integral for infinite regions
!-------------------------------------------------------------
IF(Nreg == 2) THEN
 DO m=1, Nodes
   DO n=1, Ndof
     IF(Ndest(m,n) == 0)CYCLE
     k=Ndest(m,n)
     Diag(k,n) = Diag(k,n) + 1.0
   END DO
 END DO
END IF
!-------------------------------------------------------------
!  Add Diagonal coefficients
!-------------------------------------------------------------
Collocation_points: &
DO m=1,Ndofs
 Nod=0
 DO n=1, Nodes
   DO l=1,Ndof
     IF (m == Ndest(n,l))THEN
       Nod=n
       EXIT
     END IF
   END DO
   IF (Nod /= 0)EXIT
 END DO
 DO k=1,Ndof
   DoF=Ndest(Nod,k)
   IF(DoF /= 0) THEN
     IF(NCode(DoF) == 1) THEN
       Nel=0
       Pos=0
       DO i=1,Maxe
         DO j=1,Ndofe
           IF(DoF == Ldest(i,j))THEN
             Nel=i
             Pos=j
             EXIT
           END IF
         END DO
         IF(Nel /= 0)EXIT
       END DO
       F(m) = F(m) - Diag(m,k) * Elres_u(Nel,Pos)
     ELSE
       Lhs(m,DoF)= Lhs(m,DoF) + Diag(m,k)
     END IF
```

```fortran
   END IF
 END DO
END DO &
Collocation_points
!-----------------------------------------------------------
!    Solve system of equations
!-----------------------------------------------------------
CALL Solve(Lhs,F,u1)
CLOSE(UNIT=2)
OPEN (UNIT=2,FILE='BERESULTS',FORM='FORMATTED')
!    Gather Element results from global result vector u1
Elements_2: &
DO nel=1,maxe
 D_o_F1:   &
 DO nd=1,Ndofe
   IF(Ldest(nel,nd) /= 0)THEN
     IF(NCode(Ldest(nel,nd)) == 0) THEN
       Elres_u(nel,nd) = Elres_u(nel,nd) + u1(Ldest(nel,nd))
     ELSE
       Elres_t(nel,nd) = Elres_t(nel,nd) + u1(Ldest(nel,nd))
     END IF
   END IF
 END DO &
 D_o_F1
 Elres_u(nel,:)= Elres_u(nel,:) * Scad !  Scale back
 Elres_t(nel,:)= Elres_t(nel,:) / Scat
 WRITE(2,'(24F12.5)') (Elres_u(nel,m), m=1,Ndofe)
 WRITE(2,'(24F12.5)') (Elres_t(nel,m), m=1,Ndofe)
END DO &
Elements_2
END PROGRAM
```

To make the program more readable and easier to modify, the reading of the input has been delegated to subroutines. This also gives the reader some freedom to determine the input FORMAT and implement simple meshgeneration facilities, such as those shown in Reference 1.

```fortran
SUBROUTINE Jobin(Title,Cdim,Ndof,Toa,Nreg,Ltyp,Con,E,ny &
                ,Isym,nodel,nodes,maxe)
!-----------------------------------------------------
!    Subroutine to read in basic job information
!-----------------------------------------------------
CHARACTER(LEN=80), INTENT(OUT):: Title
INTEGER, INTENT(OUT) :: Cdim,Ndof,Toa,Nreg,Ltyp,Isym,nodel
INTEGER, INTENT(OUT) :: Nodes,Maxe
REAL, INTENT(OUT)    :: Con,E,ny
READ(1,'(A80)') Title
WRITE(2,*)'Project:',Title
READ(1,*) Cdim
WRITE(2,*)'Cartesian_dimension:',Cdim
```

```fortran
READ(1,*) Ndof           !      Degrees of freedom per node
IF(NDof == 1) THEN
 WRITE(2,*)'Potential Problem'
ELSE
WRITE(2,*)'Elasticity Problem'
END IF
IF(Ndof == 2)THEN
 READ(1,*) Toa ! Analysis type (plane strain= 1,plane stress= 2)
 IF(Toa == 1)THEN
   WRITE(2,*)'Type of Analysis: Solid Plane Strain'
 ELSE
   WRITE(2,*)'Type of Analysis: Solid Plane Stress'
 END IF
END IF
READ(1,*) Nreg         !     Type of region
IF(NReg == 1) THEN
 WRITE(2,*)'Finite Region'
ELSE
 WRITE(2,*)'Infinite Region'
END IF
READ(1,*) Isym         !     Symmetry code
SELECT CASE (isym)
CASE(0)
WRITE(2,*)'No symmetry'
CASE(1)
WRITE(2,*)'Symmetry about y-z plane'
CASE(2)
WRITE(2,*)'Symmetry about y-z and x-z planes'
CASE(3)
WRITE(2,*)'Symmetry about all planes'
END SELECT                               '
READ(1,*) Ltyp           !     Element type
IF(Ltyp == 1) THEN
WRITE(2,*)'Linear Elements'
ELSE
WRITE(2,*)'Quadratic Elements'
END IF
 !     Determine number of nodes per element
IF(Cdim == 2) THEN      !     Line elements
 IF(Ltyp == 1) THEN
  Nodel= 2
 ELSE
  Nodel= 3
 END IF
ELSE                     !     Surface elements
 IF(Ltyp == 1) THEN
   Nodel= 4
 ELSE
   Nodel= 8
 END IF
END IF
 !   Read properties
```

```fortran
IF(Ndof == 1) THEN
 READ(1,*) Con
 WRITE(2,*)'Conductivity=',Con
ELSE
 READ(1,*) E,ny
 IF(ToA == 2) ny = ny/(1+ny)          ! Solid Plane Stress
 WRITE(2,*)'Modulus:',E
 WRITE(2,*)'Poisson''s ratio:',ny
END IF
READ(1,*) Nodes
WRITE(2,*)'Number of Nodes of System:',Nodes
READ(1,*) Maxe
WRITE(2,*)'Number of Elements of System:', Maxe
RETURN
END SUBROUTINE Jobin
```

```fortran
SUBROUTINE Geomin(Nodes,Maxe,xp,Inci,Nodel,Cdim)
!-------------------------------------------
!   Inputs mesh geometry
!-------------------------------------------
INTEGER, INTENT(IN) :: Nodes      !   Number of nodes
INTEGER, INTENT(IN) :: Maxe       !   Number of elements
INTEGER, INTENT(IN) :: Nodel      !   Number of Nodes of elements
INTEGER, INTENT(IN) :: Cdim       !   Cartesian Dimension
REAL, INTENT(OUT)   :: xP(:,:)    !   Node co-ordinates
REAL                :: xmax(Cdim),xmin(Cdim),delta_x(Cdim)
INTEGER, INTENT(OUT):: Inci(:,:)  !   Element incidences
INTEGER             :: Node,Nel,M,n
!----------------------------------------------------------------
!   Read Node Co-ordinates from Inputfile
!----------------------------------------------------------------
DO Node=1,Nodes
 READ(1,*) (xP(M,Node),M=1,Cdim)
 WRITE(2,'(A5,I5,A8,3F8.2)') 'Node ',Node,&
        ' Coor ',(xP(M,Node),M=1,Cdim)
END DO
!----------------------------------------------------------------
!   Read Incidences from Inputfile
!----------------------------------------------------------------
WRITE(2,*)''
WRITE(2,*)'Incidences: '
WRITE(2,*)''
Elements_1:&
DO Nel=1,Maxe
READ(1,*) (Inci(Nel,n),n=1,Nodel)
WRITE(2,'(A3,I5,A8,24I5)')'EL ',Nel,'  Inci  ',Inci(Nel,:)
END DO &
Elements_1
RETURN
END SUBROUTINE Geomin
```

```fortran
SUBROUTINE BCInput(Elres_u,Elres_t,Bcode,nodel,ndofe,ndof)
!----------------------------------------------
!    Reads boundary conditions
!----------------------------------------------
REAL,INTENT(OUT)     :: Elres_u(:,:)  ! Element results , u
REAL,INTENT(OUT)     :: Elres_t(:,:)  ! Element results , t
INTEGER,INTENT(OUT)  :: BCode(:,:)    ! Element BC´s
INTEGER,INTENT(IN)   :: nodel         ! Nodes per element
INTEGER,INTENT(IN)   :: ndofe         ! D.o.F. per Element
INTEGER,INTENT(IN)   :: ndof          ! D.o.F per Node
INTEGER :: NE_u,NE_t
WRITE(2,*)''
WRITE(2,*)'Elements with Dirichlet BC´s: '
WRITE(2,*)''
Elres_u(:,:)=0  ! Default prescribed values for u = 0.0
BCode = 0       ! Default BC= Neumann Condition
READ(1,*)NE_u
IF(NE_u > 0) THEN
Elem_presc_displ: &
DO n=1,NE_u
 READ(1,*) Nel,(Elres_u(Nel,m),m=1,Ndofe)
 BCode(Nel,:)=1
 WRITE(2,*)'Element ',Nel,'  Prescribed values: '
 Na= 1
 Nodes: &
 DO M= 1,Nodel
   WRITE(2,*) Elres_u(Nel,na:na+ndof-1)
   Na= na+Ndof
 END DO &
 Nodes
END DO &
Elem_presc_displ
END IF
WRITE(2,*)''
WRITE(2,*)'Elements with Neumann BC´s: '
WRITE(2,*)''
Elres_t(:,:)=0   !   Default prescribed values = 0.0
READ(1,*)NE_t
Elem_presc_trac: &
DO n=1,NE_t
 READ(1,*) Nel,(Elres_t(Nel,m),m=1,Ndofe)
 WRITE(2,*)'Element ',Nel,'  Prescribed values: '
 Na= 1
 Nodes1: &
 DO M= 1,Nodel
   WRITE(2,*) Elres_t(Nel,na:na+ndof-1)
   Na= na+Ndof
 END DO &
 Nodes1
END DO &
Elem_presc_trac
RETURN
END SUBROUTINE BCInput
```

7.4.1 User's manual for Program 7.1

The input data which have to be supplied in file INPUT are described below. Free field input is used, that is, numbers are separated by blanks. However, all numbers, including zero entries, must be specified.

The input is divided into two parts. First, general information about the problem is read in, then the mesh geometry is specified. The problem may consist of linear and quadratic elements, as shown in Figure 7.9. The sequence in which node numbers have to be entered when specifying incidences is also shown. Note that this order determines the direction of the outward normal, which has to point away from the material. For 3-D elements, if node numbers are entered in an anticlockwise direction, the outward normal points towards the viewer.

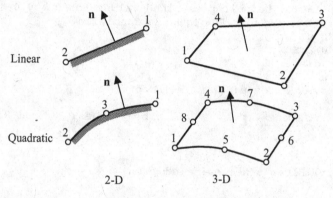

Figure 7.9   Element library

INPUT DATA SPECIFICATION FOR General_purpose-BEM program

1.0	Title specification	
	TITLE	Project Title (max 60 characters)
2.0	Cartesian dimension of problem	
	Cdim	Cartesian dimension
		2= two-dimensional problem
		3= three-dimensional problem
3.0	Problem type specification	
	Ndof	Degree of freedom per node
		1= potential problem
		2,3= elasticity problem
4.0	Analysis type (Only input for Ndof = 2 !)	
	Toa	Type of analysis
		1= Plane strain
		2= plane stress

5.0   Region type specification
    Nreg

                                             Region code
                                             1= finite region
                                           2= infinite region

6.0   Symmetry specification
    ISym

                                             Symmetry code
                                             0= no symmetry
                                           1= symmetry about y-z plane
                                           2= symmetry about y-z and x-z planes
                                           3= symmetry about all 3 planes

7.0   Element type specification
    Ltyp

                                             Element type
                                           1= linear
                                           2= quadratic

8.0   Material properties
    C1, C2

                                             Material properties
                                         $C1 = k$ (conductivity) for $Ndof=1$
                                           = E (Modulus of Elasticity) for $Ndof>1$
                                           $C2 =$ Poisson's ratio for $Ndof>1$

9.0   Node specification
    Nodes   Number of nodes

10.0   Element specification
    Maxe   Number of elements

11.0   Loop over Nodes
    x, y, (z)   Node coordinates

12.0   Loop over all elements
    Inci (1:Element nodes)   global node numbers of element nodes

13.0   *Dirichlet* boundary conditions
    NE_u   Number of elements with *Dirichlet* BC

14.0   Prescribed values for *Dirichlet* BC
    Nel, Elres_u(1 : Element D.o.F.)   Specification of boundary condition
                                               Nel = Element number to be assigned BC
                                           Elres_u = Prescribed values for all
                                           degrees of freedom of element: all d.o.F
                                           first node; all d.o.F second node etc.

15.0   *Neumann* boundary conditions
    NE_u   Number of elements with *Neumann* BC
                                               Only specify for non-zero prescribed
                                             values.

16.0   Prescribed values for *Neumann* BC
    Nel, Elres_t(1 : Element D.o.F.)   Specification of boundary condition
                                               Nel = Element number to be assigned BC
                                           Elres_t = Prescribed values for all
                                         degrees of freedom: all d.o.f. 1st node etc.

7.4.2 Sample input file

For the example of the heat flow past a cylindrical isolator, which was solved with the
Trefftz method and the direct method with constant elements, we present the input and
output files for an analysis with eight linear elements and no symmetry (Figure 7.10).

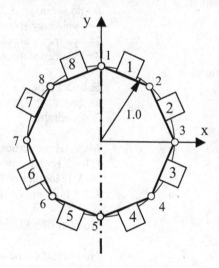

Figure 7.10 Discretisation of cylindrical isolator used for sample input file

File INPUT

```
Flow past cylindrical isolator, 8 linear elements
2
1
0
1
 1.00
 8
 8
 0.414 1.000    ! Coordinates
-0.414 1.000
-1.000 0.414
-1.000-0.414
-0.414-1.000
 0.414-1.000
 1.000-0.414
 1.000 0.414
 1 2            ! Incidences
 2 3
 3 4
 4 5
 5 6
```

```
6  7
7  8
8  1
0                    !  no Dirichlet BC's
8                    !  Neumann BC'c
1   1.000   1.000
2   0.707   0.707
3   0.000   0.000
4  -0.707  -0.707
5  -1.000  -1.000
6  -0.707  -0.707
7   0.000   0.000
8   0.707   0.707
```

7.5 CONCLUSIONS

In this chapter we have developed a general purpose program, which can be used to solve any problem in elasticity and potential flow, or if we substitute the appropriate fundamental solutions, any problem at all. This versatility has been made possible through the use of isoparametric elements and numerical integration. In essence, the boundary element method has borrowed ideas from the finite element method and, in particular, the ideas of Ergatoudis, who first suggested the use of parametric elements and numerical integration.

Indeed, there are also other similarities with the FEM in that the system of equations is obtained by assembling element contributions. In the assembly procedure we have found that the treatment of discontinous boundary conditions, as they are encountered often in practical applications, needs special attention and will change the assembly process.

The implementation of the program is far from efficient. If one does an analysis of runtime spent in each part of the program, one will realise that the computation of the element coefficient matrices will take a significant amount of time. This is because, as pointed out in Chapter 6, the order of DO loops in the numerical integration is not optimised to reduce the number of calculations. Also in the implementation, all matrices must be stored in RAM, and this may severely restrict the size of problems which can be solved.

We have noted that the system of equations obtained are fully populated, that is, the coefficient matrix contains no zero elements. This is in contrast to the FEM, where systems are sparsely populated, i.e. contain a large number of zeros. The other difference with the FEM is that the stiffness matrix is not symmetric. This has been claimed as one of the disadvantages of the method. However, this is more than compensated by the fact that the size of the system is significantly smaller.

The output from the program consists only of the values of the unknown at the boundary. The unknowns are either the temperature/displacement or the flow normal to the boundary/boundary stresses. The computation of the complete flow vectors/stress tensors at the boundary, as well as the computation of values inside the domain is discussed in the next chapter.

7.6   EXERCISES

Exercise 7.1

Using Program 7.1 compute the problem of flow past a cylindrical isolator, find out the influence of the following on the accuracy of results:

(a) when linear and quadratic boundary elements are used,
(b) when the number of elements is 8, 16 and 32.

Plot the error in the computation of maximum temperature against the number of elements.

Exercise 7.2

Modify the problem computed in Exercise 7.1 by changing the shape of the isolator, so that it has an elliptical shape, with a ratio vertical to horizontal axis of 2.0. Comment on the changes in the boundary values due to the change in shape.

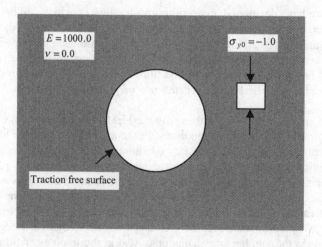

Figure 7.11   Circular excavation in an infinite domain

Exercise 7.3

Using Program 7.1, compute the problem of a circular excavation in a plane strain infinite pre-stressed domain Figure 7.11, find out the influence of the following on the accuracy of results:

(a) when linear and quadratic boundary elements are used,
(b) when the number of elements is 8, 16 and 32.

Plot the error against the number of elements.

Hint: *This is the elasticity problem equivalent to the heat flow problem in Exercise 7.1. The problem is divided into two:*

1. *Continuum with no hole and the initial stresses only.*
2. *Continuum with a hole an Neuman boundary conditions. These boundary conditionsare computed in such a way that when stresses at the boundary of problem 1 are added to those at problem 2, zero values of boundary tractions are obtained. To compute the boundary tractions equivalent to the initial stresses use equation (4.26).*

Exercise 7.4
Modify the problem computed in Exercise 7.3 by changing the shape of the excavation, so that it has an elliptical shape with a ratio vertical to horizontal axis of 2.0. Comment on the changes in the deformations due to the change in shape.

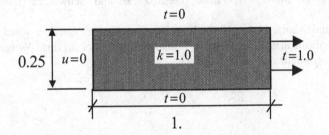

Figure 7.12 Potential problem with boundary conditions

Exercise 7.5
Using Program 7.1, compute the potential problem of the beam depicted in Figure 7.12. Assume k=1.0, length=1.0, height=0.25 and prescribed temperatures of 0.0 at the left end and a prescribed flow of 1.0 at the right. Construct two meshes, one with linear and one with quadratic boundary elements. Comment on the results.

Exercise 7.6
Using Program 7.1, compute the problem of the cantilever beam depicted in Figure 7.13. Plot the displaced shape and distribution of the normal and shear tractions at the fixed end. Construct two meshes, one with linear and the other with quadratic boundary elements. Comment on the results.

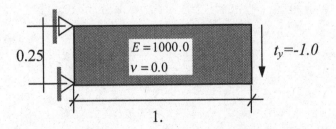

Figure 7.13 Example of a cantilever beam

Exercise 7.7

Using Program 7.1, compute the problem of the cantilever beam depicted in Figure 7.12, but apply a vertical movement of unity to the top support instead of traction at the free end. Plot the displaced shape and verify that this is just a rigid body rotation of the beam.

7.7   REFERENCES

1. Beer, G. and Watson, J.O. (1992) *Introduction to Finite and Boundary Element Methods for Engineers*. Wiley, Chichester.
2. Smith, I.M. (1999) Parallel finite element analysis. *Numerical models in geomechanics- NUMOG VII*, Pande, Pietruszczak and Schweiger (eds). Balkema, Rotterdam.
3. Payer, H.J. and Mang, H.A. (1997) İterative strategies for solving systems of linear algebraic equations arising in 3D BE-FE analyses of tunnel driving. *Numerical Linear Algebra with Applications*, **4**(3), 239-268.

8

Postprocessing

Man soll auf alles achten, denn man kann alles deuten
(You should consider everything
because you can interpret everything)
H. Hesse

8.1 INTRODUCTION

In the previous chapter, we developed a general purpose computer program for the analysis of two and three-dimensional problems in elasticity and potential flow. This program only calculates the values of unknowns (temperature/displacements or boundary flow/tractions) at the nodes of boundary elements. In this chapter, we will develop procedures for the calculation of other results which are also of interest. Among these are the flow vector or the stress tensor at the boundary and at points inside the domain.

There are two types of approximations involved in a boundary element analysis. The first is that the distribution of temperature/displacement or boundary flow/stress is approximated at the boundary by shape functions defined locally for each element. The second approximation is that the theorem by Betti is only satisfied at certain points on the boundary (collocation points).

The variation of temperature/displacements inside the domain can be determined from boundary values. It is therefore possible to compute the results at any point inside the domain as a postprocessing exercise, after the analysis has been performed. This is in contrast to the FEM, where results are only available at points inside finite elements. Indeed, in the BEM the computation of results at interior points can now be part of a graphical postprocessor with the option that the user may freely specify locations where results are required.

In Section 5.5 we have already shown how values of temperature/potential and flow inside the domain can be obtained for constant elements. However, formula (5.55) developed there cannot be applied for points which are exactly on the boundary, because a division by zero would result. On the simple test example in Chapter 5, we have seen also that when the ratio distance from the point to the nearest element (R) over the size of the element (L) is small, the quality of results deteriorates. Since exact integration is used, this cannot be attributed to precision of integration, but has to do with the approximation of boundary values. For constant boundary elements, the error is therefore attributed to poor approximation of the known and unknown boundary values by constant elements.

When isoparametric boundary elements are introduced, the approximation of boundary values is more accurate, but exact integration can no longer be used. Therefore, the quality of results will not only depend on the approximation of boundary values, but also on the precision of integration. We have already seen that when integrating functions which are singular, great care has to be taken to maintain adequate precision of integration when the point of singularity is approached. We will see that when computing internal results, the order of singularity will be one greater than for computation of the coefficient matrices, and therefore a great number of integration points will be needed, as the internal point gets close to the boundary. This means that a considerable computational effort has to be expanded for such points. There is scope, therefore, for looking at an alternative method for computing points which are very close to or even at, the boundary. We will see in the next section how the computation of results on the boundary can be achieved. This chapter is therefore divided into two parts: computation of boundary results and computation of internal results.

8.2 COMPUTATION OF BOUNDARY RESULTS

For computation of results at the boundary itself we use a procedure which is essentially the same as that used in the finite element method for computing results inside elements. Along a boundary element we know (after the solution) the variation of both u and t, because of the approximation introduced by equation (6.1). We can therefore compute fluxes or stresses tangential to the boundary by differentiation of these variations. In the following we will discuss two- and three-dimensional potential and elasticity problems separately.

8.2.1 Potential problems

The temperature distribution on a boundary element in terms of N nodal values u_n^e is:

$$u = \sum_{n=1}^{N} N_n u_n^e \qquad (8.1)$$

For two-dimensional problems we define a vector $\mathbf{V}_\xi$ in the direction tangential to the boundary (Figure 8.1), where

$$\mathbf{V}_\xi = \frac{\partial}{\partial \xi}\mathbf{x} = \sum \frac{\partial N_n}{\partial \xi}\mathbf{x}_n^e \qquad (8.2)$$

and $\mathbf{x}_n^e$ is a vector containing the coordinates of the element nodes. The flow in tangential direction is computed by:

$$q_{\bar{x}} = -k\frac{\partial u}{\partial \bar{x}} = -k\sum_{n=1}^{N}\frac{\partial N_n}{\partial \xi}\frac{\partial \xi}{\partial \bar{x}}u_n^e \tag{8.3}$$

where $\partial \xi / \partial \bar{x}$ is computed by

$$\frac{\partial \xi}{\partial \bar{x}} = \frac{1}{\sqrt{V_{\xi x}^2 + V_{\xi y}^2}} \tag{8.4}$$

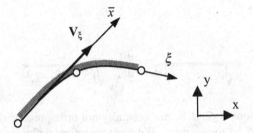

Figure 8.1   Local coordinate system for computing boundary values

For three-dimensional problems we construct, at a point inside the element, a local coordinate system x' and y' with directions as specified by vectors $\mathbf{V}_\xi$ and $\mathbf{V}_\eta$ tangential to the intrinsic coordinates ξ, η (Figure 8.2).

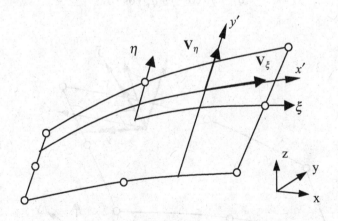

Figure 8.2   Local coordinate systems for surface element

Vector $\mathbf{V}_\xi$ has been defined previously, and $\mathbf{V}_\eta$ is given by:

$$\mathbf{V}_\eta = \frac{\partial}{\partial\eta}\mathbf{x} = \sum \frac{\partial N_n}{\partial\eta}\mathbf{x}_n^e \tag{8.5}$$

Components of the flow vector in the local directions are:

$$q_{x'} = -k\frac{\partial u}{\partial x'} = -k\sum_{n=1}^{N}\frac{\partial N_n}{\partial\xi}\frac{\partial\xi}{\partial x'}u_n^e$$

$$q_{y'} = -k\frac{\partial u}{\partial y'} = -k\sum_{n=1}^{N}\frac{\partial N_n}{\partial\eta}\frac{\partial\eta}{\partial y'}u_n^e \tag{8.6}$$

where

$$\frac{\partial\xi}{\partial x'} = \frac{1}{\sqrt{V_{\xi x}^2 + V_{\xi y}^2 + V_{\xi z}^2}}, \quad \frac{\partial\eta}{\partial y'} = \frac{1}{\sqrt{V_{\eta x}^2 + V_{\eta y}^2 + V_{\eta z}^2}} \tag{8.7}$$

Unfortunately, directions $\mathbf{V}_\xi$ and $\mathbf{V}_\eta$ are generally not orthogonal and also change with the location of the point on the element. We must therefore define an orthogonal set of axes which are independent of the directions of the intrinsic axes. The direction of these orthogonal axes is defined by unit vectors $\mathbf{v}_1$ and $\mathbf{v}_2$, which are computed using the scheme already outlined in Section 3.8 (Figure 8.3).

The flow in directions $\bar{x}, \bar{y}$ is computed by

$$q_{\bar{x}} = q_{x'}\cos^2\alpha + q_{y'}\cos^2\gamma$$

$$q_{\bar{y}} = q_{x'}\cos^2\gamma + q_{y'}\cos^2\beta \tag{8.8}$$

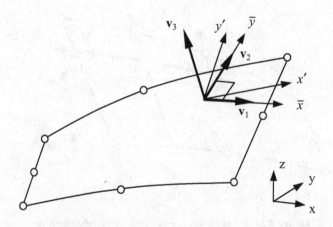

Figure 8.3 Local orthogonal coordinate system for computation of flows

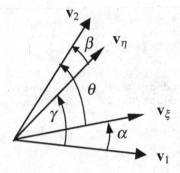

Figure 8.4   Definition of direction angles for transformation

where the direction cosines are defined as (Figure 8.4):

$$\cos\alpha = \mathbf{v}_\xi \bullet \mathbf{v}_1 \quad \cos\beta = \mathbf{v}_\eta \bullet \mathbf{v}_2 \quad \cos\gamma = \mathbf{v}_\eta \bullet \mathbf{v}_1 \quad \cos\theta = \mathbf{v}_\xi \bullet \mathbf{v}_2 \tag{8.9}$$

The theory just outlined may be programmed into a subroutine which computes boundary flows at a point on the boundary element with the intrinsic coordinates ξ, (η).

```fortran
SUBROUTINE BFLOW(Flow,xsi,eta,u,Inci,Elcor,k)
!----------------------------------------------------
!     Computes flow vectors in tangential directions
!----------------------------------------------------
IMPLICIT NONE
REAL , INTENT(OUT) :: Flow(:)  ! Flow vector
REAL , INTENT(IN)  :: xsi,eta  ! intrinsic coordinates of point
REAL , INTENT(IN)  :: u(:)     ! Nodal temperatures/potentials
INTEGER, INTENT (IN) :: Inci(:)  ! Element Incidences
REAL, INTENT (IN)    :: Elcor(:,:)  ! Element coordinates
REAL, INTENT (IN)    :: k       ! Conductivity
REAL, ALLOCATABLE    :: Vxsi(:),Veta(:),DNi(:,:),v3(:)
INTEGER :: Nodes,Cdim,Ldim
REAL :: Jxsi,Jeta,Flows(2),v1(3),v2(3),CosA,CosB,CosG,CosT
Nodes= UBOUND(ELCOR,2)     ! Number of nodes
Cdim= UBOUND(ELCOR,1)      ! Cartesian Dimension
Ldim= Cdim-1               ! Local (element) dimension
ALLOCATE (Vxsi(cdim),Dni(Nodes,Ldim),v3(Cdim))
IF(ldim > 1) ALLOCATE (Veta(cdim))
!   Compute Vector(s) tangential to boundary surface
 CALL Serendip_deriv(DNi,xsi,eta,ldim,nodes,inci)
 Vxsi(1)= Dot_Product(Dni(:,1),Elcor(1,:))
 Vxsi(2)= Dot_Product(Dni(:,1),Elcor(2,:))
IF(Cdim = 2) THEN
 CALL Vector_norm(Vxsi,Jxsi)
 Flow(1)= -k*Dot_product(Dni(:,1),u)/Jxsi
ELSE
 Vxsi(3)= Dot_Product(Dni(:,1),Elcor(3,:))
CALL Vector_norm(Vxsi,Jxsi)
```

```
Veta(1)= Dot_Product(Dni(:,2),Elcor(1,:))
Veta(2)= Dot_Product(Dni(:,2),Elcor(2,:))
Veta(3)= Dot_Product(Dni(:,2),Elcor(3,:))
CALL Vector_norm(Veta,Jeta)
!    Flows in skew coordinate system
Flows(1)= -k*Dot_product(Dni(:,1),u)/Jxsi
Flows(2)= -k*Dot_product(Dni(:,2),u)/Jeta
!    Orthoginal system
v3= Vector_ex(Vxsi,Veta)
Call Ortho(v3,v1,v2)
CosA= DOT_Product(Vxsi,v1)
CosB= DOT_Product(Veta,v2)
CosG= DOT_Product(Veta,v1)
CosT= DOT-Product(Vxsi,v2)
Flow(1)= Flows(1)*CosA**2 + Flows(2)* CosG**2
Flow(2)= Flows(1)*CosT**2 + Flows(2)* CosB**2
END IF
RETURN
END SUBROUTINE BFLOW
```

8.2.2   Elasticity problems

The computation of boundary values of stress for elasticity problems is similar to the one for potential problems. For elasticity, displacements **u** inside a boundary element are given in terms of N nodal displacements $\mathbf{u}_n^e$ by

$$\mathbf{u} = \sum_{n=1}^{N} N_n \mathbf{u}_n^e \tag{8.10}$$

For two-dimensional problems the strain in tangential direction is computed by taking the vector dot product between the derivative of the displacement vector and a unit vector in tangential direction $\mathbf{v}_\xi$, and by scaling the result (Figure 8.5):.

$$\varepsilon_{\bar{x}} = \frac{\partial u_{\bar{x}}}{\partial \bar{x}} = \left(\frac{\partial}{\partial \xi} \mathbf{u} \bullet \mathbf{v}_\xi \right) \frac{\partial \xi}{\partial \bar{x}} \tag{8.11}$$

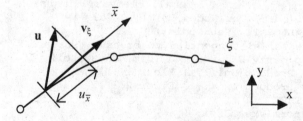

Figure 8.5   Computation of tangential strain

where

$$\frac{\partial}{\partial \xi} \mathbf{u} = \left\{ \begin{array}{c} \dfrac{\partial u_x}{\partial \xi} \\[2mm] \dfrac{\partial u_y}{\partial \xi} \end{array} \right\}$$

(8.12)

The derivatives of the displacements are given by

$$\frac{\partial u_x}{\partial \xi} = \sum_{n=1}^{N} \frac{\partial N_n}{\partial \xi} u_{xn}^e, \quad \frac{\partial u_y}{\partial \xi} = \sum_{n=1}^{N} \frac{\partial N_n}{\partial \xi} u_{yn}^e$$

(8.13)

The stresses in tangential direction are computed using Hooke's law. For plane stress conditions we have (Figure 8.6):

$$\sigma_{\bar{x}} = E \varepsilon_{\bar{x}} + \nu\, t_{\bar{y}}$$

(8.14)

where $t_{\bar{y}}$ is the traction normal to the boundary.

For plane strain we have

$$\sigma_{\bar{x}} = \frac{1}{1-\nu}\left(\frac{E}{1+\nu} \varepsilon_{\bar{x}} - \nu\, t_{\bar{y}} \right)$$

(8.15)

For three-dimensional problems we first compute the strain components in local x', y' directions. These strains are obtained in the same way as for two-dimensional problems by projecting the derivatives of displacement vector $\mathbf{u}$ onto the unit tangential vectors $\mathbf{v}_\xi$ and $\mathbf{v}_\eta$ (Figure 8.7).

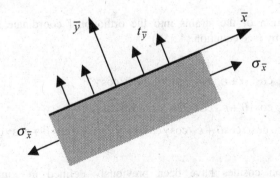

Figure 8.6   Computation of stresses for plane strain problems

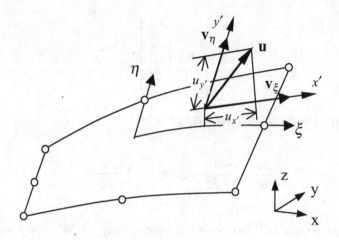

Figure 8.7   Computation of strains for three-dimensional problems

The strains in the local directions are given by

$$\varepsilon_{x'} = \frac{\partial u_{x'}}{\partial x'} = \left(\frac{\partial}{\partial \xi} \mathbf{u} \bullet \mathbf{v}_\xi \right) \frac{\partial \xi}{\partial x'}$$

$$\varepsilon_{y'} = \frac{\partial u_{y'}}{\partial y'} = \left(\frac{\partial}{\partial \eta} \mathbf{u} \bullet \mathbf{v}_\eta \right) \frac{\partial \eta}{\partial y'} \tag{8.16}$$

$$\gamma_{x'y'} = \frac{\partial u_{x'}}{\partial y'} + \frac{\partial u_{y'}}{\partial x'} = \left(\frac{\partial}{\partial \eta} \mathbf{u} \bullet \mathbf{v}_\xi \right) \frac{\partial \eta}{\partial y'} + \left(\frac{\partial}{\partial \xi} \mathbf{u} \bullet \mathbf{v}_\eta \right) \frac{\partial \xi}{\partial x'}$$

The transformation of the strains into the orthogonal coordinate system shown in Figure 8.4 is given by (see equation (4.35))

$$\varepsilon_{\bar{x}} = \varepsilon_{x'} \cos^2 \alpha + \varepsilon_{y'} \cos^2 \gamma + \gamma_{x'y'} \cos\alpha \cos\gamma$$

$$\varepsilon_{\bar{y}} = \varepsilon_{x'} \cos^2 \theta + \varepsilon_{y'} \cos^2 \beta + \gamma_{x'y'} \cos\theta \cos\beta \tag{8.17}$$

$$\gamma_{\overline{xy}} = \varepsilon_{x'} \cos\alpha \cos\theta + \varepsilon_{y'} \cos\gamma \cos\beta + \gamma_{x'y'} (\cos\alpha \cos\beta + \cos\gamma \cos\theta)$$

where the direction cosines have been previously defined in equation (8.9). The components of stress in the local orthogonal system are shown in Figure 8.8. According to Chapter 4, the stresses in the tangential plane are related to the strains by

$$\varepsilon_{\bar{x}} = \frac{1}{E}(\sigma_{\bar{x}} - v\sigma_{\bar{y}} - v\sigma_{\bar{z}})$$

$$\varepsilon_{\bar{y}} = \frac{1}{E}(\sigma_{\bar{y}} - v\sigma_{\bar{x}} - v\sigma_{\bar{z}}) \qquad (8.18)$$

$$\gamma_{\overline{xy}} = \frac{1}{G}\tau_{\overline{xy}}$$

Using equation (8.18) and considering equilibrium of stresses and the boundary tractions $t_{\bar{x}}, t_{\bar{y}}, t_{\bar{z}}$, shown in Figure 8.8, the stresses may be computed as

$$\sigma_{\bar{x}} = C_1(\varepsilon_{\bar{x}} + v\varepsilon_{\bar{y}}) + C_2 t_{\bar{z}}$$

$$\sigma_{\bar{y}} = C_1(\varepsilon_{\bar{y}} + v\varepsilon_{\bar{x}}) + C_2 t_{\bar{z}} \qquad (8.19)$$

$$\sigma_{\bar{z}} = t_{\bar{z}}$$

$$\tau_{\overline{xy}} = G\gamma_{\overline{xy}} \quad \tau_{\overline{xz}} = t_{\bar{x}} \quad \tau_{\overline{yz}} = t_{\bar{y}}$$

where

$$C_1 = \frac{E}{1-v^2}; \quad C_2 = \frac{v}{1-v} \qquad (8.20)$$

The stresses may be transformed into the global x, y, z coordinate system as outlined in Section 4.3.

The theory is translated into SUBROUTINE Bstress which computes the stress components in the tangential plane at a point with the intrinsic coordinates ξ, η on a boundary element. The subroutine is very similar to Bflow. However, in addition to vector **u**, we have to specify **t** in the list of input parameters. Also, we must provide an indicator specifying whether plane stress or plane strain is assumed for a 2-D analysis.

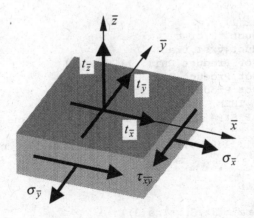

Figure 8.8   Stresses and boundary tractions at a boundary point

```fortran
SUBROUTINE BStress(Stress,xsi,eta,u,t,Inci,Elcor,E,ny,IPS)
!-------------------------------------------------
!     Computes stresses in directions tangential to a
!     Boundary Element surface
!-------------------------------------------------
REAL , INTENT(OUT)     :: Stress(:) ! Stress vector
REAL , INTENT(IN)      :: xsi,eta   ! intrinsic coordinates of point
REAL , INTENT(IN)      :: u(:,:)    ! Nodal displacements
REAL , INTENT(IN)      :: t(:,:)    ! Nodal Tractionsz
INTEGER, INTENT (IN)   :: Inci(:)   ! Element Incidences
REAL, INTENT (IN)      :: Elcor(:,:) !  Element coordinates
REAL, INTENT (IN)      :: E,ny       ! Elastic constants
INTEGER, INTENT (IN)   :: IPS !IPS= 0 plane strain; =1 plane stress
REAL, ALLOCATABLE      :: Vxsi(:),Veta(:),DNi(:,:),Ni(:),trac(:)
INTEGER :: Nodes, Cdim, Ldim
REAL :: Jxsi,Jeta,v1(3),v2(3),CosA, CosB, CosG, CosT,v3(3)
REAL :: C1,C2,G,tn,ts,ts1,ts2
REAL , ALLOCATABLE :: Dudxsi(:),Dudeta(:),Strain(:),Strains(:)
Nodes= UBOUND(ELCOR,2)  !  Number of nodes
Cdim = UBOUND(ELCOR,1)  !  Cartesian Dimension
Ldim= Cdim-1            !  Local (element) dimension
ALLOCATE (Vxsi(cdim),Veta(cdim),Dni(Nodes,Ldim),Ni(Nodes))
ALLOCATE (Dudxsi(Cdim),Dudeta(Cdim),trac(Cdim))
!   Compute Vector(s) tangential to boundary surface
CALL Serendip_deriv(DNi,xsi,eta,ldim,nodes,inci)
CALL Serendip_func(Ni,xsi,eta,ldim,nodes,inci)
trac(1)= Dot_Product(Ni,t(:,1))
trac(2)= Dot_Product(Ni,t(:,2))
Vxsi(1)= Dot_Product(Dni(:,1),Elcor(1,:))
Vxsi(2)= Dot_Product(Dni(:,1),Elcor(2,:))
IF(Cdim == 2) THEN
   ALLOCATE (Strain(1))
   CALL Vector_norm(Vxsi,Jxsi)
   V3(1)=    Vxsi(2)
   v3(2)= -  Vxsi(1)
   tn= Dot_Product(v3,trac)
   ts= Dot_Product(vxsi,trac)
   DuDxsi(1)= Dot_Product(Dni(:,1),u(1,:))
   DuDxsi(2)= Dot_Product(Dni(:,1),u(2,:))
   Strain(1)= Dot_Product(DuDxsi,Vxsi)/Jxsi
   IF(IPS /= 0) THEN
    Stress(1)= E*Strain(1) - ny*tn       !       plane stress
   ELSE
    Stress(1)= 1/(1.0-ny)*(E/(1.0-ny)*Strain(1) - ny*tn)
   END IF
    Stress(2)= tn
    Stress(3)= ts
 ELSE
   ALLOCATE (Strain(3),Strains(3))
   trac(3)= Dot_Product(Ni,t(:,3))
   Vxsi(3)= Dot_Product(Dni(:,1),Elcor(3,:))CALL
Vector_norm(Vxsi,Jxsi)
```

```
  Veta(1)= Dot_Product(Dni(:,2),Elcor(1,:))
  Veta(2)= Dot_Product(Dni(:,2),Elcor(2,:))
  Veta(3)= Dot_Product(Dni(:,2),Elcor(3,:))
  CALL Vector_norm(Veta,Jeta)
  V3= Vector_ex(Vxsi,veta)
  tn= Dot_Product(v3,trac)
  DuDxsi(1)= Dot_Product(Dni(:,1),u(1,:))
  DuDxsi(2)= Dot_Product(Dni(:,1),u(2,:))
  DuDxsi(3)= Dot_Product(Dni(:,1),u(3,:))
  DuDeta(1)= Dot_Product(Dni(:,2),u(1,:))
  DuDeta(2)= Dot_Product(Dni(:,2),u(2,:))
  DuDeta(3)= Dot_Product(Dni(:,2),u(3,:))
! Strains in skew coordinate system
  Strains(1)= Dot_product(DuDxsi,Vxsi)/Jxsi
  Strains(2)= Dot_product(DuDeta,Veta)/Jeta
  Strains(3)= Dot_product(DuDeta,Vxsi)/Jeta + &
              Dot_product(DuDxsi,Veta)/Jxsi
! Orthoginal system
  v3= Vector_ex(Vxsi,Veta)
  CALL Ortho(v3,v1,v2)
  ts1= Dot_Product(v1,trac)
  ts2= Dot_Product(v2,trac)
  CosA= DOT_Product(Vxsi,v1)
  CosB= DOT_Product(Veta,v2)
  CosG= DOT_Product(Veta,v1)
  CosT= DOT_Product(Vxsi,v2)
! Compute Strains
  Strain(1)= Strains(1)*CosA**2 + Strains(2)*CosG**2 &
                        + Strains(3)*CosA*CosG
  Strain(2)= Strains(1)*CosT**2 + Strains(2)*CosB**2 &
                        + Strains(3)*CosT*CosB
  Strain(3)= Strains(1)*CosG*CosT + Strains(2)*CosG*CosB &
                        + Strain(3)*(CosA*CosB+CosG*CosT)
! Compute stresses
  C1= E/(1.0-ny**2)  ;  C2= ny/(1.0-ny) ; G=E/(1.0-2*ny)
  Stress(1)= C1*(Strain(1)+ny*strain(2))+ C2*Tn
  Stress(2)= C1*(Strain(2)+ny*strain(1))+ C2*Tn
  Stress(3)= tn
  Stress(4)= G*Strain(3)
  Stress(5)= ts1
  Stress(6)= ts2
END IF
RETURN
End SUBROUTINE BStress
```

8.3   COMPUTATION OF INTERNAL RESULTS

The computation of internal results was already discussed briefly in Chapter 6, where flow vectors were computed inside the domain for constant elements. Here we present a more general approach for isoparametric elements. We will again discuss potential and elasticity problems separately.

8.3.1 Potential problems

To compute temperature/potential at a point P_a we rewrite equation (5.19):

$$u(P_a) = \int_S t(Q)U(P_a,Q)dS(Q) - \int_S u(Q)T(P_a,Q)dS(Q) \tag{8.21}$$

Flows at P_a in the x-, y- and z- directions are given by

$$q_x(P_a) = -k\frac{\partial u}{\partial x}(P_a) = -k\left(\int_S t(Q)\frac{\partial U}{\partial x}(P_a,Q)dS(Q) - \int_S u(Q)\frac{\partial T}{\partial x}(P_a,Q)dS(Q) \right)$$

$$q_y(P_a) = -k\frac{\partial u}{\partial y}(P_a) = -k\left(\int_S t(Q)\frac{\partial U}{\partial y}(P_a,Q)dS(Q) - \int_S u(Q)\frac{\partial T}{\partial y}(P_a,Q)dS(Q) \right) \tag{8.22}$$

$$q_z(P_a) = -k\frac{\partial u}{\partial z}(P_a) = -k\left(\int_S t(Q)\frac{\partial U}{\partial z}(P_a,Q)dS(Q) - \int_S u(Q)\frac{\partial T}{\partial z}(P_a,Q)dS(Q) \right)$$

where derivatives of U have been presented previously, and derivatives of T for two-dimensional problems are

$$\frac{\partial T}{\partial x} = \frac{\partial}{\partial x}\left[n_x\frac{\partial U}{\partial x} + n_y\frac{\partial U}{\partial y} \right] = \frac{\cos\theta}{2\pi r^2}\frac{r_x}{r}$$

$$\frac{\partial T}{\partial y} = \frac{\partial}{\partial y}\left[n_x\frac{\partial U}{\partial x} + n_y\frac{\partial U}{\partial y} \right] = \frac{\cos\theta}{2\pi r^2}\frac{r_y}{r} \tag{8.23}$$

and for three-dimensional problems

$$\frac{\partial T}{\partial x} = \frac{\partial}{\partial x}\left[n_x\frac{\partial U}{\partial x} + n_y\frac{\partial U}{\partial y} + n_z\frac{\partial U}{\partial z} \right] = \frac{\cos\theta}{4\pi r^3}\frac{r_x}{r}$$

$$\frac{\partial T}{\partial y} = \frac{\partial}{\partial y}\left[n_x\frac{\partial U}{\partial x} + n_y\frac{\partial U}{\partial y} + n_z\frac{\partial U}{\partial z} \right] = \frac{\cos\theta}{4\pi r^3}\frac{r_y}{r} \tag{8.24}$$

$$\frac{\partial T}{\partial z} = \frac{\partial}{\partial z}\left[n_x\frac{\partial U}{\partial x} + n_y\frac{\partial U}{\partial y} + n_z\frac{\partial U}{\partial z} \right] = \frac{\cos\theta}{4\pi r^3}\frac{r_z}{r}$$

The derivatives of fundamental solutions T for three-dimensional space have a singularity of $1/r^2$ for 2-D and $1/r^3$ for 3-D problems and are termed *hypersigular* in mathematical jargon.

We now extend the Laplace_lib to include the derivatives of the fundamental solution.

```fortran
FUNCTION dU(r,dxr,Cdim)
!--------------------------------
!   Derivatives of fundamental solution for potential problems
!   temperature/potential
!--------------------------------
REAL,INTENT(IN):: r      !  Distance between source and field point
REAL,INTENT(IN):: dxr(:)!  Distances in x,y directions div. by r
REAL :: dU(UBOUND(dxr,1))    !   dU is array of same dim as dxr
INTEGER ,INTENT(IN):: Cdim   !   Cartesian dimension (2-D,3-D)
REAL :: C
SELECT CASE (CDIM)
    CASE (2)            !  Two-dimensional solution
     C=1/(2.0*Pi*r)
     dU(1)= C*dxr(1)
     dU(2)= C*dxr(2)
    CASE (3)            !  Three-dimensional solution
     C=1/(4.0*Pi*r**2)
     dU(1)= C*dxr(1)
     dU(2)= C*dxr(2)
     dU(3)= C*dxr(3)
    CASE DEFAULT
END SELECT
RETURN
END FUNCTION dU
FUNCTION dT(r,dxr,Vnorm,Cdim)
!--------------------------------
!  derivatives of the fundamental solution for potential problems
!  normal gradient
!--------------------------------
INTEGER,INTENT(IN) :: Cdim    !   Cartesian dimension
REAL,INTENT(IN):: r  !   Distance between source and field point
REAL,INTENT(IN):: dxr(:)!Distances in Cartesian dir divided by R
REAL,INTENT(IN):: Vnorm(:)   !   Normal vector
REAL :: dT(UBOUND(dxr,1))     !   dT is array of same dim as dxr
REAL :: C,COSTH
COSTH= DOT_PRODUCT (Vnorm,dxr)
SELECT CASE (Cdim)
    CASE (2)            !  Two-dimensional solution
     C= 1/(2.0*Pi*r**2)
     dT(1)= C*COSTH*dxr(1)
     dT(2)= C*COSTH*dxr(2)
    CASE (3)            !  Three-dimensional solution
     C= 3/(4.0*Pi*r**3)
     dT(1)= C*COSTH*dxr(1)
     dT(2)= C*COSTH*dxr(2)
     dT(3)= C*COSTH*dxr(3)
    CASE DEFAULT

END SELECT
RETURN
END FUNCTION dT
```

The discretised form of equation (8.21) is

$$u(P_a) = \sum_{e=1}^{E} \sum_{n=1}^{N} \Delta T_n^e (P_a) u_n^e - \sum_{e=1}^{E} \sum_{n=1}^{N} \Delta U_n^e (P_a) t_n^e \qquad (8.25)$$

where u_n^e and t_n^e are the solutions obtained for the temperature/potential and boundary flow on node n on boundary element e, E is the number of elements, N the number of nodes per element and

$$\Delta T_n^e = \int_{S_e} T(P_a,Q) N_n \, dS_e(Q), \quad \Delta U_n^e = \int_{S_e} U(P_a,Q) \, N_n dS_e(Q) \qquad (8.26)$$

The discretised form of equation (8.22) is given by

$$\mathbf{q}(P_a) = -k \left(\sum_{e=1}^{E} \sum_{n=1}^{N} \Delta \mathbf{S}_n^e \, t_n^e - \sum_{e=1}^{E} \sum_{n=1}^{N} \Delta \mathbf{R}_n^e \, u_n^e \right) \qquad (8.27)$$

where

$$\mathbf{q} = \begin{Bmatrix} q_x \\ q_y \\ q_z \end{Bmatrix} \quad and \quad \Delta \mathbf{S}_n^e = \begin{Bmatrix} \Delta S_{xn}^e \\ \Delta S_{yn}^e \\ \Delta S_{zn}^e \end{Bmatrix}; \quad \Delta \mathbf{R}_n^e = \begin{Bmatrix} \Delta R_{xn}^e \\ \Delta R_{yn}^e \\ \Delta R_{zn}^e \end{Bmatrix} \qquad (8.28)$$

The components of $\mathbf{S}$ and $\mathbf{R}$ are defined as

$$\Delta S_{xn}^e = \int_{S_e} \frac{\partial U}{\partial x}(P_a,Q) N_n dS(Q); \quad \Delta S_{yn}^e = \int_{S_e} \frac{\partial U}{\partial y}(P_a,Q) N_n dS(Q) \quad etc.$$

$$\Delta R_{xn}^e = \int_{S_e} \frac{\partial T}{\partial x}(P_a,Q) N_n dS(Q); \quad \Delta R_{yn}^e = \int_{S_e} \frac{\partial T}{\partial y}(P_a,Q) N_n dS(Q) \quad etc. \qquad (8.29)$$

and can be evaluated numerically over element e using Gauss Quadrature as explained in detail in Chapter 6. For 2-D problems this is

$$\Delta U_n^e = \sum_{k=1}^{K} U(P_a,Q(\xi_k)) N_n(\xi_k) J(\xi_k) W_k$$

$$\Delta T_n^e = \sum_{k=1}^{K} T(P_a,Q(\xi_k)) N_n(\xi_k) J(\xi_k) W_k \qquad (8.30)$$

and

$$\Delta S^e_{xn} = \sum_{k=1}^{K} \frac{\partial U}{\partial x}(P_a,Q(\xi_k)) \, N_n(\xi_k)J(\xi_k)W_k$$

$$\Delta S^e_{yn} = \sum_{k=1}^{K} \frac{\partial U}{\partial y}(P_a,Q(\xi_k)) \, N_n(\xi_k)J(\xi_k)W_k$$

$$\Delta R^e_{xn} = \sum_{k=1}^{K} \frac{\partial T}{\partial y}(P_a,Q(\xi_k)) \, N_n(\xi_k)J(\xi_k)W_k \qquad (8.31)$$

$$\Delta R^e_{yn} = \sum_{k=1}^{K} \frac{\partial T}{\partial y}(P_a,Q(\xi_k)) \, N_n(\xi_k)J(\xi_k)W_k$$

For 3-D problems the equations are

$$\Delta U^e_n = \sum_{m=1}^{M}\sum_{k=1}^{K} U(P_a,Q(\xi_k,\eta_m)) \, N_n(\xi_k,\eta_m)J(\xi_k,\eta_m)W_kW_m$$

$$\Delta T^e_n = \sum_{m=1}^{M}\sum_{k=1}^{K} T(P_a,Q(\xi_k,\eta_m)) \, N_n(\xi_k,\eta_m)J(\xi_k,\eta_m)W_kW_m \qquad (8.32)$$

and

$$\Delta S^e_{xn} = \sum_{m=1}^{M}\sum_{k=1}^{K} \frac{\partial U}{\partial x}(P_a,Q(\xi_k,\eta_m)) \, N_n(\xi_k,\eta_m)J(\xi_k,\eta_m)W_kW_m \quad etc. \qquad (8.33)$$

The number of Gauss points in the ξ and η direction M,K needed for accurate integration will again depend on the proximity of P_a to the element over which the integration is carried out. For computation of displacements, kernel **T** has a singularity of $1/r$ for 2-D problems and $1/r^2$ for 3-D. Kernel **R** has a $1/r^2$ singularity for 2-D and a $1/r^3$ singularity for 3-D problems, and the number of integration points is chosen according to Table 6.1.

8.3.2 Elasticity problems

The displacements at a point P_a inside the domain can be computed by using equation (5.15):

$$\mathbf{u}(P_a) = \int_S \mathbf{U}(P_a,Q)\mathbf{t}(Q)dS - \int_S \mathbf{T}(P_a,Q)\mathbf{u}(Q)dS \qquad (8.34)$$

The strains can be computed by using equation (4.32):

$$\varepsilon = \mathbf{B}\mathbf{u}(P_a) = \int_S \mathbf{B}\mathbf{U}(P_a,Q)\mathbf{t}(Q)dS - \int_S \mathbf{B}\mathbf{T}(P_a,Q)\mathbf{u}(Q)dS \qquad (8.35)$$

Finally, stresses can be computed by pre-multiplying with the **D** matrix:

$$\sigma = \mathbf{D}\varepsilon = \int_S \mathbf{D}\mathbf{B}\mathbf{U}(P_a,Q)\mathbf{t}(Q)dS - \int_S \mathbf{D}\mathbf{B}\mathbf{T}(P_a,Q)\mathbf{u}(Q)dS \qquad (8.36)$$

Equation (8.36) can be rewritten as

$$\sigma = \int_S \mathbf{S}(P_a,Q)\mathbf{t}(Q)dS - \int_S \mathbf{R}(P_a,Q)\mathbf{u}(Q)dS \qquad (8.37)$$

where the new fundamental solutions **S** and **R** are defined as

$$\mathbf{S} = \mathbf{D}\mathbf{B}\mathbf{U}(P_a,Q), \quad \mathbf{R} = \mathbf{D}\mathbf{B}\mathbf{T}(P_a,Q) \qquad (8.38)$$

and the pseudo-stress vector σ is defined as

$$\sigma = \begin{Bmatrix} \sigma_x \\ \sigma_y \\ \sigma_z \\ \tau_{xy} \\ \tau_{yz} \\ \tau_{xz} \end{Bmatrix} \quad for \ 3-D \ and \ \sigma = \begin{Bmatrix} \sigma_x \\ \sigma_y \\ \tau_{xy} \end{Bmatrix} \quad for \ 2-D \qquad (8.39)$$

Matrices **S** and **R** which are of dimension 3 x 2 for two-dimensional problems and of dimension 6 x 3 for three-dimensional problems. Matrix **S** is given by

$$\mathbf{S} = \begin{bmatrix} S_{xxx} & S_{xxy} & S_{xxz} \\ S_{yyx} & S_{yyy} & S_{yyz} \\ S_{zzx} & S_{zzy} & S_{zzz} \\ S_{xyx} & S_{xyy} & S_{xyz} \\ S_{yzx} & S_{yzy} & S_{yzz} \\ S_{xzx} & S_{xzy} & S_{xzz} \end{bmatrix} \qquad (8.40)$$

The coefficients of **S** are computed as:

$$S_{ijk} = \frac{C_2}{r^n}\left[C_3(\delta_{ki}r_j + \delta_{kj}r_i - \delta_{ij}r_k) + (n+1)r_ir_jr_k\right]$$

where

$$\delta_{lm} = 0 \quad if \quad l \neq m$$
$$\delta_{lm} = 1 \quad if \quad l = m$$

(8.41)

Values x, y, z are substituted for i, j, k. Constants C_2, C_3 and n are defined in Table 8.1 for plane stress/strain and 3-D problems and

$$r = \sqrt{(x_P - x_Q)^2 + (y_P - y_Q)^2 + (z_P - z_Q)^2}$$

$$r_x = \frac{x_P - x_Q}{r}; r_y = \frac{y_P - y_Q}{r}; r_z = \frac{z_P - z_Q}{r}$$

(8.42)

Note that in order to simplify the equations, the definition of r_x, r_y, r_z is different from that in Chapter 4. Matrix **R** is given by:

$$\mathbf{R} = \begin{bmatrix} R_{xxx} & R_{xxy} & R_{xxz} \\ R_{yyx} & R_{yyy} & R_{yyz} \\ R_{zzx} & R_{zzy} & R_{zzz} \\ R_{xyx} & R_{xyy} & R_{xyz} \\ R_{yzx} & R_{yzy} & R_{yzz} \\ R_{xzx} & R_{xzy} & R_{xzz} \end{bmatrix}$$

(8.43)

where

$$R_{kij} = \frac{C_5}{r^{n+1}}\begin{bmatrix} (n+1)\cos\theta(C_3\delta_{ij}r_a + v(\delta_{ik}r_j + \delta_{jk}r_i) - C_6r_ir_jr_k) \\ + (n+1)v(n_ir_jr_k + n_jr_ir_k) \\ + C_3((n+1)n_kr_ir_j + n_j\delta_{ik} + n_i\delta_{jk}) - C_7n_k\delta_{ij} \end{bmatrix}$$

(8.44)

x, y, z may be substituted for the subscripts i, j, k and

$$\cos\theta = r_xn_x + r_yn_y + r_zn_z$$

(8.45)

n_x, n_y, n_z are the components of a unit vector normal to the surface at point Q. Values of the constants are given in Table 8.1.

Table 8.1 Constants for fundamental solutions **S** and **R**

	Plane strain	Plane stress	3-D
n	1	1	2
C_2	$1/4\pi(1-\nu)$	$(1+\nu)/4\pi$	$1/8\pi(1-\nu)$
C_3	$1-2\nu$	$(1-\nu)/(1+\nu)$	$1-2\nu$
C_5	$G/(2\pi(1-\nu))$	$(1+\nu)G/2\pi$	$G/(4\pi(1-\nu))$
C_6	4	4	5
C_7	$1-4\nu$	$(1-3\nu)/(1+\nu)$	$1-4\nu$

Subroutines for calculating Kernels **S** and **R** are added to the Elasticity_lib.

```fortran
SUBROUTINE SK(TS,DXR,R,C2,C3)
!-------------------------------------------------------------------
!    KELVIN SOLUTION FOR STRESS
!      TO BE MULTIPLIED WITH T

!-------------------------------------------------------------------
REAL, INTENT(OUT):: TS(:,:)  ! Fundamental solution (returned)
REAL, INTENT(IN) :: DXR(:)   ! r_x , r_y, r_z
REAL, INTENT(IN) :: R        ! r
REAL, INTENT(IN) :: C2,C3    ! Elastic constants
REAL    :: Cdim              ! Cartesian dimension
INTEGER :: NSTRES            ! No. of stress components
INTEGER :: JJ(6), KK(6)      ! sequence of stresses in pseudo-
                             ! vector

REAL    :: A,C2,C3
INTEGER :: I,N,J,K
Cdim= UBOUND(DXR,1)
IF(CDIM == 2) THEN
   NSTRES= 3
   JJ(1:3)= (/1,2,1/)
   KK(1:3)= (/1,2,2/)

ELSE
   NSTRES= 6
   JJ= (/1,2,3,1,2,3/)
   KK= (/1,2,3,2,3,1/)
END IF
Coor_directions:&
DO I=1,Cdim
   Stress_components:&
   DO N=1,NSTRES
      J= JJ(N)
      K= KK(N)
      A= 0.
      IF(I .EQ. K) A= A + DXR(J)
      IF(J .EQ. K) A= A - DXR(I)
      IF(I .EQ. J) A= A + DXR(K)
```

```
         A= A*C3
         TS(I,N)= C2/R*(A + Cdim*DXR(I)*DXR(J)*DXR(K))
         IF(Cdim .EQ. 3) TS(I,N)= TS(I,N)/2./R
       END DO &
       Stress_components
  END DO &
  Coor_directions
  RETURN
  END SUBROUTINE SK
  SUBROUTINE RK(US,DXR,R,VNORM,C3,C5,C6,C7,ny)
  !-----------------------------------------------------------
  !    KELVIN SOLUTION FOR STRESS COMPUTATION
  !    TO BE MULTIPLIED WITH U
  !-----------------------------------------------------------
  REAL, INTENT(OUT)  :: US(:,:)          ! Fundamental solution
  REAL, INTENT(IN)   :: DXR(:)           ! r_x , r_y, r_z
  REAL, INTENT(IN)   :: R                ! r
  REAL, INTENT(IN)   :: VNORM(:)         ! n_x , n_y , n_z
  REAL, INTENT(IN)   :: C3,C5,C7,ny      ! Elastic constants
  REAL :: Cdim                 ! Cartesian dimension
  INTEGER :: NSTRES            ! No. of stress components
  INTEGER :: JJ(6), KK(6) ! sequence of stresses in pseudo-
                                   vector
  REAL    :: costh, B,C
  Cdim= UBOUND(DXR,1)
  IF(CDIM == 2) THEN
     NSTRES= 3
     JJ(1:3)= (/1,2,1/)
     KK(1:3)= (/1,2,2/)
  ELSE
     NSTRES= 6
     JJ= (/1,2,3,1,2,3/)
     KK= (/1,2,3,2,3,1/)
  END IF
  COSTH= DOT_Product(dxr,vnorm)
  Coor_directions:&
  DO K=1,Cdim
     Stress_components:&

  DO N=1,NSTRES
       I= JJ(N)
       J= KK(N)
       B= 0.
       IF(I .EQ. J) B= Cdim*C3*DXR(K)
       IF(I .EQ. K) B= B + ny*DXR(J)
       IF(J .EQ. K) B= B + ny*DXR(I)
       B= COSTH *(B - C6*DXR(I)*DXR(J)*DXR(K) )
       C= DXR(J)*DXR(K)*ny
       IF(J .EQ.K) C= C + C3
       C= C*VNORM(I)
       B= B+C
```

```
        C= DXR(I)*DXR(K)*ny
        IF(I == K) C=C + C3
        C= C*VNORM(J)
        B= B+C
        C= DXR(I)*DXR(J)*Cdim*C3
        IF(I == J) C= C - C7
        C= C*VNORM(K)
        US(K,N)= (B + C)*C5/R/R
        IF(Cdim == 3) US(K,N)= US(K,N)/2./R
    END DO &
    Stress_components
END DO &
Coor_directions
RETURN
END
```

The discretised form of equation (8.34) is written as

$$\mathbf{u}(P_a) = \sum_{e=1}^{E}\sum_{n=1}^{N}\Delta\mathbf{U}_n^e\,\mathbf{t}_n^e - \sum_{e=1}^{E}\sum_{n=1}^{N}\Delta\mathbf{T}_n^e\,\mathbf{u}_n^e \tag{8.46}$$

where

$$\Delta\mathbf{U}_n^e = \int_{S_e}\mathbf{U}(P_a,Q)N_n dS(Q) \;\; ; \;\; \Delta\mathbf{T}_n^e = \int_{S_e}\mathbf{T}(P_a,Q)N_n dS(Q) \tag{8.47}$$

The discretised form of equation (8.36) is written as

$$\boldsymbol{\sigma}(P_a) = \sum_{e=1}^{E}\sum_{n=1}^{N}\Delta\mathbf{S}_n^e\,\mathbf{t}_n^e - \sum_{e=1}^{E}\sum_{n=1}^{N}\Delta\mathbf{R}_n^e\,\mathbf{u}_n^e \tag{8.48}$$

where

$$\Delta\mathbf{S}_n^e = \int_{S_e}\mathbf{S}(P_a,Q)N_n dS(Q) \;\; ; \;\; \Delta\mathbf{R}_n^e = \int_{S_e}\mathbf{R}(P_a,Q)N_n dS(Q) \tag{8.49}$$

These integrals may be evaluated using Gauss Quadrature, as explained in Chapter 6. For 2-D problems they are given by

$$\Delta\mathbf{U}_n^e = \sum_{k=1}^{K}\mathbf{U}(P_a,Q(\xi_k))\,N_n(\xi_k)J(\xi_k)W_k$$

$$\Delta\mathbf{T}_n^e = \sum_{k=1}^{K}\mathbf{T}(P_a,Q(\xi_k))\,N_n(\xi_k)J(\xi_k)W_k \quad etc. \tag{8.50}$$

For 3-D elasticity we have

$$\Delta\mathbf{U}_n^e = \sum_{m=1}^{M}\sum_{k=1}^{K}\mathbf{U}(P_a,Q(\xi_k,\eta_m))\,N_n(\xi_k,\eta_m)\,J(\xi_k,\eta_m)W_kW_m$$

(8.51)

$$\Delta\mathbf{T}_n^e = \sum_{m=1}^{M}\sum_{k=1}^{K}\mathbf{T}(P_a,Q(\xi_k,\eta_m))\,N_n(\xi_k,\eta_m)\,J(\xi_k,\eta_m)W_kW_m \quad etc.$$

The number of Gauss points in the ξ and η directions M, K needed for accurate integration, will again depend on the proximity of P_a to the element over which the integration is carried out. For computation of displacements, kernel $\mathbf{T}$ has a singularity of $1/r$ for 2-D problems and $1/r^2$ for 3-D. The number of integration points M and K are chosen according to Table 6.1.

For 3-D problems, the limiting value of R/L for the maximum number of integration points allowed for in Table 6.1 is therefore 0.22, which means that an internal point cannot be closer than about 1/5 of the size of the element adjacent to it. If we want to compute results at points closer to the element than that, then we can either subdivide the element into smaller integration zones or interpolate between the internal result at R = 0.22L and the result on a point on the boundary element, which is computed using the scheme outlined previously. The subdivision of elements is explained elsewhere[1] and will not be considered here.

8.4   PROGRAM 8.1: POSTPROCESSOR

Program Postprocessor for computing results on the boundary and inside the domain is presented. This program is run after General_purpose_BEM. It reads the INPUT file, which is the same as that read by General_Purpose_BEM, and which contains the basic job information and the geometry of boundary elements. The results of the boundary element computation are read from the file BERESULTS, which was generated by the General_purpose BEM program, and contains the values of u and t at boundary points. The coordinates of internal points are supplied in file INPUT2, and results of the program are written onto file OUTPUT.

The program first calculates fluxes/stresses at the centres of specified boundary elements and then temperatures/displacements and fluxes/stresses at specified points inside the domain.

For calculation of internal points, the integration is carried out seperately for the computation of potentials/displacements and for the flow/stresses, since the kernels have different singularities. This may not be the most efficient way and an over-integration of the first kernels may be considered to improve the efficiency, since certain computations, such as the calculation of the Jacobian, for example, need to be be carried out only once for a boundary element. Another improvement in efficiency can be made by lumping together internal points, so that only one integration loop is needed for all interior points requiring the same integration scheme. In the program there is no check made on the minimum distance an internal point may be from the boundary element, so that adequate precision is maintained.

```fortran
PROGRAM Post_processor
!---------------------------------------------------
!     General purpose postprocessor
!     for computing results at boundary and interior points
!     for elasticity and potential problems
!---------------------------------------------------
USE Utility_lib;USE Elast_lib;USE Laplace_lib
USE Integration_lib
USE Postproc_lib
IMPLICIT NONE
INTEGER, ALLOCATABLE :: Inci(:)            ! Incidences (1 elem.)
INTEGER, ALLOCATABLE :: Incie(:,:)         ! Incidences (all elem.)
INTEGER, ALLOCATABLE :: Ldest(:)           ! Destinations (1 elem.)
REAL, ALLOCATABLE    :: Elcor(:,:)         ! Element coordinates
REAL, ALLOCATABLE    :: El_u(:,:,:)        ! u and
REAL, ALLOCATABLE    :: El_t(:,:,:)        ! t results of System
REAL, ALLOCATABLE    :: El_ue(:,:)         ! Displ. of Element
REAL, ALLOCATABLE    :: El_te(:,:)         ! Tract. of Element
REAL, ALLOCATABLE    :: Disp(:)            ! Dipl. @ Nodes
REAL, ALLOCATABLE    :: Trac(:)            ! Traction @ Nodes
REAL, ALLOCATABLE    :: El_trac(:)         ! Traction @ Elements
REAL, ALLOCATABLE    :: El_disp(:)         ! Displ. @ Element
REAL, ALLOCATABLE    :: xP(:,:)            ! Cooloc. points
REAL, ALLOCATABLE    :: xPnt(:)            ! Interior points
REAL, ALLOCATABLE    :: Ni(:),GCcor(:),dxr(:),Vnorm(:)
CHARACTER (LEN=80) :: Title
REAL :: Elengx,Elenge,Rmin,Glcorx(8),Wix(8),Glcore(8),Wie(8)
REAL :: Jac
REAL, ALLOCATABLE :: Flow(:),Stress(:)!  Results boundary
REAL, ALLOCATABLE :: uPnt(:),SPnt(:) ! Results interior
REAL, ALLOCATABLE :: TU(:,:),UU(:,:) ! Kernels for u
REAL, ALLOCATABLE :: TS(:,:),US(:,:) ! Kernels for q,s
REAL, ALLOCATABLE :: Fac(:),Fac_nod(:,:)   ! for symmetry
INTEGER :: Cdim,Node,M,N,Istat,Nodel,Nel,Ndof,Cod,Nreg
INTEGER :: Ltyp,Nodes,Maxe,Ndofe,Ndofs,Ncol,ndg,ldim
INTEGER :: nod,nd,Nstres,Nsym,Isym,nsy
INTEGER :: IPS,Nan,Nen,Ios,dofa,dofe
INTEGER :: Mi,Ki,K,I
REAL    :: Con,E,ny,Fact,G,C2,C3,C5,C6,C7
REAL    :: xsi,eta,Weit,R
!    Read job information
OPEN (UNIT=1,FILE='INPUT',FORM='FORMATTED')
OPEN (UNIT=2,FILE='OUTPUT',FORM='FORMATTED')
Call Jobin(Title,Cdim,Ndof,IPS,Nreg,Ltyp,Con,E,ny,&
           Isym,nodel,nodes,maxe)
Ndofe= nodel*ndof
ldim= Cdim-1
Nsym= 2**Isym   !   number of symmetry loops
ALLOCATE(xP(Cdim,Nodes))   ! Array for node coordinates
ALLOCATE(Incie(Maxe,Nodel),Inci(Nodel),Ldest(Ndofe))
ALLOCATE(Ni(Nodel),GCcor(Cdim),dxr(Cdim),Vnorm(Cdim))
CALL Geomin(Nodes,Maxe,xp,Incie,Nodel,Cdim)
```

```
!    Compute constants
IF(Ndof == 1) THEN
 Nstres= Cdim
ELSE
 G= E/(2.0*(1+ny))
 C2= 1/(8*Pi*(1-ny))
 C3= 1.0-2.0*ny
 C5= G/(4.0*Pi*(1-ny))
 C6= 15
 C7= 1.0-4.0*ny
 Nstres= 6
 IF(Cdim == 2) THEN
  IF(IPS == 1) THEN                    ! Plane Strain
    C2= 1/(4*Pi*(1-ny))
    C5= G/(2.0*Pi*(1-ny))
    C6= 8
    Nstres= 3
  ELSE
    C2= (1+ny)/(4*Pi   )        ! Plane Stress
    C3= (1.0-ny)/(1.0+ny)
    C5= (1.0+ny)*G/(2.0*Pi)
    C6= 8
    C7= (1.0-3.0*ny)/(1.0+ny)
    Nstres= 3
  END IF
 END IF
END IF
ALLOCATE(El_u(Maxe,Nodel,ndof),El_t(Maxe,Nodel,ndof)
ALLOCATE(El_te(Nodel,ndof),El_ue(Nodel,ndof))
ALLOCATE(Fac_nod(Nodel,ndof))
ALLOCATE(El_trac(Ndofe),El_disp(Ndofe))
CLOSE(UNIT=1)
OPEN (UNIT=1,FILE='BERESULTS',FORM='FORMATTED')
WRITE(2,*) ' '
WRITE(2,*) 'Post-processed Results'
WRITE(2,*) ' '
Elements1:&
DO Nel=1,Maxe
  READ(1,*) ((El_u(nel,n,m),m=1,ndof),n=1,Nodel)
  READ(1,*) ((El_t(nel,n,m),m=1,ndof),n=1,Nodel)
END DO &
Elements1
ALLOCATE(Elcor(Cdim,Nodel))
CLOSE(UNIT=1)
OPEN (UNIT=1,FILE='INPUT2',FORM='FORMATTED')
ALLOCATE(Flow(Cdim),Stress(Nstres))
!-----------------------------------------------------------
!       Computation of boundary fluxes/stresses
!-----------------------------------------------------------
WRITE(2,*) 'Results at centres of Boundary Elements:'
READ(1,*) Nan,Nen
IF(Nan > 0) THEN
```

```
  Element_loop: &
  DO NEL= Nan,Nen
    Inci= Incie(nel,:)
    Elcor= xp(:,Inci(:))
    IF(Ndof == 1) THEN
      Flow= 0.0
      Call BFLOW(Flow,0.0,0.0,El_u(Nel,:,:),Inci,Elcor,Con)
      WRITE(2,*) 'Element',Nel,' Flux: ',Flow
    ELSE
      Stress= 0.0
      Call BStress(Stress,0.0,0.0,El_u(Nel,:,:)&
                  ,El_t(Nel,:,:),Inci,Elcor,E,ny,IPS)
      WRITE(2,*) 'Element',Nel,' Stress: ',Stress
    END IF
  END DO &
  Element_loop
END IF
ALLOCATE(uPnt(NDOF),SPnt(NSTRES),UU(NDOF,NDOF),TU(NDOF,NDOF))
ALLOCATE(TS(Nstres,Ndof),US(Nstres,Ndof))
ALLOCATE(xPnt(Cdim),Fac(Ndofe))
ALLOCATE(Disp(Cdim),Trac(Cdim))
WRITE(2,*)''
WRITE(2,*) 'Internal Results:'
WRITE(2,*)''
Internal_points: &
DO
  READ(1,*,IOSTAT=IOS) xPnt
  IF(IOS /= 0) EXIT
  WRITE(2,*) 'Coordinates: ',xPnt
!------------------------------------------------------------
! Computation of Temperatures/Displacements
!------------------------------------------------------------
  uPnt= 0.0
  Element_loop1: &
  DO NEL= 1,MAXE
    Symmetry_loop1:&
    DO nsy=1,Nsym
      Inci= Incie(nel,:)
      Elcor= xp(:,Inci(:))
      IF(ldim == 2) THEN
        ELengx= Dist((Elcor(:,3)+Elcor(:,2))/2.&
                    ,(Elcor(:,4)+Elcor(:,1))/2.,Cdim)   ! Lxsi
        ELenge= Dist((Elcor(:,2)+Elcor(:,1))/2.&
                    ,(Elcor(:,3)+Elcor(:,4))/2.,Cdim)   ! Leta
      ELSE
        Call Elength(Elengx,Elcor,nodel,ldim)
      END IF
      Ldest= 1
      Fac= 1.0
      Fac_nod=1.0
      El_ue(:,:)=El_u(Nel,:,:)
      El_te(:,:)=El_t(Nel,:,:)
```

```
    IF(Isym > 0) THEN
      DO Nod=1,Nodel
        dofa= (nod-1)*Ndof+1
        dofe= dofa+Ndof-1
        El_trac(dofa:dofe)= El_te(Nod,:)
        EL_disp(dofa:dofe)= El_ue(Nod,:)
      END DO
      CALL Mirror(Isym,nsy,Nodes,Elcor,Fac&
                     ,Inci,Ldest,El_trac,EL_disp &
                     ,nodel,ndof,Cdim)
      DO Nod=1,Nodel
        dofa= (nod-1)*Ndof+1
        dofe= dofa+Ndof-1
        El_te(Nod,:)= El_trac(dofa:dofe)
        El_ue(Nod,:)= El_disp(dofa:dofe)
        Fac_nod(Nod,:)= Fac(dofa:dofe)
      END DO
    END IF
    Rmin= Min_dist(Elcor,xPnt,Nodel,ldim,Inci)
    Mi= Ngaus(Rmin/Elengx,Cdim-1)    ! Gauss Pts. in ξ dir.
    CALL Gauss_coor(Glcorx,Wix,Mi) ! Coords/Wghts ξ dir
    Ki= 1 ; Wie(1)= 1.0 ; Glcore(1)= 0.0
    IF(Cdim == 3) THEN
      Ki= Ngaus(Rmin/Elenge,Cdim-1)    ! Gauss Pts. in η dir.
      CALL Gauss_coor(Glcore,Wie,Ki) ! Coords/Wghts η dir
    END IF
    Gauss_points_xsi: &
    DO m=1,Mi
      xsi= Glcorx(m)
      Gauss_points_eta: &
      DO k=1,Ki
        eta= Glcore(k)
        Weit= Wix(m)*Wie(k)
        CALL Serendip_func(Ni,xsi,eta,ldim,nodel,Inci)
        CALL Normal_Jac(Vnorm,Jac,xsi,eta,ldim,nodel,Inci,elcor)
        Fact= Weit*Jac
        CALL Cartesian(GCcor,Ni,ldim,elcor)  ! x,y,z of Gauss pnt
        r= Dist(GCcor,xPnt,Cdim)        ! Dist. P,Q
        dxr= (GCcor-xPnt)/r             ! rx/r , ry/r  etc
        IF(Ndof .EQ. 1) THEN
          TU= U(r,Con,Cdim)  ; UU= T(r,dxr,Vnorm,Cdim)
        ELSE
          TU= UK(dxr,r,E,ny,Cdim) ; UU= TK(dxr,r,Vnorm,ny,Cdim)
        END IF
        Node_loop1:&
        DO Node=1,Nodel
          Disp= El_ue(Node,:)* Fac_nod(Node,:)
          Trac= El_te(Node,:)* Fac_nod(Node,:)
          uPnt= uPnt + (MATMUL(TU,Trac)-&
                        MATMUL(UU,Disp))* Ni(Node)* Fact
        END DO &
```

```
      Node_loop1
    END DO &
    Gauss_points_eta
  END DO &
  Gauss_points_xsi
  END DO &
  Symmetry_loop1
END DO &
Element_loop1
WRITE(2,*) 'u: ',uPnt
!-----------------------------------------------------
!    Computation of Fluxes/Stresses
!-----------------------------------------------------
SPnt= 0.0
Element_loop2: &
DO NEL= 1,MAXE
  Symmetry_loop2: &
  DO nsy=1,Nsym
    Inci= Incie(nel,:)
    Elcor= xp(:,Inci(:))
    IF(ldim == 2) THEN
      ELengx= Dist((Elcor(:,3)+Elcor(:,2))/2. &
      ,(Elcor(:,4)+Elcor(:,1))/2.,Cdim)   ! Lxsi
      ELenge= Dist((Elcor(:,2)+Elcor(:,1))/2. &
      ,(Elcor(:,3)+Elcor(:,4))/2.,Cdim)   ! Leta
    ELSE
      Call Elength(Elengx,Elcor,nodel,ldim)
    END IF
    Ldest= 1
    Fac= 1.0
    El_ue(:,:)=El_u(Nel,:,:)
    El_te(:,:)=El_t(Nel,:,:)
    IF(Isym > 0) THEN
      DO Nod=1,Nodel
        dofa= (nod-1)*Ndof+1
        dofe= dofa+Ndof-1
        El_trac(dofa:dofe)= El_te(Nod,:)
        EL_disp(dofa:dofe)= El_ue(Nod,:)
      END DO
      CALL Mirror(Isym,nsy,Nodes,Elcor&
                ,Fac,Inci,Ldest,El_trac,El_disp&
                ,nodel,ndof,Cdim)
      DO Nod=1,Nodel
        dofa= (nod-1)*Ndof+1
        dofe= dofa+Ndof-1
        El_te(Nod,:)= El_trac(dofa:dofe)
        El_ue(Nod,:)= El_disp(dofa:dofe)
        Fac_nod(Nod,:)= Fac(dofa:dofe)
      END DO
    END IF
    Rmin= Min_dist(Elcor,xPnt,Nodel,ldim,Inci)
```

```
      Mi= Ngaus(Rmin/Elengx,Cdim)        ! Gauss Pts. in ξ dir.
      CALL Gauss_coor(Glcorx,Wix,Mi)  ! Coords/Wghts ξ dir
      Ki= 1 ; Wie(1)= 1.0 ; Glcore(1)= 0.0
      IF(Cdim == 3) THEN
       Ki= Ngaus(Rmin/Elenge,Cdim)        ! Gauss Pts. in η dir.
       CALL Gauss_coor(Glcore,Wie,Ki)  ! Coords/Wghts η dir
      END IF
      Gauss_points_xsi2: &
      DO m=1,Mi
       xsi= Glcorx(m)
       Gauss_points_eta2: &
       DO k=1,Ki
         eta= Glcore(k)
         Weit= Wix(m)*Wie(k)
         CALL Serendip_func(Ni,xsi,eta,ldim,nodel,Inci)
         CALL Normal_Jac(Vnorm,Jac,xsi,eta,ldim,nodel,Inci,elcor)
         Fact= Weit*Jac
         CALL Cartesian(GCcor,Ni,ldim,elcor)   ! x,y,z of Gauss pnt
         r= Dist(GCcor,xPnt,Cdim)         ! Dist. P,Q
         dxr= (GCcor-xPnt)/r              ! rx/r , ry/r etc
         IF(Ndof .EQ. 1) THEN
          TS(:,1)= dU(r,dxr,Cdim); US(:,1)= dT(r,dxr,Vnorm,Cdim)
         ELSE
          CALL SK(TS,DXR,R,C2,C3)
          CALL RK(US,DXR,R,VNORM,C3,C5,c6,C7,ny)
         END IF
         Node_loop2:&
         DO Node=1,Nodel
           Disp= El_ue(Node,:)* Fac_nod(Node,:)
           Trac= El_te(Node,:)* Fac_nod(Node,:)
           SPnt= SPnt + (MATMUL(TS,Trac)- &
                         MATMUL(US,Disp))* Ni(Node)* Fact
         END DO &
         Node_loop2
        END DO &
        Gauss_points_eta2
       END DO &
       Gauss_points_xsi2
     END DO &
     Symmetry_loop2
   END DO &
   Element_loop2
   IF(Ndof == 1) THEN
     WRITE(2,*) 'Flux: ',SPnt
   ELSE
     WRITE(2,*) 'Stress: ',SPnt
   END IF
   END DO &
   Internal_points
   STOP
   END PROGRAM Post_processor
```

8.4.1 Input specification for program 8.1

INPUT DATA

1.0   Boundary results

> **Nan, Nen** ... Number of first element and last element on which boundary results are to be computed

2.0   Internal point specification loop

> **x, y, (z)** ... Coordinates of internal points, one per line, specify as many as required, programs stops when end of file is reached

8.5   CONCLUSIONS

In this chapter we have discussed methods for obtaining results other than the primary ones, i.e. values of temperature/displacement and fluxes/tractions at the nodes of boundary elements. These additional results are flows/stresses acting in directions tangential to the boundary of elements and results at internal points, i.e. points where no elements exist. Results exactly on the boundary elements can be obtained by a method also known as the 'traction recovery', whereby we use the shape functions of the element to determine tangential flows/stresses. The results at internal points can be obtained with the aid of the fundamental solutions. It has been shown that such results are more accurate than comparable results from the FEM, as they satisfy the governing differential equations exactly and – for infinite domain problems – include the effect of the infinite boundary condition. The task of computing internal results can now be delegated to a postprocessor, where the user may either interactively interrogate points or define planes inside the continuum where contours are to be plotted.

It has been found that due to the high degree of singularity of the kernel functions, care must be taken that internal points are not too close to the boundary. If the proposed numerical integration scheme is used, then there is a limiting value of R/L below which the results are in error. However, since we are able to compute the results exactly on the boundary, we may use a linear interpolation between an internal point which is not too close to the boundary and a point projected onto the boundary element. There are mathematical schemes available which allow the computation of results very close to, or even exactly on the boundary without using the shape functions as done here. These methods, however, are more complicated than those outlined here and will not be discussed. The reader is referred to the literature on this subject, for example the excellent text by Banerjee [2].

8.6   EXERCISES

Exercise 8.1
For Example 7.1, plot the flow in a vertical direction along a horizontal line. Compare this with the theoretical solution and plot the error as a function of R/L as defined in 6.3.2.

Exercise 8.2

For Exercise 7.3, plot the stress in a vertical direction along a horizontal line. Compare this with the theoretical solution and plot the error as a function of R/L.

Exercise 8.3

For Exercise 7.2, plot the flow in a horizontal direction along a vertical line in the middle of the rectangular region. Compare this with the theoretical solution and plot the error as a function of R/L.

Exercise 8.4

For Exercise 7.6, plot the normal and shear stress along a vertical line, as shown in Figure 8.9. Compare this with the theoretical solution and plot the error as a function of R/L.

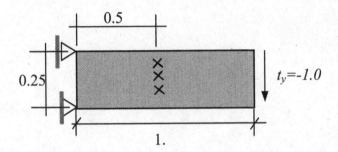

Figure 8.9 Cantilever beam with internal points

8.7   REFERENCES

1. Beer, G. and Watson, J.O. (1991) *Introduction to Finite and Boundary Element Methods for Engineers*. Wiley, Chichester.
2. Banerjee, P.K. (1994) *The Boundary Element Methods in Engineering*. McGraw-Hill, London.

9
Test Examples

Die Wahrheit wird gelebt, nicht doziert
(Truth is experienced not taught)
H. Hesse

9.1 INTRODUCTION

We have now developed all the software required to perform a boundary element analysis of problems in potential flow and elasticity. The examples which we can analyse will, however, be restricted to homogeneous domains and linear material behaviour. Before we proceed further, in an attempt to eliminate these restrictions it is opportune to pause and learn a few things about the method using test examples, especially with respect to the accuracy that can be attained.

The purpose of this chapter is twofold. Firstly, the reader will learn how problems are modelled using boundary elements with examples of simple meshes in two and three dimensions. Secondly, we show, by comparison with theory and results from finite element meshes, the accuracy which can be obtained. We will also point out possible pitfalls which must be avoided.

As with all numerical methods, examples can be presented that favour the method and others which don't. Here we find that the BEM has difficulty dealing with cantilevers with small thickness where two opposing boundaries are close to each other. On the other hand, it can deal very well with problems which involve an infinite domain. Also we will find that values at the surface are computed more accurately than with the FEM. This is, of course, an indication of the range of applications where the method is superior as compared with others: those involving a large volume to surface ratio (including infinite domains) and those where the results at the boundary are important, for example stress concentration problems.

In the following, several test examples will be presented ranging from the simple 2-D analysis of a cantilever beam to the 3-D analysis of a spherical excavation in an infinite continuum. In all cases we show the input file required to solve the problem with Program 7.1 and 8.1 and the output obtained. The results are then analysed with respect to accuracy with different discretisations. Comparison is made with theoretical results, and in some cases with finite element models.

9.2   CANTILEVER BEAM

9.2.1 Problem statement

The cantilever beam is a simple structure, which nevertheless can be used to show strengths and weaknesses of numerical methods. Here we analyse a cantilever beam with decreasing thickness and we will find that this causes some difficulties for the BEM. The problem is shown in Figure 9.1. An encastre beam is subjected to a distributed load of 10 KN/m at the end. The material properties are assumed to be: E = 10 000 MPa and ν = 0.0. We gradually decrease the thickness t of the beam and observe the accuracy of results.

9.2.2 Boundary element discretisation and input

Figure 9.2 shows the discretisation used (12 parabolic boundary elements) and the dimension of the first mesh analysed with a ratio of t/L of 0.2. The node numbering and boundary conditions are also shown.

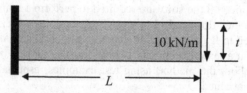

Figure 9.1   Cantilever beam: dimensions and loading assumed

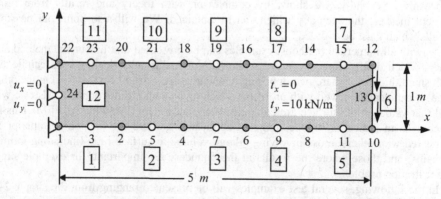

Figure 9.2   Boundary element mesh 1 (◉ ...corner node, ○ ... mid-side node)

The input file for Program 7.1 for this problem is:

```
Cantilever beam
2
2
1
1
0
2
0.1000E+05
0.0000E+00
   24
   12
     0.000       0.000
     1.000       0.000
     0.500       0.000
     2.000       0.000
     1.500       0.000
     3.000       0.000
     2.500       0.000
     4.000       0.000
     3.500       0.000
     5.000       0.000
     4.500       0.000
     5.000       1.000
     5.000       0.500
     4.000       1.000
     4.500       1.000
     3.000       1.000
     3.500       1.000
     2.000       1.000
     2.500       1.000
     1.000       1.000
     1.500       1.000
     0.000       1.000
     0.500       1.000
     0.000       0.500
      1       2       3
      2       4       5
      4       6       7
      6       8       9
      8      10      11
     10      12      13
     12      14      15
     14      16      17
     16      18      19
     18      20      21
     20      22      23
     22       1      24
      1
     12    0.0 0.0 0.0 0.0 0.0 0.0
      1
      6    0.0 -10.0 0.0 -10.0 0.0 -10.0
```

9.2.3 Results

The output obtained from Program 7.1 is:

```
Project:
 Cantilever beam
 Cartesian_dimension:  2
 Elasticity problem
 Type of analysis:   Solid plane strain
 Finite region
 No symmetry
 Quadratic elements
 Modulus:          10000.00
 Poissons ratio:  0.0
 Number of nodes of system:        24
 Number of elements of system:   12

Node     1  Coor      0.00    0.00
Node     2  Coor      1.00    0.00
Node     3  Coor      0.50    0.00
Node     4  Coor      2.00    0.00
Node     5  Coor      1.50    0.00
Node     6  Coor      3.00    0.00
Node     7  Coor      2.50    0.00
Node     8  Coor      4.00    0.00
Node     9  Coor      3.50    0.00
Node    10  Coor      5.00    0.00
Node    11  Coor      4.50    0.00
Node    12  Coor      5.00    1.00
Node    13  Coor      5.00    0.50
Node    14  Coor      4.00    1.00
Node    15  Coor      4.50    1.00
Node    16  Coor      3.00    1.00
Node    17  Coor      3.50    1.00
Node    18  Coor      2.00    1.00
Node    19  Coor      2.50    1.00
Node    20  Coor      1.00    1.00
Node    21  Coor      1.50    1.00
Node    22  Coor      0.00    1.00
Node    23  Coor      0.50    1.00
Node    24  Coor      0.00    0.50

 Incidences:
EL     1  Inci      1     2     3
EL     2  Inci      2     4     5
EL     3  Inci      4     6     7
EL     4  Inci      6     8     9
EL     5  Inci      8    10    11
EL     6  Inci     10    12    13
EL     7  Inci     12    14    15
EL     8  Inci     14    16    17
EL     9  Inci     16    18    19
```

```
EL     10    Inci      18    20    21
EL     11    Inci      20    22    23
EL     12    Inci      22     1    24


Elements with Dirichlet BC's:
Element              12  Prescribed values:
   0.00   0.00
   0.00   0.00
   0.00   0.00


Elements with Neumann BC's:
 Element               6  Prescribed values:
   0.00  -10.00
   0.00  -10.00
   0.00  -10.00


Results, Element     1
   u=    0.00000  -0.00019  -0.02685  -0.02978  -0.01409  -0.00820
   t= 298.03250   0.00000   0.00000   0.00000   0.00000   0.00000
Results, Element     2
   u=   -0.02685  -0.02978  -0.04763  -0.10710  -0.03789  -0.06325
   t=    0.00000   0.00000   0.00000   0.00000   0.00000   0.00000
Results, Element     3
   u=   -0.04763  -0.10710  -0.06258  -0.22016  -0.05575  -0.15989
   t=    0.00000   0.00000   0.00000   0.00000   0.00000   0.00000
Results, Element     4
   u=   -0.06258  -0.22016  -0.07159  -0.35718  -0.06772  -0.28642
   t=    0.00000   0.00000   0.00000   0.00000   0.00000   0.00000
Results, Element     5
   u=   -0.07159  -0.35718  -0.07457  -0.50631  -0.07377  -0.43097
   t=    0.00000   0.00000   0.00000   0.00000   0.00000   0.00000
Results, Element     6
   u=   -0.07457  -0.50631   0.07457  -0.50631   0.00000  -0.50620
   t=    0.00000 -10.00000   0.00000 -10.00000   0.00000 -10.00000
Results, Element     7
   u=    0.07457  -0.50631   0.07159  -0.35718   0.07377  -0.43097
   t=    0.00000   0.00000   0.00000   0.00000   0.00000   0.00000
Results, Element     8
   u=    0.07159  -0.35718   0.06258  -0.22016   0.06772  -0.28642
   t=    0.00000   0.00000   0.00000   0.00000   0.00000   0.00000
Results, Element     9
   u=    0.06258  -0.22016   0.04763  -0.10710   0.05575  -0.15989
   t=    0.00000   0.00000   0.00000   0.00000   0.00000   0.00000
Results, Element    10
   u=    0.04763  -0.10710   0.02685  -0.02978   0.03789  -0.06325
   t=    0.00000   0.00000   0.00000   0.00000   0.00000   0.00000
Results, Element    11
   u=    0.02685  -0.02978   0.00000  -0.00019   0.01409  -0.00820
   t=    0.00000   0.00000-298.03241   0.00000   0.00000   0.00000
Results, Element    12
   u=    0.00000  -0.00019   0.00000  -0.00019   0.00000   0.00000
```

The input file for Program 8.1 for this problem is:

```
1 12
2.5 0.1
2.5 0.2
2.5 0.3
2.5 0.4
2.5 0.5
2.5 0.6
2.5 0.7
2.5 0.8
2.5 0.9
```

The output obtained from Program 8.1 is:

```
Postprocessed results

Results at centres of boundary elements:
Element           1 Stress:    -268.5000        0.00  0.00
Element           2 Stress:    -207.8000        0.00  0.00
Element           3 Stress:    -149.5000        0.00  0.00
Element           4 Stress:    -90.10002        0.00  0.00
Element           5 Stress:    -29.80001        0.00  0.00
Element           6 Stress:      0.00000        0.00 10.00
Element           7 Stress:     29.80001        0.00  0.00
Element           8 Stress:     90.10002        0.00  0.00
Element           9 Stress:    149.5000         0.00  0.00
Element          10 Stress:    207.8000         0.00  0.00
Element          11 Stress:    268.5000         0.00  0.00
Element          12 Stress:      0.0000         0.00 10.00

Internal results:
Coordinates:      2.50000        0.10000
       u:        -0.04525       -0.14601
   Stress:    1707.69751    -3532.69019       561.74805
Coordinates:      2.50000        0.20000
       u:        -0.03347       -0.15646
   Stress:    -396.19366      435.71371       -82.49610
Coordinates:      2.50000        0.30000
       u:        -0.02200       -0.16112
   Stress:       2.74982      -84.11363        -2.52538
Coordinates:      2.50000        0.40000
       u:        -0.01106       -0.15886
   Stress:     -63.19255       43.75769       -18.32864
Coordinates:      2.50000        0.50000
       u:         0.00000       -0.15926
   Stress:       0.00000        0.00006       -16.43005
Coordinates:      2.50000        0.60000
       u:         0.01106       -0.15886
   Stress:      63.19250      -43.75760       -18.32874
```

```
Coordinates:       2.50000        0.70000
          u:       0.02200       -0.16112
    Stress:       -2.74981       84.11368      -2.52534
Coordinates:       2.50000        0.80000
          u:       0.03347       -0.15646
    Stress:      396.19373     -435.71402     -82.49619
Coordinates:       2.50000        0.90000
          u:       0.04525       -0.14601
    Stress:    -1707.69714     3532.68799     561.74780
```

A total of three analyses were carried out, gradually reducing the value of t to 0.5 and 0.2m. The results of the analyses are summarised in Table 9.1 and compared with results obtained from the classical beam theory[1] (Bernoulli hypothesis). The maximum deflection at the free end and the bending stresses at the fixed end are compared. Note that the bending stresses are obtained directly from the analysis (these are equal to the tractions t_x on element 12).

Table 9.1 Summary of results for cantilever beam with parabolic boundary elements

t/L	Mesh	Max. deflection *(mm)*		Max. stress *(MPa)*	
		Computed	Beam theory	Computed	Beam theory
0.2	1	0.5063	0.500	0.298	0.300
0.1	1	3.6970	4.000	1.103	1.200
0.04	1	36.497	62.50	4.359	7.500

Note that the results obtained for a ratio t/L of 0.2 include the additional effect of shear displacement and are probably more accurate than the results computed by beam theory. It can be seen that the results deteriorate rapidly with decreasing value of t/L, and that for a ratio t/L of 0.04 the error is unacceptable. Note that for an equivalent finite element mesh, no problems arises if the thickness is reduced.

The reasons for this deterioration in accuracy are twofold: firstly, as we already pointed out in Chapter 6, our integration scheme is fairly limited in that no element subdivision is allowed for, and that the maximum number of Gauss points is limited to eight. Therefore, as collocation points are getting very close to the elements over which we integrate, the integration error becomes rapidly unacceptable.

While this error can be rectified by allowing more Gauss points or element subdivision, another source of error will still persist. This is the error introduced by the fact that in the collocation points are very close to each other in one direction and far away in the other direction. Apparently, this causes a lack of diagonal dominance and the system of equations to become unstable. Numerical experiments have shown that even if the precision of integration is increased significantly by using element subdivision, errors still persist if the ratio t/L is decreased to a very low value. A relatively simple remedy to this problem is to increase the number of elements as the ratio t/L is decreased. Such adaptive meshes for $t/L = 0.1$ and 0.04 are shown in Figure 9.3.

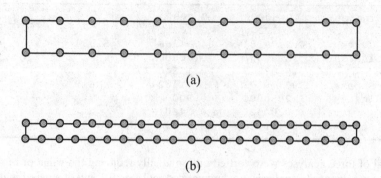

(a)

(b)

Figure 9.3   Adaptive meshes: (a) mesh 2 and (b) mesh 3 (only corner nodes shown)

As can be seen from Table 9.2, the accuracy of the results is now greatly improved.

Table 9.2   Results for refined meshes

t/L	Mesh	Max. deflection (mm)		Max. stress (MPa)	
		Computed	Beam theory	Computed	Beam theory
0.2	1	0.5063	0.500	0.298	0.300
0.1	2	3.9930	4.000	1.192	1.200
0.04	3	62.234	62.50	7.459	7.500

If we compute the stresses at the centre of the beam (i.e. at 2.5m from the fixed end using the postprocessor (Program 8.1) for mesh 2 with $t/L = 0.2$), then we get the distribution in Figure 9.4 for the bending stresses and in Figure 9.5 for the shear stresses.

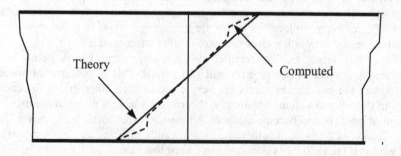

Theory

Computed

Figure 9.4   Bending stresses computed by postprocessing

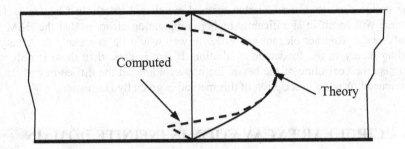

Figure 9.5   Shear stresses computed by postprocessing

It can be clearly seen that the limited accuracy of the integration which we allowed for in the postprocessing program leads to large errors as we get close to the boundary element, but that values computed directly on the boundary are very accurate. The results near the boundary elements can be easily improved if we use subregions of integration.

9.2.4 Comparison with FEM

If we compare this example with the finite element method, then we can see that an equivalent FE mesh with parabolic elements exactly represents the bending stress, but will only be able to approximate the parabolic distribution of the shear stress by a constant distribution.

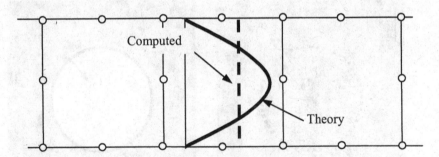

Figure 9.6   Part of FE mesh and computed shear stress distribution

9.2.5 Conclusions

The conclusion from this example is that there is almost no difference in the discretisation effort and computing time between the FEM and BEM for this example. We find that the postprocessing program which we developed has limitations with respect to the computations of interior results if points are close to the boundary. Also, we found that for slender beams the program leads to significant errors for coarse discretisations. However, there are ways in which we can significantly improve these results, for example by either improving the precision of integration or by using schemes other than point collocation, (e.g. the Galerkin method mentioned briefly in Chapter 6). However, all these will result in significantly higher computation effort so that the BEM will lose its advantage. Another elegant and efficient way would be to include the classical beam bending theory in the fundamental solution. If this is done then there is only a need to discretise the centreline of the beam, thereby avoiding all the difficulties which we have experienced. A good description of this method is given by Hartmann[2].

9.3   CIRCULAR EXCAVATION IN INFINITE DOMAIN

9.3.1 Problem statement

Consider an excavation made in an infinite, homogeneous, elastic space. The elastic space is assumed to have a modulus of elasticity of 10 000 MPa, a Poisson's ratio of zero and to have been pre-stressed with a stress field of $\sigma_x = 0.0$ MPa, $\sigma_y = -3.0$ MPa, (compression) and $\tau_{xy} = 0.0$ MPa. The displacements and the changed stress distribution due to excavation are required. To obtain this solution, we actually have to solve two problems (Figure 9.7): one (trivial) one where no excavation exists; and one where the supporting tractions t_0 computed in the first step are released, i.e. applied in the opposite direction.

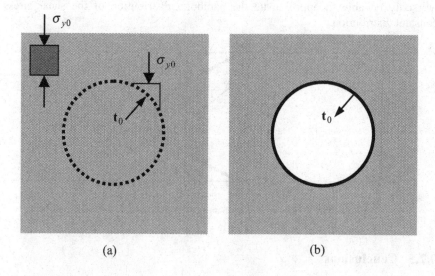

$$(a) \qquad\qquad\qquad\qquad (b)$$

Figure 9.7   Problems to be solved: infinite space (a) without (b) with excavation

We can use equation (4.26) to solve problem (a), i.e. to compute the tractions t_0 as

$$t_0 = \left\{ \begin{array}{c} 0.0 \\ n_y \sigma_{y0} \end{array} \right\} \qquad\qquad (9.1)$$

9.3.2 Boundary element discretisation and input

To solve problem (b) we use the BEM with two planes of symmetry. For the first mesh, a single parabolic element (Figure 9.8) is used. Subsequently, two (mesh 2) and four elements (mesh 3) are used for a quarter of the boundary. The mesh is subjected to *Neumann* boundary conditions with values of t_0 applied, as shown in Figure 9.8.

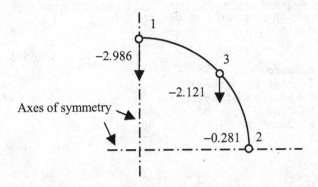

Figure 9.8   Boundary element mesh with *Neumann* boundary conditions

If the element were able to describe an exact circle then the values of the traction t_{y0} should be exactly −3.0 at node 1 and 0.0 at node 2. However, since the element can only describe a parabola, the y-component of the normal vector will not be exactly −1.0 at node 1 and not exactly 0.0 at node 2. Therefore a small geometrical error occurs due to the coarse discretisation.

The input file for Program 7.1 for this problem is:

```
Circular hole
2
2
1
2
2
2
0.1000E+05
0.0000E+00
       3
       1
         0.000        1.000
         1.000        0.000
         0.707        0.707
       1     2     3
       0
       1
       1     0.00000 -2.98681    0.00000 -0.28103    0.00000 -2.12132
```

The output obtained from Program 7.1 is:

```
Project:
 Circular hole
 Cartesian_dimension:                              2
 Elasticity problem
 Type of analysis:
 Solid plane                                 Strain
 Infinite region
 Symmetry about y-z and x-z planes
 Quadratic elements
 Modulus:
                                             10000.00
 Poisson's ratio:                            0
 Number of nodes of system:           3
 Number of elements of system:        1
Node     1  Coor      0.00     1.00
Node     2  Coor      1.00     0.00
Node     3  Coor      0.71     0.71

 Incidences:
EL      1  Inci       1     2     3

 Elements with Dirichlet BC's:

 Elements with Neumann BC's:

 Element             1  Prescribed values:
  0.00  -2.986810
  0.00  -0.281030
  0.00  -2.121320
 Results, Element     1
   u=    0.00000  -0.00060   0.00029   0.00000   0.00021  -0.00041
   t=    0.00000  -2.98681   0.00000  -0.28103   0.00000  -2.12132
```

The input file for Program 8.1 for this problem is:

```
1 1
1.1   0
1.2   0
1.3   0
1.4   0
1.5   0
```

The output obtained from Program 8.1 is:

```
Postprocessed results
Results at the boundary:
Element 1            ξ=-1     Stress:      2.900716        0.00   0.00
                     ξ= 0     Stress:     -1.550000        0.00   0.00
                     ξ=+1     Stress:     -5.624045        0.00   0.00

Internal results:
Coordinates:      1.10000        0.00000
       u:         0.00029        0.00000
   Stress:       -1.70867       -3.62508        0.00000
Coordinates:      1.20000        0.00000
       u:         0.00028        0.00000
   Stress:       -0.95274       -3.07557        0.00000
Coordinates:      1.30000        0.00000
       u:         0.00027        0.00000
   Stress:       -1.03789       -2.38682        0.00000
Coordinates:      1.40000        0.00000
       u:         0.00026        0.00000
   Stress:       -1.07541       -1.87793        0.00000
Coordinates:      1.50000        0.00000
       u:         0.00025        0.00000
   Stress:       -1.06400       -1.51019        0.00000
```

9.3.3 Results

In Table 9.3 the results at the boundary for the three meshes are compared. It can be seen that even the coarse mesh with only one element gives acceptable results for this problem. The internal results along a horizontal line are shown in Figures 9.9 and 9.10.

Table 9.3   Results for meshes with parabolic boundary elements

Mesh	No. Elem	Max. deflection (mm)	Max. stress (MPa)	Min. stress (MPa)
1	1	0.60	-8,62	2.90
2	2	0.60	-8.99	2.99
3	4	0.60	-9.00	3.00
Theory		0.60	-9.00	3.00

Note that in the figures the number of elements given is the total number including the mirrored elements. This means that for mesh 1 only one element is specified by the user with 3mirrored elements automatically generated by the program. It can be seen that, due to the integration scheme employed, the results cannot be computed very close to the boundary for the coarse mesh.

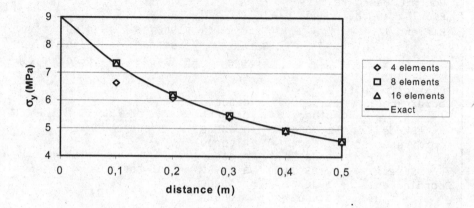

Figure 9.9   Distribution of vertical stress along a horizontal line

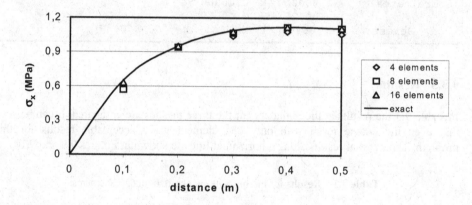

Figure 9.10   Distribution of horizontal stress along a horizontal line

9.3.4 Comparison with FEM

The problem was analysed with the FEM with three different meshes of finite elements with quadratic shape function as shown in Figure 9.11. Symmetry was considered by fixing the corresponding displacements at the symmetry planes. This discretisation has the same variation of displacements along the excavation boundary as the boundary element mesh. Note that in the FEM we have to truncate the mesh at some distance away from the excavation. A truncation of two diameters away from the excavation is often used. At the truncation surface all displacements are assumed to be fixed. In order to eliminate the truncation error-coupled analyses were also made where boundary elements were used at the truncated surface of the FE mesh (see chapter 14).

Table 9.4 shows the results of the analysis. It can be seen that they are less accurate than those obtained for the BEM, and that the truncation error is still significant with a truncation distance of two diameters of the circle.

Table 9.4   Results for meshes with parabolic finite elements

		FEM only		Coupled with BEM	
Mesh	No. Elem	u_{max} *(mm)*	σ_{max} *(MPa)*	u_{max} *(mm)*	σ_{max} *(MPa)*
1	2	0.480	-6.840	0.535	-8.48
2	6	0.494	-7.189	0.583	-9.01
3	16	0.506	-8.165	0.598	-8.97
Theory		0.600	-9.000	0.600	-9.00

9.3.5 Conclusions

In contrast to the previous example, this one favours the boundary element method. We see that with the FEM we have two sources of error: one associated with the truncation of the mesh that is necessary because the method is unable to model infinite domains, the other one is that in the FEM the variation of the unknown has to be approximated by shape functions in the directions inside the continuum as well as along the boundary surface. It can be seen that without much additional effort, the first error can be virtually eliminated by using the coupled method, and by specifying boundary elements at the truncated boundary. This example demonstrates that the BEM is most efficient when the ratio boundary surface to volume is very small. For problems in geomechanics, where the soil/rock mass can be assumed to have infinite extent, this ratio actually approaches zero.

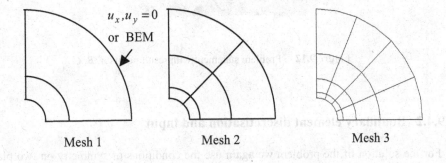

$u_x, u_y = 0$
or BEM

Mesh 1   Mesh 2   Mesh 3

Figure 9.11   Finite element meshes used

9.4   SQUARE EXCAVATION IN INFINITE ELASTIC SPACE

9.4.1 Problem statement

This example was chosen to demonstrate the ability of the BEM to model stress concentrations. The problem is identical to the previous one except that the shape of the excavation is square instead of circular. The exact solution for this problem is not known, but according to the theory of elasticity, a singularity of the vertical stress occurs as the corner is approached.

It is known[3] that for a corner with a subtended angle of 180° (crack), the displacements tend to zero with $\sqrt{r}$ and the stresses tend to infinity with $1/\sqrt{r}$. A boundary element with quadratic variation of displacements will not be able to model this variation so we expect some loss of accuracy for coarse meshes. While there is obviously no point in trying to compute an infinite value of stress, the variation of the displacement can be used to compute stress intensity factors[2]. So there is some scope to be able to accurately predict the variation of displacements.

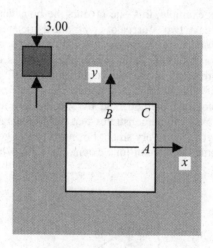

Figure 9.12   Problem statement with result points A, B, C

9.4.2 Boundary element discretisation and input

For the solution of the problem we again use the conditions of symmetry on two planes and consider four meshes, three of which are shown in Figure 9.13. Meshes 1 to 3 simply represent a uniform subdivision into 2, 4 and 16 elements. Mesh 4 is a graded mesh where the element size has been reduced near the corner.

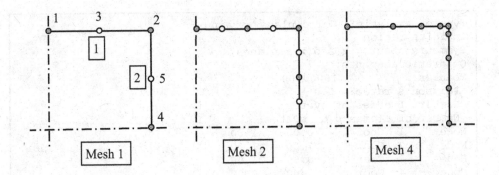

Figure 9.13   Meshes used (mid-side nodes not shown for graded mesh 4)

The input file for Program 7.1 for mesh 1 is:

```
Square excavation - Mesh 1
2
2
1
2
2
2
0.1000E+05
0.0000E+00
     5
     2
     0.000      1.000
     1.000      1.000
     0.500      1.000
     1.000      0.000
     1.000      0.500
     1      2      3
     2      4      5
     0
     2
     1  0.00000  -3.00000  0.00000  -3.00000  0.00000  -3.00000
     2  0.00000   0.00000  0.00000   0.00000  0.00000   0.00000
```

The output obtained from Program 7.1 is:

```
Project:
 Square excavation - Mesh 1
 Cartesian_dimension:      2
 Elasticity problem
```

```
Type of analysis:  Solid plane strain
Infinite region
Symmetry about y-z and x-z planes
Quadratic elements
Modulus:            10000.00
 Poisson's ratio:  0.0
 Number of nodes of system:        5
 Number of elements of system:     2
Node      1  Coor      0.00    1.00
Node      2  Coor      1.00    1.00
Node      3  Coor      0.50    1.00
Node      4  Coor      1.00    0.00
Node      5  Coor      1.00    0.50

 Incidences:
EL      1  Inci      1    2    3
EL      2  Inci      2    4    5

 Elements with Dirichlet BC's:

 Elements with Neumann BC's:

 Element            1  Prescribed values:
  0.00  -3.00
  0.00  -3.00
  0.00  -3.00
 Element            2  Prescribed values:
  0.00   0.00
  0.00   0.00
  0.00   0.00

 Results, Element    1
 u=   0.00000  -0.00072   0.00018  -0.00031   0.00014  -0.00063
 t=   0.00000  -3.00000   0.00000  -3.00000   0.00000  -3.00000
 Results, Element    2
 u=   0.00018  -0.00031   0.00020   0.00000   0.00021  -0.00012
 t=   0.00000   0.00000   0.00000   0.00000   0.00000   0.00000
```

The input file for Program 8.1 for mesh 1 is:

```
1 2
1.1   0
1.2   0
1.3   0
1.4   0
1.5   0
```

The output obtained from Program 8.1 is:

```
Postprocessed results
Results at centres of boundary elements:
Element                    1 Stress:        1.800000      0.00  0.00
Element                    2 Stress:       -3.100000      0.00  0.00
Internal results:
Coordinates:      1.10000        0.00000
         u:       0.00020        0.00000
     Stress:      0.09520       -2.03256        0.00000
Coordinates:      1.20000        0.00000
         u:       0.00021        0.00000
     Stress:      0.05180       -2.09529        0.00000
Coordinates:      1.30000        0.00000
         u:       0.00021        0.00000
     Stress:     -0.06386       -1.96108        0.00000
Coordinates:      1.40000        0.00000
         u:       0.00020        0.00000
     Stress:     -0.15494       -1.80109        0.00000
Coordinates:      1.50000        0.00000
         u:       0.00020        0.00000
     Stress:     -0.23919       -1.62653        0.00000
```

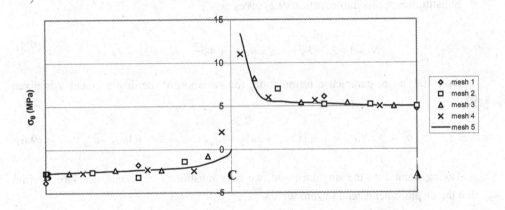

Figure 9.14   Distribution of tangential stress on boundary B, C, A

Boundary stresses obtained from Program 8.1 are plotted in Figure 9.14. It can be seen that the magnitude of the stress concentration depends on the fineness of the boundary element mesh near the corner, and that a fine graded mesh can reasonably approximate the theoretical stress distribution at the corner.

9.4.3 'Quarter point' elements

A numerical trick can be used to simulate a singularity: if we move the 'midside' node of an element to the 'quarter point' on one side, then it will be shown that the Jacobian tends to zero as the nearest corner node is approached. The following derivation has been suggested by Trevor Davies.[4]

Consider the simple element in Figure 9.15, which is located along the x-axis with one point at the origin. The derivation is made simpler if the intrinsic coordinate ξ is transformed into $\bar{\xi}$ which ranges from 0 to 1.0.

Expressed in this coordinate system, the three shape functions are given by

$$N_1 = 2\left(\bar{\xi} - 1/2\right)\left(\bar{\xi} - 1\right), \quad N_3 = -4\bar{\xi}\left(\bar{\xi} - 1\right), \quad N_2 = 2\bar{\xi}\left(\bar{\xi} - 1/2\right) \tag{9.2}$$

The coordinate x of a point with local coordinate $\bar{\xi}$ can be computed by the interpolation

$$x = \sum N_i\left(\bar{\xi}\right) x_i \tag{9.3}$$

Substituting for the coordinates of the nodes ($x_1 = 0.0, x_2 = 0.25, x_3 = 1.0$), we obtain simply

$$x = \bar{\xi}^2 \tag{9.4}$$

Substitution of this into equation (9.2) gives

$$N_1 = 1 + 2x - 3\sqrt{x}; \quad N_3 = -4x + 4\sqrt{x}; \quad N_2 = 2x - \sqrt{x} \tag{9.5}$$

Assuming an isoparametric formulation, the variation of the displacement u is given by

$$u_x = \sum N_i u_{xi} = u_1 + x(2u_{x1} + 2u_{x2} - 4u_{x3}) + \sqrt{x}(3u_{x1} + 4u_{x2} - u_{x3}) \tag{9.6}$$

Taking point 1 as the singular point, we may substitute $r = x$, and therefore we find that the displacements tend to zero with $\sqrt{r}$.

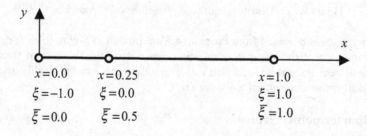

Figure 9.15   'Quarter point' boundary element

The strains are computed by taking the derivative of the displacements, and are given by

$$\varepsilon_x = \frac{\partial u_x}{\partial x} = c + \frac{d}{\sqrt{r}} \qquad (9.7)$$

where c and d are constants. Since for elastic material the stresses σ_x are proportional to the strains, we see that they go to infinity with $o(1/\sqrt{r})$.

In Figure 9.14, is shown that, with the simple expedient of moving the third node point of the element at the corner, we can obtain similar or slightly better results for mesh 3 with a 'quarter point' element (mesh 5) than with the graded mesh 4. Such elements have been successfully used for the computation of stress intensity factors[5].

9.4.4 Comparison with finite elements

If we compare with finite element results, we find that the results are influenced not only by the discretisation along the boundary, but also the subdivision inside the elastic space. In fact any result for the stress concentration at the edge can be obtained depending on the element subdivision. A reasonably fine graded mesh is shown in Figure 9.16. It consists of 40 elements with quadratic variation of displacements. The distribution of the tangential stress along a vertical line, however, shows that the general trend of the theoretical distribution cannot be obtained.

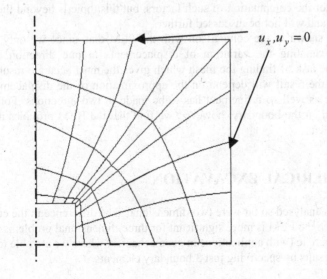

$u_x , u_y = 0$

Figure 9.16   Finite element mesh

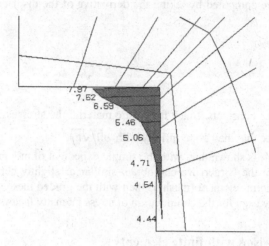

Figure 9.17   Distribution of vertical stress along a nearly vertical line through the Gauss points

9.4.5 Conclusions

In this example we have shown how the boundary element method deals with singularities as they sometimes arise when we have corners. These singularities are of course only theoretical, since there is no such a thing as a perfect corner in nature. Also, the stresses cannot reach an infinite value because they will be limited by a maximum value that a material can sustain. However, in fracture mechanics we may compute stress intensity factors based on the variation of the displacements near the crack. The BEM is well suited for the computation of such factors, but this topic is beyond the limited scope of this book and will not be discussed further.

We have shown in the comparison with the FEM that, since we only have to worry about approximating the variation of displacements in one direction, i.e. along the boundary, the task of finding the mesh which gives the most accurate result is simplified. In the FEM the result will depend on the approximation of the displacements inside the elastic space as well so refinement has to be made in two directions. For a comparable discretisation on the boundary, however, we find that the BEM provides a better answer for this problem.

9.5   SPHERICAL EXCAVATION

All problems analysed so far were two-dimensional. The difference in the computing effort between the FEM is more significant for three-dimensional problems. Here we show one example (with applications in geomechanics) where we are able to obtain reasonable results by specifying just 3 boundary elements.

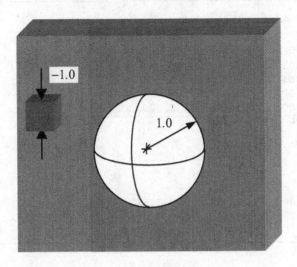

Figure 9.18   Problem statement

9.5.1 Problem statement

The example is similar to the example of a circular excavation in an infinite domain, except that the excavation is now spherical, and the virgin stress is given by

$$\boldsymbol{\sigma}_0 = \begin{bmatrix} 0 & 0 & -1 & 0 & 0 & 0 \end{bmatrix}^T \tag{9.8}$$

9.5.2 Boundary element discretisation and input

Two fairly coarse discretisations are used. Both meshes consist of three boundary elements; one consists of linear the other of parabolic elements. Three planes of symmetry are assumed so only one octant of the problem had to be considered. The meshes are shown in Figure 9.19.

The input file for Program 7.1 for mesh 1 is:

```
Spherical excavation
3
3
2
3
1
0.1000E+05
0.0000E+00
```

```
7
3
 1.000        0.000        0.000
      0.707        0.000        0.707
      0.577        0.577        0.577
      0.707        0.707        0.000
      0.000        0.707        0.707
      0.000        1.000        0.000
      0.000        0.000        1.000
      5     1     2     3
      3     2     4     6
      2     1     7     4
      0
      3
1 0.0 0.0 -0.3574 0.0 0.0 -0.3677 0.0 0.0 -0.2683 0.0 0.0 -0.2818
2 0.0 0.0 -0.2818 0.0 0.0 -0.2683 0.0 0.0 -0.3671 0.0 0.0 -0.3574
3 0.0 0.0 -0.9251 0.0 0.0 -0.8864 0.0 0.0 -0.8628 0.0 0.0 -0.8864
```

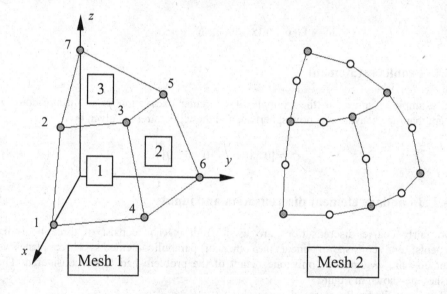

Figure 9.19   Boundary meshes used

The output obtained from Program 7.1 is:

```
Project:
 Spherical excavation
 Cartesian_dimension:           3
 Elasticity problem
 Infinite region
 Symmetry about all planes
 Linear elements
```

```
 Modulus:            10000.00
 Poisson's ratio:  0.0
 Number of nodes of system:             7
 Number of elements of system:          3
Node      1  Coor       1.00      0.00      0.00
Node      2  Coor       0.71      0.00      0.71
Node      3  Coor       0.58      0.58      0.58
Node      4  Coor       0.71      0.71      0.00
Node      5  Coor       0.00      0.71      0.71
Node      6  Coor       0.00      1.00      0.00
Node      7  Coor       0.00      0.00      1.00


 Incidences:
EL      1  Inci       1      2      3      4
EL      2  Inci       4      3      5      6
EL      3  Inci       3      2      7      5


 Elements with Dirichlet BC's:

 Elements with Neumann BC's:

 Element              1  Prescribed values:
  0.00   0.00  -0.3575400
  0.00   0.00  -0.3672700
  0.00   0.00  -0.2689700
  0.00   0.00  -0.2824100
 Element              2  Prescribed values:
  0.00   0.00  -0.2824100
  0.00   0.00  -0.2689700
  0.00   0.00  -0.3672700
  0.00   0.00  -0.3575400
 Element              3  Prescribed values:
  0.00   0.00  -0.9248300
  0.00   0.00  -0.8862000
  0.00   0.00  -0.8627400
  0.00   0.00  -0.8862000


Results, Element     1
   u=       0.00002   0.00000   0.00000   0.00001   0.00000  -0.00005
            0.00001   0.00001  -0.00004   0.00001   0.00001   0.00000
   t=       0.00000   0.00000  -0.35754   0.00000   0.00000  -0.36727
            0.00000   0.00000  -0.26897   0.00000   0.00000  -0.28241
Results, Element     2
   u=       0.00001   0.00001   0.00000   0.00001   0.00001  -0.00004
            0.00000   0.00001  -0.00005   0.00000   0.00002   0.00000
   t=       0.00000   0.00000  -0.28241   0.00000   0.00000  -0.26897
            0.00000   0.00000  -0.36727   0.00000   0.00000  -0.35754
Results, Element     3
   u=       0.00001   0.00001  -0.00004   0.00001   0.00000  -0.00005
            0.00000   0.00000  -0.00007   0.00000   0.00001  -0.00005
   t=       0.00000   0.00000  -0.92483   0.00000   0.00000  -0.88620
            0.00000   0.00000  -0.86274   0.00000   0.00000  -0.88620
```

The input file for Program 8.1 for mesh 1 is:

```
 1 3
 1.1     0      0
 1.2     0      0
 1.3     0      0
 1.4     0      0
 1.5     0      0
 2.0     0      0
 4.0     0      0
 6.0     0      0
10.0     0      0
```

The output obtained from Program 8.1 is:

```
Postprocessed results

 Results at centres of boundary elements:
   Element   1  Stress:    -0.02713   -0.06991     0.07922
                            0.00000    0.00000     0.00000
   Element   2  Stress:     0.11654   -0.52780     0.67830
                            0.00000    0.00000     0.00000
   Element   3  Stress:     0.07749    0.05937     0.08258
                            0.00000    0.00000     0.00000

 Internal results:
Coordinates:       1.10000      0.00000      0.00000
          u:       0.00002      0.00000      0.00000
     Stress:      -0.20158      0.15633     -0.39412
                   0.00000      0.00000      0.00000
Coordinates:       1.20000      0.00000      0.00000
          u:       0.00001      0.00000      0.00000
     Stress:      -0.16961      0.12147     -0.25647
                   0.00000      0.00000      0.00000
Coordinates:       1.30000      0.00000      0.00000
          u:       0.00001      0.00000      0.00000
     Stress:      -0.14644      0.09865     -0.18208
                   0.00000      0.00000      0.00000
Coordinates:       1.40000      0.00000      0.00000
          u:       0.00001      0.00000      0.00000
     Stress:      -0.12693      0.08157     -0.13503
                   0.00000      0.00000      0.00000
Coordinates:       1.50000      0.00000      0.00000
          u:       0.00001      0.00000      0.00000
     Stress:      -0.11019      0.06824     -0.10311
                   0.00000      0.00000      0.00000
Coordinates:       2.00000      0.00000      0.00000
          u:       0.00001      0.00000      0.00000
     Stress:      -0.05632      0.03147     -0.03531
                   0.00000      0.00000      0.00000
```

```
Coordinates:        4.00000        0.00000        0.00000
           u:        0.00000        0.00000        0.00000
      Stress:       -0.00832        0.00427       -0.00340
Coordinates:        6.00000        0.00000        0.00000
           u:        0.00000        0.00000        0.00000
      Stress:       -0.00254        0.00128       -0.00095
                     0.00000        0.00000        0.00000
Coordinates:       10.00000        0.00000        0.00000
           u:        0.00000        0.00000        0.00000
      Stress:       -0.00056        0.00028       -0.00020
                     0.00000        0.00000        0.00000
```

9.5.3 Results

For the mesh with parabolic boundary elements the maximum displacement at the crown is computed as 0.09 mm compared with the theoretical value of 0.09. The maximum value of stress at the meridian of the sphere is computed as 2.09 MPa compared with the theoretical solution of 2.0. The results for the stresses are summarised in Figure 9.20, where the vertical stress is plotted along a horizontal line. It can be seen that to get more accurate results inside the region, a finer mesh is necessary.

9.5.4 Comparison with FEM

To be able to model this problem with the FEM we have to truncate the mesh in the same way as with the 2-D example. A fairly coarse mesh is shown in Figure 9.21. The mesh has 135 degrees of freedom as compared with 16 degrees of freedom of the boundary element mesh 2. The maximum displacements obtained from this analysis of 0.084, the maximum stress is 1.53 MPa which represents a poor agreement with the theory. A finer mesh would be required.

9.6   CONCLUSIONS

The purpose of this chapter was to show on a number of simple examples how the method works and what sort of accuracy can be expected by comparing the results with the theoretical solution. The examples have been chosen in such a way that they also include some examples where the BEM as implemented in this book has some difficulties. In some cases we highlight the fact that with the relatively unsophisticated integration scheme which we implemented in this book, care has to be taken when computing interior results close to the boundary.

After reading this chapter, the reader should have learned how to generate boundary element meshes and the input files for Programs 7.1 General_purpose_BEM and 8.1 Post_processor. A good appreciation of the method, the accuracy that can be obtained and the pitfalls that should be avoided should also have been gained. The reader may now proceed to learn more about more advanced topics.

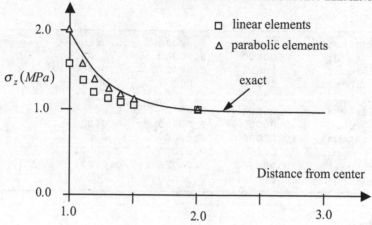

Figure 9.20   Distribution of vertical stress

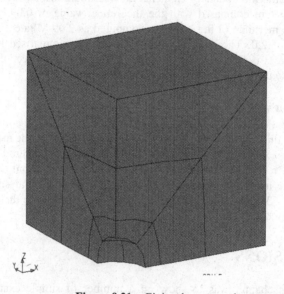

Figure 9.21   Finite element mesh

9.7   REFERENCES

1. Krätzig, W.B. and Wittek, U. (1995) *Tragwerke 1*. Springer, Berlin.
2. Hartmann, F. (1987) *Methode der Randelemente*. Springer, Berlin.
3. Kanninen, M.F. and Popelar, C.H. (1985) *Advanced Fracture Mechanics*. Oxford. Science Series 15, Oxford University Press, New York.
4. Davies, T. (2000) Personal communication.
5. Ingraffea, A.R. and Manu, C. (1980) Stress intensity factor computation in three dimensions with quarter point elements, *Int J. Numer. Meth. in Eng.,* **15**, 1427-1445.

10

Multiple Regions

Die Tiefe ist im Klaren und Heiteren
(The depth lies in the clear, the pleasing)
H. Hesse

10.1 INTRODUCTION

The solution procedures described so far are only applicable to homogeneous domains, as the fundamental solution used assumes that material properties do not change inside the domain being analysed. There are many instances, however, where this assumption does not hold. For example, in a soil mass, the modulus of elasticity may change with depth or there might be various soil layers with different properties. For some special types of non-homogeneity it is possible to derive fundamental solutions, for example, if the material properties vary linearly with depth. However, such fundamental solutions are often complicated and the programming effort significant.

In cases when we have layers or zones of different materials, however, we can develop special solution methods based on the kernels discussed in Chapter 4. The basic idea is to consider a number of regions which are connected to each other, much like pieces of a puzzle. Each region is treated in the same way as discussed previously, but can now be assigned different material properties. Since at the interfaces between the regions both **t** and **u** are not known, the number of unknowns is increased and additional equations are required to be able to solve the problem. These equations can be obtained from the conditions of equilibrium and compatibility at the region interfaces. There are two approaches which can be taken in the implementation of the method.

In the first[1], we modify the assembly procedure, so that a larger system of equations is now obtained including the additional unknowns at the interfaces. The second method[2] is similar to the approach taken by the finite element method. Here we construct a 'stiffness matrix' **K** of each region, the coefficients of which are the fluxes or tractions due to unit temperatures/displacements. The matrices **K** for all regions are then assembled in the same way as with the FEM. The second method is more efficient and more amenable to implementation on parallel computers. The method may also be used for coupling boundary with finite elements, as outlined in Chapter 14).

10.2 MULTI-REGION ASSEMBLY

For the purpose of explaining multi-region assembly, consider the example in Chapter 7 (Figure 7.3) to which we add another region (Figure 10.1). For the purpose of the explanation, we assume there is only one degree of freedom per node and a *Neumann* BC (*t* known) is assumed everywhere, except on element 4 of region I, where a *Dirichlet* BC is assumed (*u* known). At the interface between the regions, however, both *u* and *t* are unknown giving an additional two unknowns for the example, that is, a total of eight unknowns. These are numbered in parentheses in Figure 10.1.

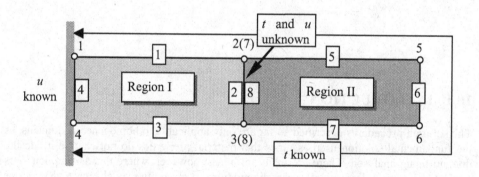

Figure 10.1   A simple two-region problem for explaining assembly

Each region is defined by linear boundary elements describing its boundary and a code which specifies whether the region is finite or infinite. In our example, region I consists of elements 1, 2, 3, 4 and region II of elements 5, 6, 7, 8. Both are finite regions. The idea is now to compute the coefficient matrices for each region separately and then to assemble them. At the interface between regions we have two boundary elements, one belonging to region I (element 2) the other to region II (element 8). These elements are identical, except that the sequence of node numbers and therefore the outward normal is reversed.

We introduce two different numbering systems, a local one, specified for each region, as shown in Figure 10.2, which forms the basis for computation of element contributions to the coefficient matrices for each region, and a global one, as shown in Figure 10.1, used for the region assembly. Incidences of elements in the global and local (region) numbering are shown in Table 10.1.

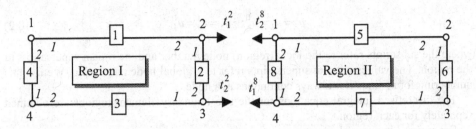

Figure 10.2   Local (region) numbering of nodes and local (element) numbering (italics)

Table 10.1 Incidences of boundary elements in global and local numbering

Element	Global		Local (region)	
	1	2	1	2
1	1	2	1	2
2	2	3	2	3
3	3	4	3	4
4	4	1	4	1
5	2	5	1	2
6	5	6	2	3
7	6	3	3	4
8	3	2	4	1

For assembly we need to expand the destination vector to include destinations for the additional unknowns at the interface elements. This means that for element 2 of region I, where *Ldest* was previously defined as (/2,3/), this now becomes (/2,3,7,8/), whereas for element 8 of region II, *Ldest* is expanded to (/3,2,8,7/).

Additional equations are needed in order to determine the additional unknown. These are obtained from the conditions of equilibrium (or heat balance) and compatibility (or uniqueness of temperature/potential). For the specific case considered here, the equations for the preservation of heat balance at the interface are written as:

$$t_1^2 + t_2^8 = 0; \quad t_2^2 + t_1^8 = 0 \tag{10.1}$$

Note that in our numbering scheme, the superscript refers to the boundary element number and the subscript to the local (element) node number. We use this numbering scheme for t since it may be discontinuous at a node.

The conditions of compatibility, or in our case uniqueness of temperature /potential, are written as:

$$u_2 = u_2^{\mathrm{I}} = u_1^{\mathrm{II}}; \quad u_3 = u_3^{\mathrm{I}} = u_4^{\mathrm{II}} \tag{10.2}$$

where the subscripts refer to the local (region) node number and the roman superscripts to the region. The values without superscripts refer to a global node numbering. We use this numbering for u, since it is always continuous at a node.

The discretised integral equations for the two boundary element regions are obtained separately for each region.

For region I known values are:

$$t_1^1; \quad t_2^1; \quad t_1^2; \quad t_2^2; \quad u_1^I; \quad u_4^I \tag{10.3}$$

Unknown values are:

$$u_2^I; \quad u_3^I; \quad t_1^4; \quad t_2^4; \quad t_1^2; \quad t_2^2 \tag{10.4}$$

The discretised integral equations are written as:

$$
\begin{aligned}
&cu(P_i) + (\Delta T_{2i}^1 + \Delta T_{1i}^2) \, u_2^I + (\Delta T_{2i}^2 + \Delta T_{1i}^3) \, u_3^I \\
&- \Delta U_{1i}^4 \, t_1^4 - \Delta U_{2i}^4 \, t_2^4 - \Delta U_{1i}^2 \, t_1^2 - \Delta U_{2i}^2 t_2^2 = \qquad\qquad\text{for} \quad i = 1,2,3,4 \\
&\Delta U_{1i}^1 \, t_1^1 + \Delta U_{2i}^1 t_2^1 + \Delta U_{1i}^2 \, t_1^2 + \Delta U_{2i}^2 \, t_2^2 - \Delta T_{2i}^4 \, u_1^I - \Delta T_{1i}^4 \, u_4^I
\end{aligned}
\tag{10.5}
$$

For region II, known values are:

$$t_1^5; \quad t_2^5; \quad t_1^6; \quad t_2^6; \quad t_1^7; \quad t_2^7 \tag{10.6}$$

and unknown values are specified as:

$$t_1^8; \quad t_2^8; \quad u_1^{II}; \quad u_2^{II}; \quad u_3^{II}; \quad u_4^{II} \tag{10.7}$$

Discretised equations for region II are given by:

$$
\begin{aligned}
&cu(P_i) + (\Delta T_{2i}^8 + \Delta T_{1i}^5) \, u_1^{II} + (\Delta T_{2i}^5 + \Delta T_{1i}^6) \, u_2^{II} + (\Delta T_{2i}^6 + \Delta T_{1i}^7) \, u_3^{II} \\
&+ (\Delta T_{2i}^7 + \Delta T_{1i}^8) \, u_4^{II} - \Delta U_{1i}^8 \, t_1^8 - \Delta U_{2i}^8 t_2^8 + = \qquad\qquad\text{for} \quad i = 1,2,3,4 \\
&\Delta U_{1i}^5 \, t_1^5 + \Delta U_{2i}^5 t_2^5 + \Delta U_{1i}^6 \, t_1^6 + \Delta U_{2i}^6 \, t_2^6 + \Delta U_{1i}^7 \, t_1^7 + \Delta U_{2i}^7 \, t_2^7
\end{aligned}
\tag{10.8}
$$

The system of equations for each region can be written in matrix form. For region I we have:

P_i 1   2   3   4   7   8 $\leftarrow$ *unknown*

$$
\begin{bmatrix}
-\Delta U_{21}^4 & \Delta T_{21}^1 + \Delta T_{11}^2 & \Delta T_{21}^2 + \Delta T_{11}^3 & -\Delta U_{11}^4 & -\Delta U_{11}^2 & -\Delta U_{21}^2 \\[2ex]
-\Delta U_{22}^4 & \left[\Delta T_{22}^1 + \Delta T_{12}^2\right] & \Delta T_{22}^2 + \Delta T_{12}^3 & -\Delta U_{12}^4 & -\Delta U_{12}^2 & -\Delta U_{22}^2 \\[2ex]
-\Delta U_{23}^4 & \Delta T_{23}^1 + \Delta T_{13}^2 & \left[\Delta T_{23}^2 + \Delta T_{13}^3\right] & -\Delta U_{13}^4 & -\Delta U_{13}^2 & -\Delta U_{23}^2 \\[2ex]
-\Delta U_{24}^4 & \Delta T_{24}^1 + \Delta T_{14}^2 & \Delta T_{24}^2 + \Delta T_{14}^3 & -\Delta U_{14}^4 & -\Delta U_{14}^2 & -\Delta U_{24}^2
\end{bmatrix}^I
\begin{Bmatrix}
t_2^4 \\ u_2^{II} \\ u_3^{II} \\ t_1^4 \\ t_1^2 \\ t_2^2
\end{Bmatrix} = \{F\}^I \quad (10.9)
$$

where

$$
F_i^I = \Delta U_{1i}^1\, t_1^1 + \Delta U_{2i}^1 t_2^1 + \Delta U_{1i}^3\, t_1^3 + \Delta U_{2i}^3\, t_2^3 - \Delta T_{1i}^4\, u_1^I - \Delta T_{2i}^4\, u_4^I \quad (10.10)
$$

For region II the equations are:

2   5   6   3   7   8

$$
\begin{bmatrix}
\left[\Delta T_{21}^8 + \Delta T_{11}^5\right] & \Delta T_{21}^5 + \Delta T_{11}^6 & \Delta T_{21}^6 + \Delta T_{11}^7 & \Delta T_{21}^7 + \Delta T_{11}^8 & -\Delta U_{21}^8 & -\Delta U_{11}^8 \\[2ex]
\Delta T_{22}^8 + \Delta T_{12}^5 & \left[\Delta T_{22}^5 + \Delta T_{12}^6\right] & \Delta T_{22}^6 + \Delta T_{12}^7 & \Delta T_{22}^7 + \Delta T_{12}^8 & -\Delta U_{22}^8 & -\Delta U_{12}^8 \\[2ex]
\Delta T_{23}^8 + \Delta T_{13}^5 & \Delta T_{23}^5 + \Delta T_{13}^6 & \left[\Delta T_{23}^6 + \Delta T_{13}^7\right] & \Delta T_{23}^7 + \Delta T_{13}^8 & -\Delta U_{23}^8 & -\Delta U_{13}^8 \\[2ex]
\Delta T_{24}^8 + \Delta T_{14}^5 & \Delta T_{24}^5 + \Delta T_{14}^6 & \Delta T_{24}^6 + \Delta T_{14}^7 & \left[\Delta T_{24}^7 + \Delta T_{14}^8\right] & -\Delta U_{24}^8 & -\Delta U_{14}^8
\end{bmatrix}^{II}
\begin{Bmatrix}
u_1^{II} \\ u_2^{II} \\ u_3^{II} \\ u_4^{II} \\ t_2^8 \\ t_1^8
\end{Bmatrix} = \{F\}^{II} \quad (10.11)
$$

and

$$F_i^{II} = \Delta U_{1i}^5\, t_1^5 + \Delta U_{2i}^5 t_2^5 + \Delta U_{1i}^6\, t_1^6 + \Delta U_{2i}^6\, t_2^6 + \Delta U_{1i}^7\, t_1^7 + \Delta U_{2i}^7\, t_2^7 \tag{10.12}$$

Using equations (10.1) and (10.2), both matrix equations can now be written as one system of equations:

$$
\begin{array}{cc}
unknown \rightarrow & 1\quad 2\quad 3\quad 4\quad 5\quad 6\quad 7\quad 8
\end{array}
$$

$$
\begin{array}{ccc}
1 & 1 \\
2 & 2 \\
3 & 3 \\
n_r\quad 4\ P_i & 4 \\
\downarrow\quad 5\ \downarrow & 1 \\
6 & 2 \\
7 & 3 \\
8 & 4
\end{array}
\begin{bmatrix}
a_{11}^I & a_{21}^I & a_{31}^I & a_{41}^I & 0 & 0 & a_{51}^I & a_{61}^I \\
a_{12}^I & a_{22}^I & a_{32}^I & a_{42}^I & 0 & 0 & a_{52}^I & a_{62}^I \\
a_{13}^I & a_{23}^I & a_{33}^I & a_{43}^I & 0 & 0 & a_{53}^I & a_{63}^I \\
a_{14}^I & a_{24}^I & a_{34}^I & a_{44}^I & 0 & 0 & a_{54}^I & a_{64}^I \\
0 & a_{11}^{II} & a_{41}^{II} & 0 & a_{21}^{II} & a_{31}^{II} & a_{51}^{II} & a_{61}^{II} \\
0 & a_{12}^{II} & a_{42}^{II} & 0 & a_{22}^{II} & a_{32}^{II} & a_{52}^{II} & a_{62}^{II} \\
0 & a_{13}^{II} & a_{43}^{II} & 0 & a_{23}^{II} & a_{33}^{II} & a_{53}^{II} & a_{63}^{II} \\
0 & a_{14}^{II} & a_{44}^{II} & 0 & a_{24}^{II} & a_{34}^{II} & a_{54}^{II} & a_{64}^{II}
\end{bmatrix}
\begin{bmatrix}
t_1^4 \\ u_2 \\ u_3 \\ t_2^4 \\ u_5 \\ u_6 \\ t_1^2 \\ t_2^2
\end{bmatrix}
=
\begin{Bmatrix}
\mathbf{F}^I \\ \mathbf{F}^{II}
\end{Bmatrix}
\tag{10.13}
$$

where a_{ji}^N are the coefficients of the matrices, as specified in equations (10.9) and (10.11). The first subscript j refers to the column number of the region matrix, the second, i, to the collocation point number P_i and the roman superscript N refers to the region number. In the assembled system of equations, the columns are numbered according to the global numbering of unknowns. For the row numbering n_r, the collocation points of each region are entered sequentially. Note that the coefficient matrix of the system of equations is no longer fully populated. In equation (10.13) the unknown fluxes are referred to as t_1^2, t_2^2. The fluxes t_1^8, t_2^8 can be determined from equation (10.1). The programming of the method is fairly straightforward, but we will not show an example of implementation here, as we believe that the method discussed next has more potential for an efficient implementation.

10.3 STIFFNESS MATRIX ASSEMBLY

The multi-region assembly is not very efficient in cases where sequential excavation/construction (for example, in tunnelling) is to be modelled, since the coefficient matrices of all regions have to be computed and assembled every time a region is added or removed. Also, if hardware with parallel processing capabilities is used, then it would be convenient if each region matrix could be assembled and computed completely separately. Also, as we will see, the method just discussed is not very efficient for problems where only some of the region nodes are connected.

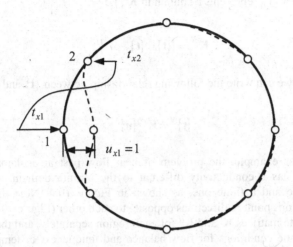

Figure 10.3   Example of computation of stiffness coefficients (elasticity problem)

The alternative to multi-region assembly, here referred to as stiffness matrix assembly, utilises a philosophy similar to that used by the finite element method. The idea is to compute a 'stiffness matrix' $\mathbf{K}^N$ for each region N. Coefficients of $\mathbf{K}^N$ are values of t due to unit values of u at all region nodes. In potential flow problems these would correspond to fluxes due to unit temperatures, while in elasticity they would be tractions due to unit displacements. To obtain the 'stiffness matrix' $\mathbf{K}^N$ of a region, we simply solve the *Dirichlet* problem M times, where M is the number of degrees of freedom of the BE region. For example, to get the first column of $\mathbf{K}^N$, we apply a unit value of temperature or of displacement in the x-direction, as shown in Figure 10.3.

For computation of *Dirichlet* problems we use equation (7.3), with a modified right-hand side

$$[\Delta U]\{t\}_1 = [\Delta T]\{u\}_1 \tag{10.14}$$

Here $[\Delta T], [\Delta U]$ are the assembled coefficient matrices, $\{t\}_1$ is the first column of the stiffness matrix $\mathbf{K}^M$ and $\{u\}_1$ is a vector with a unit value in the first row, i.e.

$$\{u\}_1 = \begin{Bmatrix} 1 \\ 0 \\ 0 \\ \vdots \end{Bmatrix} \tag{10.15}$$

If we perform the multiplication of $[\Delta T]\{u\}_1$ it can easily be seen that the right-hand side of equation (10.14) is simply the first column of matrix $[\Delta T]$. Computation of the region 'stiffness matrix' is therefore basically a solution of $[\Delta U]\{t\}_i = \{F\}_i$, with N right-hand sides $\{F\}_i$, where each right-hand side corresponds to a column in $[\Delta T]$.

Each solution vector $\{\mathbf{t}\}_i$ represents a column in $\mathbf{K}$, i.e.

$$\mathbf{K}^N = [\{\mathbf{t}\}_1 \ \{\mathbf{t}\}_2 \ \cdots] \tag{10.16}$$

For each region N we can write the following relationship between $\{\mathbf{t}\}$ and $\{\mathbf{u}\}$:

$$\{\mathbf{t}\}^N = \mathbf{K}^N \{\mathbf{u}\}^N \tag{10.17}$$

To compute, for example, the problem of heat flow past an isolator, which is not impermeable, but has a conductivity different to the infinite domain, we specify two regions, an infinite and a finite one, as shown in Figure 10.4. Note that the outward normals of the regions point in directions opposite to each other (Figure 10.5).

First we compute matrices $\mathbf{K}^I$ and $\mathbf{K}^{II}$ for each region separately, and then we assemble the regions using the conditions for flow balance and uniqueness of temperature. These conditions are written as

$$\{\mathbf{t}\}^I + \{\mathbf{t}\}^{II} = \{\mathbf{t}\}; \quad \{\mathbf{u}\}^I = \{\mathbf{u}\}^{II} = \{\mathbf{u}\} \tag{10.18}$$

where $\{\mathbf{t}\}$ is a vector of applied flux at the interface.

The assembled system of equations for the example in Figure 10.4 is simply:

$$\{\mathbf{t}\} = \left(\mathbf{K}^I + \mathbf{K}^{II}\right)\{\mathbf{u}\} \tag{10.19}$$

which can be solved for $\{\mathbf{u}\}$ if $\{\mathbf{t}\}$ is known.

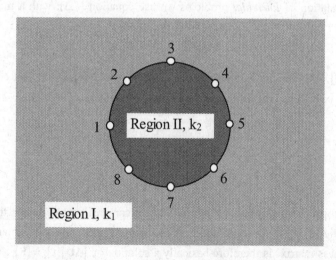

Figure 10.4   Example of a multi-region analysis: inclusion with different conductivity in an infinite domain

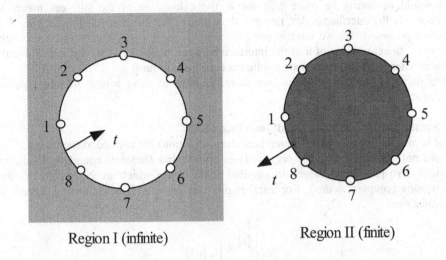

Region I (infinite)   Region II (finite)

Figure 10.5   The two regions of the problem

10.3.1 Partially coupled problems

In many cases we have problems where not all nodes of the regions are connected. We refer to these problems as partially coupled. Consider, for example, the modified heat flow problem in Figure 10.6, where we have added another circular hole (or impermeable isolator) where *Neumann* boundary conditions are specified. Here only some of the nodes of region I are connected to region II.

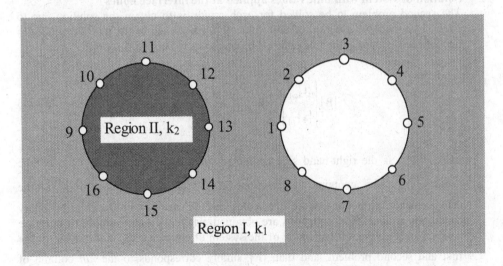

Figure 10.6   Problem with a circular inclusion and a hole

It would obviously be more efficient in the calculation of the stiffness matrix to consider only the interface nodes, i.e. only those nodes that are connected to a region. It is therefore proposed that we modify our procedure in such a way that we first solve the problem with zero values of **u** at the interface between region I and II, and then solve the problem where unit values of **u** are applied at each node in turn.

For partially coupled problems, we therefore have to solve a total of four types of problems:

1. **Solution of system with 'fixed' interface nodes**
   The first problem is where given boundary conditions are applied at the nodes which are not connected to other regions (free nodes) and *Dirichlet* boundary conditions with zero prescribed values are applied at the nodes which are connected to other regions (coupled nodes). For each region we can write the following system of equations:

$$[\mathbf{B}]\begin{Bmatrix} \{\mathbf{t}\}_{c0}^{N} \\ \{\mathbf{x}\}_{f0}^{N} \end{Bmatrix} = \{\mathbf{F}\}_{0}^{N} \tag{10.20}$$

where $[\mathbf{B}]$ is the assembled left-hand side and $\{\mathbf{F}\}_{o}^{N}$ contains the right-hand side due to given boundary conditions. The unknown vector $\{\mathbf{t}\}_{co}^{N}$ then contains the tractions at the coupled nodes and the vector $\{\mathbf{x}\}_{fo}^{N}$ either displacements or tractions at the free nodes, depending on the boundary conditions specified.

2. **Solution of system with unit values applied at the interface nodes**
   The second problem to be solved for each region is to obtain the solution due to *Dirichlet* boundary condition of unit value applied at each of the interface nodes in turn and zero values at the free nodes. The equations to be solved are

$$[\mathbf{B}]\begin{Bmatrix} \{\mathbf{t}\}_{cn}^{N} \\ \{\mathbf{x}\}_{fn}^{N} \end{Bmatrix} = \{\mathbf{F}\}_{n} \qquad n=1,2\ldots N_{c} \tag{10.21}$$

where $\{\mathbf{F}\}_{n}^{N}$ is the right-hand side computed for a unit value of u at node n. The vector $\{\mathbf{t}\}_{cn}^{N}$ now contains the tractions at the coupled nodes and $\{\mathbf{x}\}_{fn}^{N}$ the displacements or tractions at the free nodes, for the case of unit *Dirichlet* boundary conditions at node n. N_{c} equations are obtained, where N_{c} is the number of interface nodes. Note that the left-hand side of the system of equations $[\mathbf{B}]$ is the same for the first and second problem, and that $\{\mathbf{F}\}_{n}$ simply corresponds to the *nth* column of $[\Delta\mathbf{T}]$.

After the solution of the first two problems $\{\mathbf{t}\}_c^N$ and $\{\mathbf{x}\}_f^N$ can be expressed in terms of $\{\mathbf{u}\}_c^N$ by:

$$\begin{Bmatrix} \{\mathbf{t}\}_c^N \\ \{\mathbf{x}\}_f^N \end{Bmatrix} = \begin{Bmatrix} \{\mathbf{t}\}_{c0}^N \\ \{\mathbf{x}\}_{f0}^N \end{Bmatrix} + \begin{bmatrix} \mathbf{K}^N \\ \mathbf{A}^N \end{bmatrix} \{\mathbf{u}\}_c^N \tag{10.22}$$

where the matrices $\mathbf{K}$ and $\mathbf{A}$ are defined by:

$$\mathbf{K} = \begin{bmatrix} \{\mathbf{t}\}_{c1} & \cdots & \{\mathbf{t}\}_{cN_c} \end{bmatrix} \; ; \quad \mathbf{A} = \begin{bmatrix} \{\mathbf{x}\}_{c1} & \cdots & \{\mathbf{x}\}_{cN_c} \end{bmatrix} \tag{10.23}$$

3. **Assembly of regions, calculation of interface unknowns**
   After all the region stiffness matrices $\mathbf{K}^N$ have been computed, they are assembled into a system of equations which can be solved for the unknown $\{\mathbf{u}\}_c$. For the assembly we use conditions of heat balance and uniqueness of temperature, as explained previously. This results in the following system of equations:

$$[\mathbf{K}]\{\mathbf{u}\}_c = \{\mathbf{F}\} \tag{10.24}$$

where $[\mathbf{K}]$ is the assembled 'stiffness matrix' of the interface nodes, and $\{\mathbf{F}\}$ is the assembled right-hand side. This system is solved for the unknown $\{\mathbf{u}\}_c$ at the nodes of all interfaces of the problem.

4. **Calculation of other unknowns of region**
   After the interface unknowns have been determined, the values of $\mathbf{t}$ at the interface and the value of $\mathbf{u}$ at the free nodes are determined for each region by the application of

$$\begin{Bmatrix} \{\mathbf{t}\}_c^N \\ \{\mathbf{x}\}_f^N \end{Bmatrix} = \begin{Bmatrix} \{\mathbf{t}\}_{c0}^N \\ \{\mathbf{x}\}_{f0}^N \end{Bmatrix} + \begin{bmatrix} \mathbf{K}^N \\ \mathbf{A}^N \end{bmatrix} \{\mathbf{u}\}_c^N \tag{10.25}$$

Note that $\{\mathbf{u}\}_c^N$ is obtained by gathering values from the vector of unknowns at all the interface nodes $\{\mathbf{u}\}_c$.

10.3.2 Example

Consider the example, in Figure 10.7, of a potential problem with two regions, *Dirichlet* boundary conditions with prescribed zero values applied on the left side and *Neumann*

BCs on the right side, as shown. All other boundaries are assumed to have *Neumann* BC with zero prescribed values. The interface only involves nodes 2 and 3, and therefore only 2 interface unknowns exist.

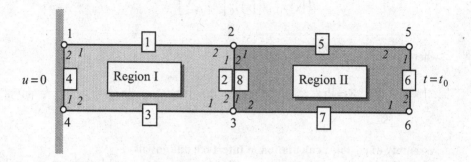

Figure 10.7   Example for stiffness assembly, partially coupled problem. Global node numbering and local (element) numbering (italics) shown

Figure 10.8 shows the local (region) numbering and the problems which have to be solved for obtaining vector $\{\mathbf{t}\}_{co}$ and the two rows of matrices $\mathbf{K}$ and $\mathbf{A}$. The sequence in which the nodes of the region are numbered in local numbering is such that the interface nodes are numbered first. Note that when there is a roman superscript indicating the region number, then the subscript refers to the local (region) numbering system.

It is obvious that for the first problem to be solved for region I, $\{\mathbf{t}\}_{io}$ will be zero. Following the procedure in Chapter 7 and referring to the element numbering of Figure 10.7, we obtain the following integral equations for the second and third problem:

$$\Delta U_{1i}^2\, t_{11}^I + \Delta U_{2i}^2\, t_{21}^I + \Delta U_{1i}^4\, t_{31}^I + \Delta U_{2i}^4\, t_{41}^I = \left(\Delta T_{2i}^1 + \Delta T_{1i}^2\right) 1$$

$$\Delta U_{1i}^2\, t_{12}^I + \Delta U_{2i}^2\, t_{22}^I + \Delta U_{1i}^4\, t_{32}^I + \Delta U_{2i}^4\, t_{42}^I = \left(\Delta T_{1i}^3 + \Delta T_{2i}^2\right) 1 \qquad (10.26)$$

This gives the following system of equations, with two right-hand sides:

$$\begin{bmatrix} \Delta U_{11}^2 & \Delta U_{21}^2 & \Delta U_{11}^4 & \Delta U_{21}^4 \\ \Delta U_{12}^2 & \Delta U_{22}^2 & \Delta U_{12}^4 & \Delta U_{22}^4 \\ \Delta U_{13}^2 & \Delta U_{23}^2 & \Delta U_{13}^4 & \Delta U_{23}^4 \\ \Delta U_{14}^2 & \Delta U_{24}^1 & \Delta U_{14}^4 & \Delta U_{24}^4 \end{bmatrix} \begin{Bmatrix} t_{11}^I & t_{12}^I \\ t_{21}^I & t_{22}^I \\ t_{31}^I & t_{32}^I \\ t_{41}^I & t_{42}^I \end{Bmatrix} = \begin{Bmatrix} \Delta T_{21}^1 + \Delta T_{11}^2 & \Delta T_{21}^2 + \Delta T_{11}^3 \\ \Delta T_{22}^1 + \Delta T_{12}^2 & \Delta T_{22}^1 + \Delta T_{12}^2 \\ \Delta T_{23}^1 + \Delta T_{13}^2 & \Delta T_{23}^1 + \Delta T_{13}^2 \\ \Delta T_{24}^1 + \Delta T_{14}^2 & \Delta T_{24}^1 + \Delta T_{14}^2 \end{Bmatrix} \qquad (10.27)$$

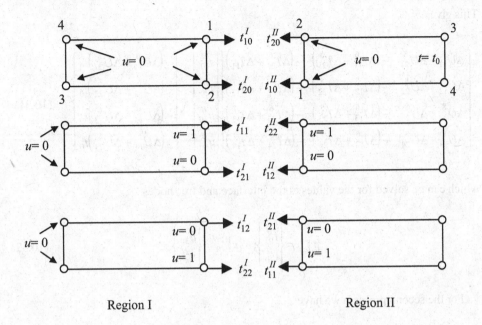

Figure 10.8   The different problems to be solved for regions I and II (potential problem)

After solving the system of equations, we obtain

$$\{ \mathbf{t} \}_c^I = \mathbf{K}^I \{ \mathbf{u} \}_c^I; \quad \{ \mathbf{x} \}_f^I = \mathbf{A}^I \{ \mathbf{u} \}_c^I \tag{10.28}$$

where

$$\{ \mathbf{t} \}_c^I = \begin{Bmatrix} t_1^I \\ t_2^I \end{Bmatrix}; \quad \mathbf{K}^I = \begin{bmatrix} t_{11}^I & t_{12}^I \\ t_{21}^I & t_{22}^I \end{bmatrix}; \quad \{ \mathbf{u} \}_c^I = \begin{Bmatrix} u_1^I \\ u_2^I \end{Bmatrix}$$

$$\{ \mathbf{x} \}_f^I = \begin{Bmatrix} t_3^I \\ t_4^I \end{Bmatrix}; \quad \mathbf{A}^I = \begin{bmatrix} t_{31}^I & t_{32}^I \\ t_{41}^I & t_{42}^I \end{bmatrix} \tag{10.29}$$

For region II we have for the case of zero u at the interface nodes

$$\Delta U_{2i}^8 \, t_{20}^{II} + \Delta U_{1i}^8 \, t_{10}^{II} - \left(\Delta T_{1i}^6 + \Delta T_{2i}^5 \right) u_{30}^{II} - \left(\Delta T_{2i}^6 + \Delta T_{1i}^7 \right) u_{40}^{II} = -\left(\Delta U_{1i}^6 + \Delta U_{2i}^6 \right) t_0 \tag{10.30}$$

This gives

$$
\begin{bmatrix}
\Delta U_{11}^8 & \Delta U_{21}^8 & -\left(\Delta T_{11}^6 + \Delta T_{21}^5\right) & -\left(\Delta T_{21}^6 + \Delta T_{11}^7\right) \\
\Delta U_{12}^8 & \Delta U_{22}^8 & -\left(\Delta T_{12}^6 + \Delta T_{22}^5\right) & -\left(\Delta T_{22}^6 + \Delta T_{12}^7\right) \\
\Delta U_{13}^8 & \Delta U_{23}^8 & -\left(\Delta T_{13}^6 + \Delta T_{23}^5\right) & -\left(\Delta T_{23}^6 + \Delta T_{13}^7\right) \\
\Delta U_{14}^8 & \Delta U_{24}^8 & -\left(\Delta T_{14}^6 + \Delta T_{24}^5\right) & -\left(\Delta T_{24}^6 + \Delta T_{14}^7\right)
\end{bmatrix}
\begin{Bmatrix}
t_{10}^{II} \\
t_{20}^{II} \\
u_{30}^{II} \\
u_{40}^{II}
\end{Bmatrix}
=
\begin{Bmatrix}
-\left(\Delta U_{11}^6 + \Delta U_{21}^6\right)t_0 \\
-\left(\Delta U_{12}^6 + \Delta U_{22}^6\right)t_0 \\
-\left(\Delta U_{13}^6 + \Delta U_{23}^6\right)t_0 \\
-\left(\Delta U_{14}^6 + \Delta U_{24}^6\right)t_0
\end{Bmatrix}
\tag{10.31}
$$

which can be solved for the values at the interface and free nodes

$$
\{t\}_{co}^{II} = \begin{Bmatrix} t_{10}^{II} \\ t_{20}^{II} \end{Bmatrix}; \quad \{x\}_{f0}^{II} = \begin{Bmatrix} u_{30}^{II} \\ u_{40}^{II} \end{Bmatrix}
\tag{10.32}
$$

For the second problem we have

$$
\Delta U_{1i}^8 t_{12}^{II} + \Delta U_{2i}^8 t_{22}^{II} - \left(\Delta T_{1i}^6 + \Delta T_{2i}^5\right)u_{32}^{II} - \left(\Delta T_{2i}^6 + \Delta T_{1i}^7\right)u_{42}^{II} = \left(\Delta T_{1i}^5 + \Delta T_{2i}^8\right) 1
$$
$$
\Delta U_{1i}^8 t_{11}^{II} + \Delta U_{2i}^8 t_{21}^{II} - \left(\Delta T_{1i}^6 + \Delta T_{2i}^5\right)u_{31}^{II} - \left(\Delta T_{2i}^6 + \Delta T_{1i}^7\right)u_{41}^{II} = \left(\Delta T_{2i}^7 + \Delta T_{1i}^8\right) 1
\tag{10.33}
$$

Note that the left-hand side of the systems of equations is the same as for the previous problem. The solutions can be written as

$$
\{t\}_c^{II} = \{t\}_{c0}^{II} + K^{II}\{u\}_c^{II}; \quad \{x\}_f^{II} = \{x\}_{f0}^{II} + A^{II}\{u\}_c^{II}
\tag{10.34}
$$

where

$$
\{t\}_c^{II} = \begin{Bmatrix} t_1^{II} \\ t_2^{II} \end{Bmatrix}; \quad K^{II} = \begin{bmatrix} t_{11}^{II} & t_{12}^{II} \\ t_{21}^{II} & t_{22}^{II} \end{bmatrix}; \quad \{t\}_{co}^{II} = \begin{Bmatrix} t_{10}^{II} \\ t_{20}^{II} \end{Bmatrix}; \quad \{u\}_c^{II} = \begin{Bmatrix} u_1^{II} \\ u_2^{II} \end{Bmatrix}
$$
$$
\{x\}_c^{II} = \begin{Bmatrix} u_3^{II} \\ u_4^{II} \end{Bmatrix}; \quad A^{II} = \begin{bmatrix} u_{31}^{II} & u_{32}^{II} \\ u_{41}^{II} & u_{42}^{II} \end{bmatrix}; \quad \{x\}_{co}^{II} = \begin{Bmatrix} u_{30}^{II} \\ u_{40}^{II} \end{Bmatrix}
\tag{10.35}
$$

The equations of compatibility and preservation of heat at the interface can be written as

$$\left\{\begin{matrix} t_{1c}^I \\ t_{2c}^I \end{matrix}\right\} + \left\{\begin{matrix} t_{2c}^{II} \\ t_{1c}^{II} \end{matrix}\right\} = \{\mathbf{t}\}_c ; \quad \left\{\begin{matrix} u_{1c}^I \\ u_{2c}^I \end{matrix}\right\} = \left\{\begin{matrix} u_{2c}^{II} \\ u_{1c}^{II} \end{matrix}\right\} = \{\mathbf{u}\}_c \tag{10.36}$$

Substituting equations (10.29) and (10.35), we obtain

$$\mathbf{K}\,\{\mathbf{u}\}_c + \{\mathbf{t}\}_c = 0 \tag{10.37}$$

where

$$\mathbf{K} = \begin{bmatrix} t_{11}^I + t_{22}^{II} & t_{21}^I + t_{21}^{II} \\ t_{21}^I + t_{12}^{II} & t_{22}^I + t_{12}^{II} \end{bmatrix}; \quad \{\mathbf{u}\}_c = \left\{\begin{matrix} u_2 \\ u_3 \end{matrix}\right\}; \quad \{\mathbf{t}\}_c = \left\{\begin{matrix} t_{10}^{II} \\ t_{20}^{II} \end{matrix}\right\} \tag{10.38}$$

This system can be solved for the interface unknowns. The calculation of the other unknowns is done separately for each region. For region I we have

$$\{\mathbf{t}\}_c^{II} = \mathbf{K}^I\{\mathbf{u}\}_c^{I}; \quad \{\mathbf{x}\}_f^{II} = \mathbf{A}^I\{\mathbf{u}\}_c^{I} \tag{10.39}$$

whereas for region II

$$\{\mathbf{t}\}_c^{II} = \{\mathbf{t}\}_{c0}^{II} + \mathbf{K}^{II}\{\mathbf{u}\}_c^{II}; \quad \{\mathbf{x}\}_f^{II} = \{\mathbf{x}\}_{f0}^{II} + \mathbf{A}^{II}\{\mathbf{u}\}_c^{II} \tag{10.40}$$

Consider the equivalent elasticity problem of a cantilever beam. We can see (Figure 10.8) that for region II the problem where the interface displacements are fixed, tractions at the interface corresponding to a shortened cantilever beam are obtained. If $u_x = 1$ is applied only a rigid body motion results and therefore no resulting tractions at the interface occur.

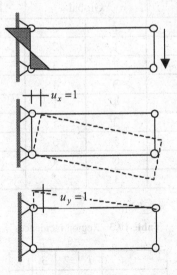

Figure 10.9 Effect of application of *Dirichlet* boundary conditions on region II of cantilever beam (elasticity problem)

The application of $u_y = 1$ will result in shear tractions at the interface, however.

10.4   COMPUTER IMPLEMENTATION

We now consider the computer implementation of the stiffness matrix assembly method. We divide this into two tasks. First we develop a **SUBROUTINE** Stiffness_BEM for calculation of the matrix **K**. If the problem is not fully coupled, then this subroutine will also determine the matrix **A**, and the solutions for zero values of **u** at the interface. Secondly, we develop a program General_purpose_BEM_Multi.

For the calculation we have to consider three different numbering systems, each one is related to the global numbering system as shown in Figure 10.7:

1. Local (element) numbering. This is the sequence in which the nodes to which an element connect are entered in the element incidence vector. The incidences for the simple example problem just discussed are given in Table 10.2
2. Local (region) numbering. This is the sequence in which the nodes to which a region connects are entered in the region incidence vector. This numbering is used for computing the 'stiffness matrix' of a region. The region incidence vector for the example problem is given in Table 10.3.
3. Local (interface) numbering. This is basically the sequence in which the interface nodes are entered in the interface incidence vector. For the example problem the interface incidences are given in Table 10.4. This sequence is determined in such a way that the first node of the first interface element will start the sequence.

Table 10.2   Incidences of boundary elements in global and local numbering

Element	Global		Local (Region)	
	1	*2*	*1*	*2*
1	1	2	4	1
2	2	3	1	2
3	3	4	2	3
4	4	1	3	4
5	2	5	2	3
6	5	6	3	4
7	6	3	4	1
8	3	2	1	2

Table 10.3   Region incidences

	1	*2*	*3*	*4*
Region I	2	3	4	1
Region II	3	2	5	6

Table 10.4   Interface incidences

	1	*2*
Region I	2	3
Region II	3	2

Note that the interface incidences are simply the first two values of the region incidence vector. For problems involving more than one unknown per node the incidence vectors have to be expanded to *destination* vectors, as explained in Chapter 7.

10.4.1   Subroutine Stiffness_BEM

The tasks for the **SUBROUTINE** Stiffness_BEM are essentially the same as for the General_Purpose_BEM, except that the boundary conditions considered are expanded. We add a new code (2) which is used for the nodes at the interface.

The input parameters for **SUBROUTINE** Stiffness_BEM are the incidence vectors of the boundary elements which describe the boundary of the region, the coordinates of the nodes and the boundary conditions. Note that the vector of incidences as well as the coordinates has to be in the local (region) numbering. **SUBROUTINE** AssemblySTIFF is basically the same as **SUBROUTINE** Assembly except that a boundary code 2 for interface conditions has been added. Boundary code 2 is treated the same as code 1 (*Dirichlet*) except that columns of $[\Delta T]^e$ are assembled into the array RhsM (multiple right-hand sides).

SUBROUTINE Solve is modified into Solve_Multi which can handle both single (Rhs) and multiple (RhsM) right-hand sides.

The output parameter of the **SUBROUTINE** is the stiffness matrix **K**, and for partially coupled problems in addition the matrix **A** as well **as** $\{t\}_c$. The rows and columns of these matrices will be numbered in a local (interface) numbering. The values of $\{u\}_{f0}$ and $\{t\}_{c0}$ are stored in the array El_res, which contains the element results. They can be added at element level.

We show below the library module Stiffness_lib which contains all the necessary declarations and subroutines for the computation of the stiffness matrix. The symmetry option has been left out in the implementation shown to simplify the coding.

```
MODULE Stiffness_lib
USE Utility_lib ; USE Integration_lib ; USE Geometry_lib
CONTAINS
SUBROUTINE Stiffness_BEM(nr,xP,Nodel,Ndof,Ndofe,NodeR&
,Ncode,NdofR,Ndofc,KBE,A,tc,Cdim,Elres_u,Elres_t,&
IncieR,LdesteR,Nbel,ListR,TypeR,Bcode,Con,E,ny,Ndest,Isym)
!--------------------------------------------------
!    Computes the stiffness matrix
!    of a boundary element region
!--------------------------------------------------
IMPLICIT NONE
REAL, INTENT(INOUT):: xP(:,:)      ! Array of node coordinates
INTEGER, INTENT(IN):: nr           ! region number
```

```fortran
INTEGER, INTENT(IN):: Ncode(:)      ! Global restraint code
INTEGER, INTENT(IN):: NdofR          ! D.o.F. of region
INTEGER, INTENT(IN):: Ndofc          ! No of interface D.o.F.
INTEGER, INTENT(IN):: NodeR(:)       ! Nodes/region
INTEGER, INTENT(IN):: TypeR(:)       ! Types of region
INTEGER, INTENT(IN):: Cdim           ! Cartesian dimension
INTEGER, INTENT(IN):: IncieR(:,:)    ! Incidences (region numbering)
INTEGER, INTENT(IN):: LdesteR(:,:)   ! Destinations
INTEGER, INTENT(INOUT):: Bcode(:,:)  ! Bounadry code
INTEGER, INTENT(IN):: Nbel(:)        ! Number of elements/region
INTEGER, INTENT(IN):: ListR(:,:)     ! List of elements/region
INTEGER, INTENT(IN):: Nodel
INTEGER, INTENT(IN):: Ndof
INTEGER, INTENT(IN):: Ndofe
INTEGER, INTENT(IN):: Isym
INTEGER, INTENT(IN):: Ndest(:,:)     ! Node destinations
REAL, INTENT(INOUT):: Elres_u(:,:),Elres_t(:,:) ! Elem. results
REAL, INTENT(INOUT):: E,ny,Con
REAL(KIND=8), INTENT(OUT)   :: KBE(:,:)  ! Stiffness matrix
REAL(KIND=8), INTENT(OUT)   :: A(:,:)    ! u due to unit values ui
REAL(KIND=8), INTENT(OUT)   :: tc(:)     ! interface tractions
!    temporal arrays :
REAL(KIND=8), ALLOCATABLE :: dUe(:,:),dTe(:,:),Diag(:,:)
REAL(KIND=8), ALLOCATABLE :: Lhs(:,:)
REAL(KIND=8), ALLOCATABLE :: Rhs(:),RhsM(:,:) ! right-hand sides
REAL(KIND=8), ALLOCATABLE :: u1(:),u2(:,:)      ! results
REAL, ALLOCATABLE :: Elcor(:,:)
REAL    :: Scat,Scad
INTEGER :: NdofF
INTEGER :: Dof,k,l,nel
INTEGER :: n,m,Pos,i,j,nd,ne
ALLOCATE(dTe(NdofR,Ndofe),dUe(NdofR,Ndofe))
ALLOCATE(Diag(NdofR,Ndof))
ALLOCATE(Lhs(NdofR,NdofR),Rhs(NdofR),RhsM(NdofR,NdofR))
ALLOCATE(u1(NdofR),u2(NdofR,NdofR))
ALLOCATE(Elcor(Cdim,Nodel))
!---------------------------------------------
!     Scaling
!---------------------------------------------
CALL Scal(E,xP,Elres_u,Elres_t,Cdim,Scad,Scat)
!-------------------------------------------------------------------
! Compute and assemble element coefficient matrices
!-------------------------------------------------------------------
Lhs= 0.0
Diag= 0.0
Rhs= 0.0
RhsM= 0.0
Elements_1:&
DO Nel=1,Nbel(nr)
  ne= ListR(nr,Nel)
  Elcor(:,:)= xP(:,IncieR(ne,:))          !   gather element coords
```

```
    IF(Cdim == 2) THEN
      IF(Ndof == 1) THEN
        CALL Integ2P(Elcor,IncieR(ne,:),Nodel,NodeR(nr)&
                    ,xP,Con,dUe,dTe)
      ELSE
        CALL Integ2E(Elcor,IncieR(ne,:),Nodel,NodeR(nr)&
                    ,xP,E,ny,dUe,dTe,Ndest,Isym)
      END IF
    ELSE
      CALL Integ3(Elcor,IncieR(ne,:),Nodel,NodeR(nr)&
                 ,xP,Ndof,E,ny,Con,dUe,dTe,Ndest,Isym)
    END IF
    CALL AssemblyMR(ne,Ndof,Ndofe,Nodel,Lhs&
                   ,Rhs,RhsM,DTe,DUe,LdesteR(ne,:),Ncode,Bcode&
                   ,Diag,Elres_u,Elres_t,Scad)
END DO &
Elements_1
!------------------------------------------------------------
!  Add azimuthal integral for infinite regions
!------------------------------------------------------------
IF(TypeR(nr) == 2) THEN
 DO m=1, NodeR(nr)
   DO n=1, Ndof
     k=Ndof*(m-1)+n
     Diag(k,n) = Diag(k,n) + 1.0
   END DO
 END DO
END IF
!------------------------------------------------------------
!  Add Diagonal coefficients
!------------------------------------------------------------
Nodes_global: &
DO m=1,NodeR(nr)
 Degrees_of_Freedoms_node: &
 DO n=1, Ndof
   DoF = (m-1)*Ndof + n     !  global degree of freedom no.
   k = (m-1)*Ndof + 1       !  address in coeff. matrix (row)
   l = k + Ndof - 1         !  address in coeff. matrix (column)
   IF (NCode(DoF) == 1 .or. NCode(DoF) == 2) THEN
     Pos = 0
     Nel = 0
!    get local degree of freedom no corresponding to global one
     Elements_all: &
     DO i=1,Nbel(nr)
       ne= ListR(nr,i)
       Degrees_of_freedom_elem: &
       DO j=1,Ndofe
         IF (DoF == LdesteR(ne,j)) THEN
           Nel = ne
           Pos = j
           EXIT
```

```
        END IF
    END DO &
     Degrees_of_freedom_elem
    IF (Nel /= 0) EXIT
    END DO &
    Elements_all
    Rhs(k:l) = Rhs(k:l) - Diag(k:l,n)*Elres_u (Nel,Pos)
    IF(NCode(DoF) == 2)THEN
      RhsM(k:l,DoF) = RhsM(k:l,DoF) - Diag(k:l,n) / Scad
    END IF
  ELSE
    Lhs(k:l,Dof)= Lhs(k:l,Dof) + Diag(k:l,n)
  END IF
END DO &
Degrees_of_Freedoms_node
END DO &
Nodes_global
!    Solve problem
CALL Solve_Multi(Lhs,Rhs,RhsM,u1,u2)
!-------------------------------------------
!   Scale back
!-------------------------------------------
Interface_tractions: &
DO N=1,NdofC
u1(N)= u1(N) / Scat
u2(N,:)= u2(N,:) / Scat
END DO &
Interface_tractions
M=NdofC
NdofF= NdofR-NdofC
Free_D_o_F: &
DO N=1,NdofF
M=M+1
 IF(NCode(M) == 0) THEN
   u1(M)= u1(M) * Scad
   u2(M,:)= u2(M,:) * Scad
 ELSE
   u1(M)= u1(M) / Scat
   u2(M,:)= u2(M,:) / Scat
 END IF
END DO &
Free_D_o_F
Elres_u(:,:)= Elres_u(:,:) * Scad
Elres_t(:,:)= Elres_t(:,:) / Scat
!-------------------------------------
!  Gather element results due to
!  zero Dirichlet conditions at the interface
!-------------------------------------
Elements2: &
DO nel=1,Nbel(nr)
ne= ListR(nr,nel)
D_o_F1:   &
```

```
DO nd=1,Ndofe
  IF(Ncode(LdesteR(ne,nd)) == 0) THEN
    Elres_u(ne,nd) =  u1(LdesteR(ne,nd))
   ELSE IF(Bcode(ne,nd) == 1 .or. Bcode(ne,nd) == 2) THEN
    Elres_t(ne,nd) =  u1(LdesteR(ne,nd))
  END IF
 END DO &
 D_o_F1
END DO &
Elements2
!---------------------------------------
!   Gather stiffness matrix KBE and matrix A
!---------------------------------------
Interface_DoFs: &
DO N=1,Ndofc
    KBE(N,:)= u2(N,:)
    tc(N)= u1(N)
END DO &
Interface_DoFs
A= 0.0
M=NdofC
Free_DoFs: &
DO N=1,NdofF
    M= M+1
    A(N,1:NdofC)= u2(M,:)
END DO &
Free_DoFs
DEALLOCATE (dUe,dTe,Diag,Lhs,Rhs,RhsM,u1,u2,Elcor)
RETURN
END SUBROUTINE Stiffness_BEM


SUBROUTINE Solve_Multi(Lhs,Rhs,RhsM,u,uM)
!--------------------------------------------------
!    Solution of system of equations
!    by Gauss Elimination
!    for multple right-hand sides
!--------------------------------------------------
REAL(KIND=8) ::    Lhs(:,:)      !    Equation Left-hand side
REAL(KIND=8) ::    Rhs(:)        !    Equation right-hand side 1
REAL(KIND=8) ::    RhsM(:,:)     !    Equation right-hand sides 2
REAL(KIND=8) ::    u(:)          !    Unknowns 1
REAL(KIND=8) ::    uM(:,:)       !    Unknowns 2
REAL(KIND=8) ::    FAC
INTEGER  M,Nrhs                  !    Size of system
INTEGER  i,n,nr
M= UBOUND(RhsM,1) ; Nrhs= UBOUND(RhsM,2)
! Reduction
Equation_n: &
DO n=1,M-1
   IF(ABS(Lhs(n,n)) < 1.0E-10) THEN
     CALL Error_Message('Singular Matrix')
   END IF
```

```fortran
  Equation_i: &
   DO i=n+1,M
      FAC= Lhs(i,n)/Lhs(n,n)
      Lhs(i,n+1:M)= Lhs(i,n+1:M) - Lhs(n,n+1:M)*FAC
      Rhs(i)= Rhs(i) - Rhs(n)*FAC
     RhsM(i,:)= RhsM(i,:) - RhsM(n,:)*FAC
     END DO &
   Equation_i
 END DO &
 Equation_n
 !      Backsubstitution
 Unknown_1: &
 DO n= M,1,-1
  u(n)= -1.0/Lhs(n,n)*(SUM(Lhs(n , n+1:M)*u(n+1:M)) - Rhs(n))
 END DO &
 Unknown_1
 Load_case: &
 DO Nr=1,Nrhs
  Unknown_2: &
  DO n= M,1,-1
   uM(n,nr)= -1.0/Lhs(n,n)&
            *(SUM(Lhs(n , n+1:M)*uM(n+1:M , nr)) - RhsM(n,nr))
   END DO &
  Unknown_2
 END DO &
 Load_case
 RETURN
 END SUBROUTINE Solve_Multi

 SUBROUTINE AssemblyMR(Nel,Ndof,Ndofe,Nodel,Lhs&
 ,Rhs,RhsM,DTe,DUe,Ldest,Ncode&
 ,Bcode,Diag,Elres_u,Elres_t,Scad)
 !-------------------------------------------------
 !  Assembles element contributions DTe , DUe
 !  into global matrix Lhs, vector Rhs
 !  and matrix RhsM
 !-------------------------------------------------
 INTEGER,INTENT(IN)        :: NEL    ! Element no.
 REAL(KIND=8)              :: Lhs(:,:) ! Eq.left-hand side
 REAL(KIND=8)              :: Rhs(:)  ! Right-hand side
 REAL(KIND=8)              :: RhsM(:,:) ! Matrix of right-hand sides
 REAL(KIND=8), INTENT(IN):: DTe(:,:),DUe(:,:)   ! Element arrays
 REAL, INTENT(INOUT)    :: Elres_u(:,:),Elres_t(:,:)
 INTEGER , INTENT(IN)     :: LDest(:) ! Element destinations
 INTEGER , INTENT(IN) :: NCode(:) ! Boundary code (global)
 INTEGER , INTENT(IN) :: BCode(:,:) ! Boundary code (global)
 INTEGER , INTENT(IN) :: Ndof
 INTEGER , INTENT(IN) :: Ndofe
 INTEGER , INTENT(IN) :: Nodel
 REAL(KIND=8) :: Diag(:,:) ! Array with diagonal coeff of DT
 INTEGER :: n,Ncol,m,k,l
```

```
DoF_per_Element:&
DO m=1,Ndofe
 Ncol=Ldest(m)           !    Column number
 IF(BCode(nel,m) == 0) THEN!    Neumann BC
   Rhs(:) = Rhs(:) + DUe(:,m)*Elres_t(nel,m)
   !    The assembly of dTe depends on the global BC
   IF (NCode(Ldest(m)) == 0) THEN
     Lhs(:,Ncol)= Lhs(:,Ncol) + DTe(:,m)
   ELSE
     Rhs(:) = Rhs(:) - DTe(:,m) * Elres_u(nel,m)
   END IF
 ELSE IF(BCode(nel,m) == 1) THEN    !    Dirichlet BC
   Lhs(:,Ncol) = Lhs(:,Ncol) - DUe(:,m)
   Rhs(:)= Rhs(:) - DTe(:,m) * Elres_u(nel,m)
 END IF
 IF(BCode(nel,m) == 2) THEN    !    Interface
   Lhs(:,Ncol) = Lhs(:,Ncol) - DUe(:,m)
 END IF
 IF(NCode(Ldest(m)) == 2) THEN    !    Interface
   RhsM(:,Ncol)= RhsM(:,Ncol) - DTe(:,m) / Scad
 END IF
END DO &
DoF_per_Element
 !     Sum of off-diagonal coefficients
DO n=1,Nodel
 DO k=1,Ndof
   l=(n-1)*Ndof+k
   Diag(:,k)= Diag(:,k) - DTe(:,l)
 END DO
END DO
RETURN
END SUBROUTINE AssemblyMR

END MODULE Stiffness_lib
```

10.5   PROGRAM 10.1: GENERAL PURPOSE PROGRAM, DIRECT METHOD, MULTIPLE REGIONS

Using the library for the stiffness matrix computation, we now develop a general purpose program for the computation of multi-region problems. The input to the program is the same as for one region, except that we must now specify additional information about the regions. A region is specified by a list of elements that describe its boundary, a region code that indicates if the region is finite or infinite and (if symmetry is considered) the symmetry code. In order to simplify the code, symmetry will not be considered here. The various tasks to be carried out are:

1. **Detect interface elements, number interface nodes/degrees of freedom**
    The first task of the program will be to determine which elements belong to an

interface between regions, and to establish a local interface numbering. Interface elements can be detected by the fact that two boundary elements connect to exactly the same nodes, although not in the same sequence since the outward normals will be different. The number of interface degrees of freedom will determine the size of matrices **K** and **A**.

2.   **For each region do**

   (a) *Establish local (region) numbering for element incidences*
       For the treatment of the individual regions we have to renumber the nodes/degrees of freedom for each region into a local (region) numbering system as explained previously. The incidence and destination vectors of boundary elements as well as coordinate vector are modified accordingly.

   (b) *Determine **K** and **A** and results due to 'fixed' interface nodes*
       The next task is to determine the matrix **K** and to assemble it into the global system of equations using the interface destination vector. For partially coupled problems we calculate and store at the same time the results for the elements due to zero values of $\{\mathbf{u}\}_c$ at the interface. These values are stored in the element result vectors Elres_u and Elres_t. Finally the matrix **A** and the vector $\{\mathbf{t}\}_c$ are determined and stored.

3.   **Solve global system of equations**
     The global system of equations is solved for the interface unknowns $\{\mathbf{u}\}_c$

4.   **For each region determine $\{\mathbf{t}\}_c$ and $\{\mathbf{u}\}_f$**

     Using equation (10.25), the values for the fluxes/tractions at the interface and (for partially coupled problems) the temperatures/displacements at the free nodes are determined and added to the values already stored in Elres_u and Elres_t. Note that before equation (10.25) can be used, the interface unknowns which are in global interface numbering have to be gathered.

```
PROGRAM General_purpose_MRBEM
!-----------------------------------------------------------
!      General purpose BEM program
!      for solving elasticity and potential problems
!      with multiple regions
!-----------------------------------------------------------
USE Utility_lib; USE Elast_lib; USE Laplace_lib
USE Integration_lib; USE Stiffness_lib
IMPLICIT NONE
INTEGER, ALLOCATABLE        :: NCode(:,:)    ! Element BC´s
INTEGER, ALLOCATABLE        :: Ldest_KBE(:)  ! Interface destinations
INTEGER, ALLOCATABLE        :: TypeR(:)      ! Type of BE-regions
REAL, ALLOCATABLE           :: Elcor(:,:)    ! Element coordinates
REAL, ALLOCATABLE           :: xP(:,:)       ! Node co-ordinates
REAL, ALLOCATABLE           :: Elres_u(:,:)  ! Element results
REAL, ALLOCATABLE           :: Elres_t(:,:)  ! Element results
```

```
REAL(KIND=8),  ALLOCATABLE :: KBE(:,:,:) ! Region stiffness
REAL(KIND=8),  ALLOCATABLE :: A(:,:,:)    ! Results due to ui=1
REAL(KIND=8),  ALLOCATABLE :: Lhs(:,:),Rhs(:) ! global matrices
REAL(KIND=8),  ALLOCATABLE :: uc(:) !    interface unknown
REAL(KIND=8),  ALLOCATABLE :: ucr(:)! interface unknown(region)
REAL(KIND=8),  ALLOCATABLE :: tc(:)   ! interface tractions
REAL(KIND=8),  ALLOCATABLE :: xf(:)   ! free unknown
REAL(KIND=8),  ALLOCATABLE :: tcxf(:) ! unknowns of region
REAL, ALLOCATABLE   :: XpR(:,:) ! Region node coordinates
REAL, ALLOCATABLE   :: ConR(:)   ! Conductivity of regions
REAL, ALLOCATABLE   :: ER(:)     ! Youngs modulus of regions
REAL, ALLOCATABLE   :: nyR(:)    ! Poisson's ratio of regions
REAL                :: E,ny,Con
INTEGER,ALLOCATABLE:: InciR(:,:)! Incidences (region)
INTEGER,ALLOCATABLE:: Incie(:,:)! Incidences (global)
INTEGER,ALLOCATABLE:: IncieR(:,:) ! Incidences (local)
INTEGER,ALLOCATABLE:: ListC(:)   ! List of interface nodes
INTEGER,ALLOCATABLE:: ListEC(:,:) ! List of interface Elem.
INTEGER,ALLOCATABLE:: ListEF(:,:) ! List of free Elem.
INTEGER,ALLOCATABLE:: LdestR(:,:) ! Destinations(local numbering)
INTEGER,ALLOCATABLE:: Nbel(:)     ! Number of BE per region
INTEGER,ALLOCATABLE:: NbelC(:)   ! Number of Interf. Elem./reg.
INTEGER,ALLOCATABLE:: NbelF(:)   ! Number of free elem./region
INTEGER,ALLOCATABLE:: Bcode(:,:) ! Boundary code for all elements
INTEGER,ALLOCATABLE:: Ldeste(:,:) ! Destinations (global)
INTEGER,ALLOCATABLE:: LdesteR(:,:)!Destinations (local)
INTEGER,ALLOCATABLE:: NodeR(:)   ! No. of nodes of Region
INTEGER,ALLOCATABLE:: NodeC(:)   ! No. of nodes on Interface
INTEGER,ALLOCATABLE:: ListR(:,:) ! List of Elements/region
INTEGER,ALLOCATABLE:: Ndest(:,:)
INTEGER    :: Cdim   ! Cartesian dimension
INTEGER    :: Nodes  ! No. of nodes of System
INTEGER    :: Nodel  ! No. of nodes per element
INTEGER    :: Ndofe  ! D.o.F's of Element
INTEGER    :: Ndof   ! No. of degrees of freedom per node
INTEGER    :: Ndofs  ! D.o.F's of System
INTEGER    :: NdofR  ! Number of D.o.F. of region
INTEGER    :: NdofC  ! Number of interface D.o.F. of region
INTEGER    :: NdofF  ! Number D.o.F. of free nodes of region
INTEGER    :: NodeF  ! Number of free Nodes of region
INTEGER    :: NodesC ! Total number of interface nodes
INTEGER    :: NdofsC ! Total number of interface D.o.F.
INTEGER    :: Toa    ! Type of analysis (plane strain/stress)
INTEGER    :: Nregs  ! Number of regions
INTEGER    :: Ltyp   ! Element type(linear = 1, quadratic = 2)
INTEGER    :: Isym   ! Symmetry code
INTEGER    :: Maxe   ! Number of Elements of System
INTEGER    :: nr,nb,ne,ne1,nel
INTEGER    :: n,node,is,nc,no,ro,co
INTEGER    :: k,m,nd,nrow,ncln,DoF_KBE,DoF
CHARACTER(LEN=80) :: Title
```

```fortran
!-----------------------------------------------------------
!   Read job information
!-----------------------------------------------------------
OPEN (UNIT=1,FILE='INPUT',FORM='FORMATTED') ! Input
OPEN (UNIT=2,FILE='OUTPUT',FORM='FORMATTED')! Output
Call JobinMR(Title,Cdim,Ndof,Toa,Ltyp,Isym;nodel,nodes,maxe)
Ndofs= Nodes * Ndof            ! D.O.F's of System
Ndofe= Nodel * Ndof            ! D.O.F's of Element
Isym= 0  !   no symmetry considered here
ALLOCATE(Ndest(Nodes,Ndof))
Ndest= 0
READ(1,*)Nregs    !    read number of regions
ALLOCATE(TypeR(Nregs),Nbel(Nregs),ListR(Nregs,Maxe))
IF(Ndof == 1)THEN
 ALLOCATE(ConR(Nregs))
ELSE
 ALLOCATE(ER(Nregs),nyR(Nregs))
END IF
CALL Reg_Info(Nregs,ToA,Ndof,TypeR,ConR,ER,nyR,Nbel,ListR)
ALLOCATE(xP(Cdim,Nodes))   !  Array for node coordinates
ALLOCATE(Incie(Maxe,Nodel)) !  Array for incidences
CALL Geomin(Nodes,Maxe,xp,Incie,Nodel,Cdim)
ALLOCATE(BCode(Maxe,Ndofe))
ALLOCATE(Elres_u(Maxe,Ndofe),Elres_t(Maxe,Ndofe))
CALL BCinput(Elres_u,Elres_t,Bcode,nodel,ndofe,ndof)
!--------------------------------------------------
!     Determine element destination vector for assembly
!--------------------------------------------------
ALLOCATE(Ldeste(Maxe,Ndofe))
Elements_of_region2:&
DO Nel=1,Maxe
 k=0
 DO n=1,Nodel
   DO m=1,Ndof
     k=k+1
     IF(Ndof > 1) THEN
       Ldeste(Nel,k)= ((Incie(Nel,n)-1)*Ndof + m)
     ELSE
       Ldeste(Nel,k)= Incie(Nel,n)
     END IF
   END DO
 END DO
END DO &
Elements_of_region2
!--------------------------------------------------
!    Detect interface elements,
!    assign interface boundary conditions
!    Determine number of interface nodes
!--------------------------------------------------
ALLOCATE(ListC(Nodes))
NodesC=0
```

```
ListC=0
Elements_loop: &
DO ne=1,Maxe
 Elements_loop1: &
 DO ne1=ne+1,Maxe
   IF(Match(Incie(ne1,:),Incie(ne,:))) THEN
     BCode(ne,:)= 2 ; BCode(ne1,:)= 2    !  assign interface BC
     Element_nodes: &
     DO n=1,nodel
       Node= Incie(ne,n)
       is= 0
       Interface_nodes: &
       DO nc=1,NodesC
         IF(Node == ListC(nc)) is= 1
       END DO &
       Interface_nodes
       IF(is == 0) THEN
         NodesC= NodesC + 1
         ListC(NodesC)= Node
       END IF
     END DO &
     Element_nodes
     EXIT
   END IF
 END DO &
 Elements_loop1
END DO &
Elements_loop
NdofsC= NodesC*Ndof
ALLOCATE(InciR(Nregs,Nodes),IncieR(Maxe,Nodel))
ALLOCATE(KBE(Nregs,NdofsC,NdofsC),A(Nregs,Ndofs,Ndofs))
ALLOCATE(Lhs(NdofsC,NdofsC),Rhs(NdofsC),uc(NdofsC),tc(NdofsC))
ALLOCATE(NodeR(Nregs),NodeC(Nregs))
ALLOCATE(ListEC(Nregs,maxe))
ALLOCATE(ListEF(Nregs,maxe))
ALLOCATE(LdesteR(Maxe,Ndofe))
ALLOCATE(Ldest_KBE(Ndofs))
ALLOCATE(NCode(Nregs,Ndofs))
ALLOCATE(LdestR(Nregs,Ndofs))
ALLOCATE(NbelC(Nregs))
ALLOCATE(NbelF(Nregs))
LdesteR= 0
Ncode= 0
NbelF= 0
NbelC= 0
!-------------------------------------------------
!    Assign local (region) numbering
!    and incidences of BE in local numbering
!-------------------------------------------------
ListEC= 0
ListEF= 0
```

```
DoF_KBE= 0
Regions_loop_1: &
DO nr=1,Nregs
 node= 0
 Elements_of_region: &
 DO nb=1,Nbel(nr)
  ne= ListR(nr,nb)
  Interface_elements: &
  IF(Bcode(ne,1) == 2) THEN
   NbelC(nr)= NbelC(nr) + 1
   ListEC(nr,NbelC(nr))= ne
   Nodes_of_Elem: &
   DO n=1,Nodel
!   check if node has already been entered
        is=0
     DO no=1,node
       IF(InciR(nr,no) == Incie(ne,n)) THEN
         is= 1
         EXIT
       END IF
     END DO
     IF(is == 0) THEN
        node=node+1
        InciR(nr,node)= Incie(ne,n)
        IncieR(ne,n)= node
     ELSE
        IncieR(ne,n)= no
     END IF
   END DO &
   Nodes_of_Elem
  END IF &
  Interface_elements
END DO &
Elements_of_region
NodeC(nr)= Node        ! No of interface nodes of Region nr
NdofC= NodeC(nr)*Ndof  ! D.o.F. at interface of Region nr
Elements_of_region1: &
DO nb=1,Nbel(nr)
 ne= ListR(nr,nb)
 Free_elements: &
 IF(Bcode(ne,1) /= 2) THEN
   NbelF(nr)= NbelF(nr) + 1
   ListEF(nr,NbelF(nr))= ne
   Nodes_of_Elem1: &
   DO n=1,Nodel
        is=0
     DO no=1,node
       IF(InciR(nr,no) == Incie(ne,n)) THEN
         is= 1
         EXIT
       END IF
```

```
       END DO
       IF(is == 0) THEN
         node=node+1
         InciR(nr,node)= Incie(ne,n)
         IncieR(ne,n)= node
       ELSE
         IncieR(ne,n)= no
       END IF
     END DO &
     Nodes_of_Elem1
   END IF &
   Free_elements
END DO &
Elements_of_region1
NodeR(nr)= node                    !   number of nodes per region
!----------------------------------------------
!      Determine local element destination vector
!----------------------------------------------
Elements:&
DO Nel=1,Nbel(nr)
  k=0
  ne= ListR(nr,Nel)
  DO n=1,Nodel
    DO m=1,Ndof
      k=k+1
      IF(Ndof > 1) THEN
        LdesteR(ne,k)= ((IncieR(ne,n)-1)*Ndof + m)
      ELSE
        LdesteR(ne,k)= IncieR(ne,n)
      END IF
    END DO
  END DO
END DO &
Elements
  !----------------------------------------------
  !      Determine local node destination vector
  !----------------------------------------------
n= 0
DO no=1, NodeR(nr)
  DO m=1, Ndof
    n= n + 1
    LdestR(nr,n)= (InciR(nr,no)-1) * Ndof + m
  END DO
END DO
!----------------------------------------------
!      Determine global boundary code vector for assembly
!----------------------------------------------
NdofR= NodeR(nr)*Ndof ! Total degrees of freedom of region
DoF_o_System: &
DO nd=1,NdofR
  DO Nel=1,Nbel(nr)
```

```
    ne=ListR(nr,Nel)
    DO m=1,Ndofe
       IF (nd == LdesteR(ne,m) .and. NCode(nr,nd) == 0) THEN
            NCode(nr,nd)= NCode(nr,nd)+BCode(ne,m)
       END IF
    END DO
  END DO
END DO &
DoF_o_System
END DO &
Regions_loop_1
Regions_loop_2: &
DO nr=1,Nregs
!------------------------------------
!   Allocate coordinates in local(region) numbering
!------------------------------------
ALLOCATE(XpR(Cdim,NodeR(nr)))
Region_nodes: &
DO Node=1,NodeR(nr)
  XpR(:,Node)= Xp(:,InciR(nr,node))
END DO &
Region_nodes
!------------------------------------------------------------
!    Determine interface destination vector for region assembly
!------------------------------------------------------------
No_o_Interfaceelements:&
DO n=1, NbelC(nr)
  ne= ListEC(nr,n)
  DoF_o_Element:&
  DO m=1, Ndofe
    DoF= Ldeste(ne,m)
    IF(Ldest_KBE(DoF) == 0)THEN
      DoF_KBE= DoF_KBE + 1
      Ldest_KBE(DoF)= DoF_KBE
    END IF
  END DO &
   DoF_o_Element
END DO &
No_o_Interfaceelements
NdofR= NodeR(nr)*Ndof ! Total degrees of freedom of region
NdofC= NodeC(nr)*Ndof ! D.o.F. of interface of Region nr
E=ER(nr)
ny=nyR(nr)
CALL Stiffness_BEM(nr,XpR,Nodel,Ndof,Ndofe&
     ,NodeR,Ncode(nr,:),NdofR,NdofC,KBE(nr,:,:)&
     ,A(nr,:,:),tc,Cdim,Elres_u,Elres_t,IncieR&
     ,LdesteR,Nbel,ListR,TypeR,Bcode,Con,E,ny,Ndest,Isym)
DO ro=1,NdofC
  DoF= LdestR(nr,ro)
  Nrow= Ldest_KBE(DoF)
  Rhs(Nrow)= Rhs(Nrow) + tc(ro)
```

```
  DO co=1, NdofC
    DoF= LdestR(nr,co)
    Ncln= Ldest_KBE(DoF)
    Lhs(Nrow,Ncln)= Lhs(Nrow,Ncln) - KBE(nr,ro,co)
  END DO
 END DO
 DEALLOCATE (XPR)
END DO &
Regions_loop_2
DEALLOCATE(tc)
!---------------------------------
!   Solve for interface unknown
!---------------------------------
CALL Solve(Lhs,Rhs,uc)
!---------------------------------
! Compute and add effect of interface displ.
!---------------------------------
Regions_loop_3: &
DO nr=1,Nregs
! gather region interface displacements
  NdofC= NodeC(nr)*Ndof
 ALLOCATE(ucr(NdofC))
 Interface_dof: &
 DO n=1,NdofC
    DoF= LdestR(nr,n)
    ucr(n)= uc(Ldest_KBE(DoF))
 END DO &
 Interface_dof
!--------------------------------------------------------------------
! Store interface displacements into Elres_u
!--------------------------------------------------------------------
 Interface_DoF1:&
 DO nd=1, NdofC
   DO n=1, Nbel(nr)
     ne=ListR(nr,n)
     DO m=1,Ndofe
       IF(nd == LdesteR(ne,m))THEN
         Elres_u(ne,m)= Elres_u(ne,m) + ucr(nd)
       END IF
     END DO
   END DO
 END DO &
 Interface_DoF1
 !   effects of interface displacement in local (region) numbering
 NdofR= NodeR(nr)*Ndof
 NdofF= (NodeR(nr) - NodeC(nr))*Ndof  !   d.o.F , free nodes
 ALLOCATE(tc(NdofC),xf(NdofF),tcxf(NdofR))
 tc= 0.0; xf= 0.0; tcxf= 0.0
 tc= MATMUL(KBE(nr,1:NdofC,1:NdofC),ucr)
 xf= MATMUL(A(nr,1:NdofF,1:NdofC),ucr)
 tcxf(1:NdofC)= tc
```

```
tcxf(NdofC+1:NdofR)= xf
!--------------------------------------------------------------
! Store interface tractions into Elres_t
!--------------------------------------------------------------
DO nd=1, NdofC
  DO n=1, NbelC(nr)
    ne=ListEC(nr,n)
    DO m=1, Ndofe
      IF(nd == LdesteR(ne,m))THEN
        Elres_t(ne,m)= Elres_t(ne,m) + tcxf(nd)
      END IF
    END DO
  END DO
END DO
!--------------------------------------------------------------
! Store results of free nodes into Elres_u or Elres_t
!--------------------------------------------------------------
DO nd=NdofC+1, NdofR
  DO n=1, NbelF(nr)
    ne=ListEF(nr,n)
    DO m=1, Ndofe
      IF(nd == LdesteR(ne,m))THEN
        IF(Ncode(nr,nd) == 0)THEN
          Elres_u(ne,m)= Elres_u(ne,m) + tcxf(nd)
        ELSE IF(Bcode(ne,m) == 1)THEN
          Elres_t(ne,m)= Elres_t(ne,m) + tcxf(nd)
        END IF
      END IF
    END DO
  END DO
END DO
DEALLOCATE(tc,xf,tcxf,ucr)
END DO &
Regions_loop_3
!---------------------------
!    Print out results
!---------------------------
CLOSE(UNIT=2)
OPEN(UNIT=2,FILE= 'BERESULTS',FORM='FORMATTED')
Elements_all:&
DO nel=1,Maxe
  WRITE(2,*) ' Results, Element ',nel
  WRITE(2,*) 'u=' , (Elres_u(nel,m), m=1,Ndofe)
  WRITE(2,*) 't=' , (Elres_t(nel,m), m=1,Ndofe)
END DO &
Elements_all
END PROGRAM General_purpose_MRBEM
```

10.5.1 User's manual for Program 10.1

The input data which have to be supplied in the data file INPUT are described below. Free field input is used, that is, numbers are separated by blanks. However, all numbers including the zero entries must be specified.

The input is divided into three parts. First general information about the problem is read in. Next the mesh geometry is specified. The problem may consist of linear and quadratic elements, as shown in Figure 10.10. The sequence in which the node numbers have to be entered when specifying the incidences is also shown. Note that this order determines the direction of the outward normal, which has to point away from the material. For 3-D elements, if the node numbers are entered in an anticlockwise sense the outward normal points towards the viewer.

Finally, information about regions has to be specified. For each region we must input the number of boundary elements that describe the region, the region code (finite or infinite) and the material properties.

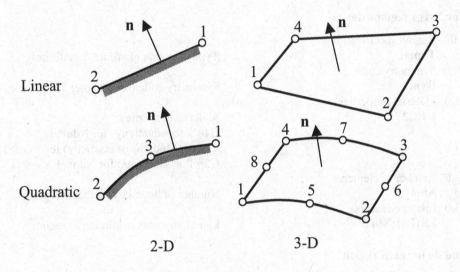

Figure 10.10 Element library

INPUT DATA SPECIFICATION

1.0 Title specification
     TITLE   Project title (max 60 characters)
2.0 Cartesian dimension of problem
     Cdim   Cartesian dimension
     2= two-dimensional problem
     3= three-dimensional problem
3.0 Problem type specification
     Ndof   Degree of freedom per node
                                                  1   = potential problem

2,3 = elasticity problem

4.0   Type of analysis
      Toa   2-D analysis type:
                                           1= plane strain
                                           2= plane stress

5.0   Element type specification
      Ltyp   Element type
                                           1= linear
                                           2= quadratic

6.0   Node specification
      Nodes   Number of nodes
7.0   Element specification
      Maxe   Number of elements
8.0   Region specification
      Nregs   Number of regions

For Nregs regions do:

9.0   Region specification
      TypeR   Type of region (1=finite, 2=infinite)
10.0 Symmetry code
      Isym   Symmetry code (see Chapter 7)
11.0 Material properties
      C1,C2   Material properties :
                                           C1= k (conductivity) for Ndof=1
                                               = E (Modulus of elasticity) for Ndof>1
                                           C2= Poisson's ratio for Ndof>1

12.0 Number of elements
      Nbel   Number of boundary elements/region
13.0 List of elements
      ListR(1:Nbel)   List of elements belonging to region

End do for each region

14.0 Loop over all elements
      Inci (1:Element nodes)   Global node numbers of element nodes
15.0 *Dirichlet* boundary conditions
      NE_u   Number of elements with *Dirichlet* BC
16.0 Prescribed values for *Dirichlet* BC
      Nel, Elres_u(1 : Element D.o.F.) Specification of boundary condition
                                           Nel = Element number to be assigned BC
                                           Elres_u = Prescribed values for all
                                           degrees of freedom of element: all d.o.F
                                           first node; all d.o.F second node, etc.

17.0 *Neumann* boundary conditions
      NE_u   Number of elements with *Neumann* BC
                                           Only specify for non-zero prescribed
                                           values

18.0   Prescribed values for *Neumann* BC
       Nel, Elres_t(1 : Element D.o.F.)   Specification of boundary condition
                                                      Nel = Element number to be assigned BC
                                                      Elres_t = Prescribed values for all
                                                      degrees of freedom of element: all d.o.F
                                                      first node; all d.o.F second node etc.

10.5.2   Sample problem

The example problem is the same as the cantilever problem in Chapter 9, except that two
regions are specified instead of one. The mesh is shown in Figure 10.11.

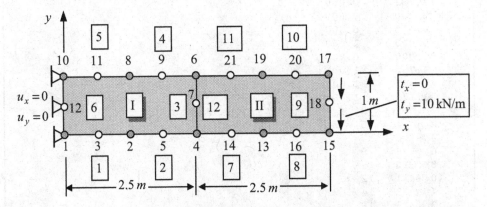

Figure 10.11   Cantilever beam multi-region mesh

The input file for this problem is:

```
Cantilever beam multi-region    !   Title
2                        !   Cartesian dimension
2                        !   Elasticity problem
1                        !   T.o.A.= plane strain
2                        !   Parabolic elements
   21                    !   Nodes
   12                    !   Elements
    2                    !   Number of regions
1                        !   Region 1: Type of region= Finite
0                        !   No symmetry
0.1000E+05 0.0000E+00    !   E , v
   6                     !   Number of elements
   1 2 3 4 5 6           !   List of elements
1                        !   Region 2: Type of region= Finite
0                        !   No symmetry
0.1000E+05 0.0000E+00    !   E , v
```

```
    6                   !  Number of elements
  7 8 9 10 11 12  !  List of elements
      0.000       0.000   !       Node coordinates
      1.250       0.000
      0.625       0.000
      2.500       0.000
      1.875       0.000
      2.500       1.000
      2.500       0.500
      1.250       1.000
      1.875       1.000
      0.000       1.000
      0.625       1.000
      0.000       0.500
      3.750       0.000
      3.125       0.000
      5.000       0.000
      4.375       0.000
      5.000       1.000
      5.000       0.500
      3.750       1.000
      4.375       1.000
      3.125       1.000
      1      2      3  !      Element incidences
      2      4      5
      4      6      7
      6      8      9
      8     10     11
     10      1     12
      4     13     14
     13     15     16
     15     17     18
     17     19     20
     19      6     21
      6      4      7
      1
  6 0.0 0.0 0.0 0.0 0.0 0.0    !  Dirichlet BC
    1
    9 0.0 -10.0 0.0 -10.0 0.0 -10.0   !  Neumann BC
```

The output from Program 10.1 is:

```
Project:
 Cantilever beam multi-region
 Cartesian_dimension:              2
 Elasticity Problem
 Type of Analysis: Solid Plane Strain
 Quadratic Elements
Number of Nodes of System:             21
Number of Elements of System:          12
Region
```

```
 Finite region
 No symmetry
 Youngs modulus:    10000.00
 Poisson's ratio:   0.0000000E+00
 List of boundary elements:
 1              2              3              4              5              6
Region           2
 Finite region
 No symmetry Youngs modulus:    10000.00
 Poissons ratio:  0.0000000E+00
 List of boundary elements:
 7              8              9             10             11            12
Node     1  Coor      0.00     0.00
Node     2  Coor      1.25     0.00
Node     3  Coor      0.63     0.00
Node     4  Coor      2.50     0.00
Node     5  Coor      1.88     0.00
Node     6  Coor      2.50     1.00
Node     7  Coor      2.50     0.50
Node     8  Coor      1.25     1.00
Node     9  Coor      1.88     1.00
Node    10  Coor      0.00     1.00
Node    11  Coor      0.63     1.00
Node    12  Coor      0.00     0.50
Node    13  Coor      3.75     0.00
Node    14  Coor      3.13     0.00
Node    15  Coor      5.00     0.00
Node    16  Coor      4.38     0.00
Node    17  Coor      5.00     1.00
Node    18  Coor      5.00     0.50
Node    19  Coor      3.75     1.00
Node    20  Coor      4.38     1.00
Node    21  Coor      3.13     1.00


 Incidences:
EL      1  Inci       1      2      3
EL      2  Inci       2      4      5
EL      3  Inci       4      6      7
EL      4  Inci       6      8      9
EL      5  Inci       8     10     11
EL      6  Inci      10      1     12
EL      7  Inci       4     13     14
EL      8  Inci      13     15     16
EL      9  Inci      15     17     18
EL     10  Inci      17     19     20
EL     11  Inci      19      6     21
EL     12  Inci       6      4      7
 Elements with Dirichlet BCs:

 Element             6  Prescribed values:
  0.0000000E+00   0.0000000E+00
```

```
    0.0000000E+00   0.0000000E+00
    0.0000000E+00   0.0000000E+00

Elements with Neumann BC´s:

Element              9  Prescribed values:
    0.0000000E+00  -10.00000
    0.0000000E+00  -10.00000
    0.0000000E+00  -10.00000
Results, Element    1
u=   0.0000E+00   0.0000E+00
    -3.2348E-02  -4.4511E-02
    -1.7147E-02  -1.2137E-02
t=   0.0000E+00   0.0000E+00
     0.0000E+00   0.0000E+00
     0.0000E+00  0.0000E+00
    .


    .

Results, Element     9
u= -7.3849E-02  -0.5012
    7.3849E-02  -0.5012
    1.0140E-09  -0.5011
    .

    .


    .

Results, Element    12
u=   5.5386E-02  -0.1582
    -5.5386E-02  -0.1582
     1.0792E-09  -0.1582
t=  -147.7         5.933
     147.7         5.933
    -1.2626E-06   12.060
```

It can be seen that the maximum displacement is 0.5012 as compared with the theoretical value of 0.500, and that the multi-region method results in a negligible loss of accuracy.

10.6 CONCLUSIONS

In this chapter we have extended the capabilities of program 7.1 so that problems with non-homogeneous material properties can be handled. The 'stiffness matrix assembly' approach taken is quite different from the methods usually published in the literature and uses some ideas of the finite element method. There are several advantages: since each region can be treated completely separately the method is well suited to parallel processing because each processor could be assigned to the computation of the stiffness matrix of one region. Furthermore, with this method it is possible to model sequential excavation and construction as is required, for example, in the field of tunnelling[3]. By choosing to implement the method we have also laid the groundwork for the coupling with the finite element method so that there is not much more theory to discuss in Chapter 14. The multi-region method extends the capability of the BEM not only to handle non-homogeneous domains but also, as will be demonstrated later, can be applied to contact and crack propagation problems.

10.7   REFERENCES

1. Butterfield, R. and Tomlin, G.R. (1972) Integral techniques for solving zoned anisotropic continuum problems. *Int. Conf. Variational Methods in Engineering*, Southampton University, pp. 9/31-9/53.
2. Beer, G. (1993) An efficient numerical method for modelling initiation and propagation of cracks along a material interface. *Int. J. Numer. Methods Eng.*, **36** (21), 3579-3594.
3 Beer, G. and Dünser, Ch. (2000) Boundary element analysis of problems in tunnelling. *Developments in Theoretical Geomechanics* (D.W. Smith and J.P. Carter eds.) AA. Balkema, Rotterdam, 103-122.

11

Edges and Corners

He who goes beneath the surface
does this at his own risk
O. Wilde

11.1   INTRODUCTION

The multi-region method outlined in the previous chapter works well if the interfaces between the regions are smooth, that is, each interface point has a unique tangent. If the boundary is not smooth but has corners and edges, i.e. the outward normals are different on each side of a point, then normal flow t or normal traction $\mathbf{t}$ are also different on each side. Such a case would arise, for example, if the shape of the inclusion in the example of the previous chapter is square instead of circular (Figure 11.1). In this case, two values of normal flow would have to be computed at the corner node instead of one. However, the integral equations allows us to compute one value of t at a node. Therefore, additional equations are needed.

The problem also arises if *Dirichlet* boundary conditions are specified on both sides of a corner node. If a *Dirichlet* boundary condition is specified on one side and a *Neumann* condition on the other, then the solution can be obtained, as was previously shown with the two region cantilever beam example.

This chapter deals in some detail with the treatment of corners in the boundary element method. This is of particular relevance to the multi-region method, because in many applications it is not possible to avoid interfaces that are not smooth.

A number of schemes for dealing with this problem have been proposed in the past. The following are the methods that have been suggested:

- Numerically round off the corner by using an average outward normal, i.e. an average of all normal vectors of elements connecting to the node. This is not really correct, as the geometry of the element should be rounded off too.

- Use two or more nodes on a corner, each one moved slightly away form the edge[1]. It is necessary to use non-conformal interpolation[2], but the additional nodes can be utilised as collocation points and additional equations can be obtained. Although this scheme is fairly straightforward to implement, it has the disadvantage that, because collocation points are very close to each other, the equation system may become unstable. Also, the use of non-conforming elements results in a larger system of equations.

- The unknown values of fluxes/tractions are computed by extrapolation from the nodes adjacent to the corner node[3]. This method is also not difficult to implement, but its accuracy would greatly depend on the size of the boundary elements adjacent to the corner.

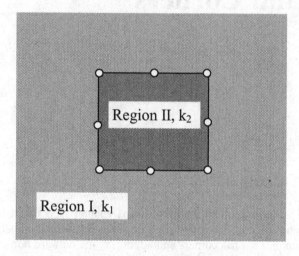

Figure 11.1   Example of a multi-region problem with corners

Here we present a mechanically consistent procedure as was recently published by Gao and Davies[4]. We start with potential problems and then proceed to elasticity.

11.2   POTENTIAL PROBLEMS

Consider, for example, the potential problem in Figure 11.2 which is divided into three regions. We note that at node 8 there are two values of t, depending on which side of the node we are. We isolate region III in Figure 11.3 and show local (region) as well as the local (element) node numbering. If we solve for the *Dirichlet* boundary condition ($u_1=0$, $u_2=0$, $u_3=0$), the flux normal to the boundary t at node 2 is different on the horizontal and vertical edge.

For this problem the unknown fluxes are:

$$t_1^1, \quad t_2^1, \quad t_1^4, \quad t_2^4, \quad u_4 \qquad\qquad (11.1)$$

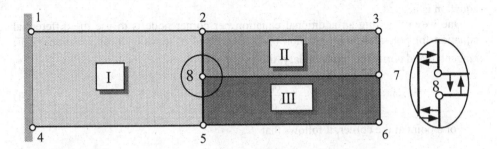

Figure 11.2 2-D potential problem showing discontinuous boundary fluxes at corner

Following the procedure in Chapter 7, the system of discretised integral equations can be written as:

$$\Delta U_{1i}^1\, t_1^1 + \Delta U_{2i}^1 t_2^1 + \Delta U_{1i}^4\, t_1^4 + \Delta U_{2i}^4\, t_2^4 - \left(\Delta T_{2i}^2 + \Delta T_{1i}^3\right) u_4 =$$

$$-\left(\Delta U_{1i}^2\, t_1^2 + \Delta U_{2i}^2 t_2^2 + \Delta U_{1i}^3\, t_1^3 + \Delta U_{2i}^3\, t_2^3\right) \qquad for \quad i = 1,2,3,4 \tag{11.2}$$

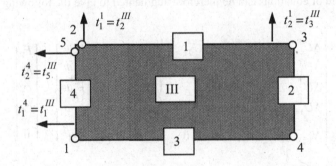

Figure 11.3 Isolated view of region III showing notation for t

The system of equations may be recast in matrix form as:

$$\begin{bmatrix} \Delta U_{11}^1 & \Delta U_{21}^1 & \Delta U_{11}^4 & \Delta U_{21}^4 & -\left(\Delta T_{21}^2 + \Delta T_{11}^3\right) \\ \Delta U_{12}^1 & \Delta U_{22}^1 & \Delta U_{12}^4 & \Delta U_{22}^4 & -\left(\Delta T_{22}^2 + \Delta T_{12}^3\right) \\ \Delta U_{13}^1 & \Delta U_{23}^1 & \Delta U_{13}^4 & \Delta U_{23}^4 & -\left(\Delta T_{23}^2 + \Delta T_{13}^3\right) \\ \Delta U_{14}^1 & \Delta U_{24}^1 & \Delta U_{14}^4 & \Delta U_{24}^4 & -\left(\Delta T_{24}^2 + \Delta T_{14}^3\right) \end{bmatrix} \begin{Bmatrix} t_1^1 \\ t_2^1 \\ t_1^4 \\ t_2^4 \\ u_4 \end{Bmatrix} = \begin{Bmatrix} F_1 \\ F_2 \\ F_3 \\ F_4 \end{Bmatrix} \tag{11.3}$$

The system of four equations cannot be solved for five unknowns and an additional equation is needed.

One way of getting an additional equation per corner node is to use the differential equation for potential flow (equation (4.7). Expressed in terms of local directions $\bar{x}, \bar{y}$, tangential and normal to the boundary, this is given by

$$\frac{\partial q_{\bar{x}}}{\partial \bar{x}} + \frac{\partial q_{\bar{y}}}{\partial \bar{y}} = 0 \tag{11.4}$$

For a point at the corner, it follows that

$$\frac{\partial q_{\bar{y}}}{\partial \bar{x}} = \frac{\partial t}{\partial \bar{x}} = 0 \tag{11.5}$$

Substituting the interpolation for t (Equation 6.1) we obtain

$$\frac{\partial t}{\partial \bar{x}} = \frac{1}{J} \sum_{n=1}^{2(3)} \frac{\partial N_n}{\partial \xi} t_n^e = 0 \tag{11.6}$$

For linear elements with two nodes, we can write with $N_{n,\xi} = \partial N_n / \partial \xi$

$$N_{1,\xi}\, t_1^e + N_{2,\xi}\, t_2^e = 0 \tag{11.7}$$

The system of equations can be therefore augmented to give the following:

$$\begin{bmatrix} \Delta U_{11}^1 & \Delta U_{21}^1 & \Delta U_{11}^4 & \Delta U_{21}^4 & -\left(\Delta T_{21}^2 + \Delta T_{11}^3\right) \\ \Delta U_{12}^1 & \Delta U_{22}^1 & \Delta U_{12}^4 & \Delta U_{22}^4 & -\left(\Delta T_{22}^2 + \Delta T_{12}^3\right) \\ \Delta U_{13}^1 & \Delta U_{23}^1 & \Delta U_{13}^4 & \Delta U_{23}^4 & -\left(\Delta T_{23}^2 + \Delta T_{13}^3\right) \\ \Delta U_{14}^1 & \Delta U_{24}^1 & \Delta U_{14}^4 & \Delta U_{24}^4 & -\left(\Delta T_{24}^2 + \Delta T_{14}^3\right) \\ N_{1,\xi} & N_{2,\xi} & 0 & 0 & 0 \end{bmatrix} \begin{Bmatrix} t_2^{III} \\ t_3^{III} \\ t_1^{III} \\ t_5^{III} \\ u_4^{III} \end{Bmatrix} = \begin{Bmatrix} F_1 \\ F_2 \\ F_3 \\ F_4 \\ 0 \end{Bmatrix} \tag{11.8}$$

where $t_2^{III} = t_1^1$, $t_3^{III} = t_2^1$ has been substituted, as shown in Figure 11.3. Similar equations can be obtained for region II. Note that for region I, no special attention has to be given to any corners.

11.3   TWO-DIMENSIONAL ELASTICITY

Considering the same problem in 2-D elasticity, we have a traction vector $\mathbf{t}$ instead of a scalar. As shown in Figure 11.4, we now have four unknown per corner, two of which can be computed by the integral equations and therefore there are two additional unknowns for corner node 1.

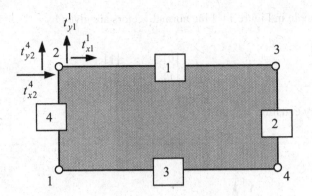

Figure 11.4 2-D elasticity problem showing discontinuous tractions at corner 1

One auxiliary equation can be obtained from the assumption that the stress tensor is unique at a point. The relationship between tractions and stresses has been defined in Chapter 4. In matrix form this is given by

$$\mathbf{t}_j^e = [\sigma]\mathbf{n}^e \quad for \quad e = 1,4 \tag{11.9}$$

where j is the local node number of the corner point and matrix $[\sigma]$ is defined as

$$[\sigma] = \begin{bmatrix} \sigma_x & \tau_{xy} \\ \tau_{yx} & \sigma_y \end{bmatrix} \tag{11.10}$$

Pre-multiplying the first equation with $\mathbf{n}^1$ and the second with $\mathbf{n}^4$, we obtain

$$\mathbf{n}^1 \bullet \mathbf{t}_j^4 = \mathbf{n}^1 \bullet [\sigma]\mathbf{n}^4 \quad ; \quad \mathbf{n}^4 \bullet \mathbf{t}_j^1 = \mathbf{n}^4 \bullet [\sigma]\mathbf{n}^1 \tag{11.11}$$

Because of the symmetry of $[\sigma]$, it follows that

$$\mathbf{n}^1 \bullet \mathbf{t}_j^4 = \mathbf{n}^4 \bullet \mathbf{t}_j^1 \tag{11.12}$$

The auxiliary equation for corner node 1 is therefore given by

$$\begin{bmatrix} n_x^1 & n_y^1 & -n_x^4 & -n_y^4 \end{bmatrix} \dots \begin{Bmatrix} t_{xj}^4 \\ t_{yj}^4 \\ t_{xj}^1 \\ t_{yj}^1 \end{Bmatrix} = 0 \tag{11.13}$$

For the example in Figure 11.4 the normal vectors are given by

$$\mathbf{n}^1 = \begin{Bmatrix} 0 \\ 1 \end{Bmatrix}; \quad \mathbf{n}^4 = \begin{Bmatrix} 1 \\ 0 \end{Bmatrix} \tag{11.14}$$

The auxiliary equation is simply

$$t_{yj}^4 = t_{xj}^1 \tag{11.15}$$

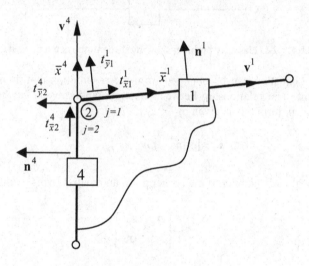

Figure 11.5   Definitions of local tractions

One additional equation is needed. Gao and Davies[4] suggest an elegant way of obtaining auxiliary equations using the differential equations of equilibrium. The result is that the shear stress approaches a constant value near the corner node but a detailed proof of this is given in Reference 4. Now for each node two additional auxiliary equations can be written:

$$\frac{\partial t_{\bar{x}j}^e}{\partial \bar{x}^e} = 0 \quad \text{for} \quad e = 1,4 \tag{11.16}$$

The local directions $\bar{x}^e$ for each element are defined by unit vectors $\mathbf{v}^e$ tangential to the boundary (see Figure. 11.5, where one boundary of element 4 has been inclined slightly for the purpose of explaining the method for general geometries). $\mathbf{v}^e$ is computed as follows:

$$\mathbf{V}^e = \frac{\partial}{\partial \xi} \mathbf{x}^e = \sum_{n=1}^{2(3)} \frac{\partial N_n}{\partial \xi} \mathbf{x}_n^e; \quad \mathbf{v}^e = \mathbf{V}^e \frac{1}{J^e} \tag{11.17}$$

where

$$\frac{1}{J^e} = \frac{\partial \xi}{\partial \bar{x}^e} = \frac{1}{\sqrt{\left(V_x^e\right)^2 + \left(V_y^e\right)^2}} \tag{11.18}$$

The traction in local $\bar{x}$-directions in terms of the global tractions $\mathbf{t}_j^e$ are computed by:

$$t_{\bar{x}j}^e = \mathbf{t}_j^e \bullet \mathbf{v}^e \tag{11.19}$$

Equations (11.16) can be rewritten as

$$\frac{\partial t_{\bar{x}j}^e}{\partial \bar{x}^e} = \frac{1}{J^e} \frac{\partial \mathbf{t}_j^e}{\partial \xi} \bullet \mathbf{v}^e = 0 \tag{11.20}$$

Using the interpolation function for the traction, we have

$$\frac{\partial \mathbf{t}_j^e}{\partial \xi} = \sum_{n=1}^{2(3)} \frac{\partial N_n(\xi_j)}{\partial \xi} \mathbf{t}_n^e \tag{11.21}$$

and

$$\frac{\partial t_{\bar{x}j}^e}{\partial \bar{x}^e} = \frac{1}{|J|^e} \sum_{n=1}^{2(3)} \frac{\partial N_n(\xi_j)}{\partial \xi} \mathbf{t}_n^e \bullet \mathbf{v}^e = 0 \tag{11.22}$$

For a linear element $\dfrac{\partial N_1}{\partial \xi} = -\dfrac{1}{2}$ and $\dfrac{\partial N_1}{\partial \xi} = \dfrac{1}{2}$, and the auxiliary equations are written as

$$\left(\mathbf{t}_1^e - \mathbf{t}_2^e\right) \bullet \mathbf{v}^e = 0 \quad for \quad e = 1,4 \tag{11.23}$$

or in matrix form as

$$\begin{bmatrix} v_x^1 & v_y^1 & -v_x^1 & -v_y^1 & 0 & 0 & 0 & 0 \\ 0 & 0 & 0 & 0 & v_x^4 & v_y^4 & -v_x^4 & -v_y^4 \end{bmatrix} \begin{Bmatrix} t_{x1}^1 \\ t_{y1}^1 \\ t_{x2}^1 \\ t_{y2}^1 \\ t_{x1}^4 \\ t_{y1}^4 \\ t_{x2}^4 \\ t_{y2}^4 \end{Bmatrix} = 0 \tag{11.24}$$

For the example in Figure 11.4, the normal vectors are given by

$$\mathbf{v}^1 = \begin{Bmatrix} 1 \\ 0 \end{Bmatrix} \quad ; \quad \mathbf{v}^4 = \begin{Bmatrix} 0 \\ 1 \end{Bmatrix} \tag{11.25}$$

The auxiliary equation becomes

$$\begin{bmatrix} 1 & 0 & -1 & 0 & 0 & 0 & 0 & 0 \\ 0 & 0 & 0 & 0 & 0 & 1 & 0 & -1 \end{bmatrix} \begin{Bmatrix} t_{x1}^1 \\ t_{y1}^1 \\ t_{x2}^1 \\ t_{y2}^1 \\ t_{x1}^4 \\ t_{y1}^4 \\ t_{x2}^4 \\ t_{y2}^4 \end{Bmatrix} = 0 \tag{11.26}$$

or simply

$$t_{x1}^1 = t_{x2}^1 \quad ; \quad t_{y1}^4 = t_{y2}^4 \tag{11.27}$$

There are two possibilities as to the way auxiliary equations can be used. We can either take equations (11.12) and (11.16), or we can use only equation (11.16). Gao and Davies[4] suggest that it is better to use the latter, as they do not imply continuity of the stress tensor, an assumption which is no longer true if there is a change in material properties between regions. We note, however, that in special cases, as for the example in Figure 11.4 where the boundaries are parallel to the x and y-axes, we cannot freely choose which components of the traction are to be computed by the auxiliary equations. If the edges are directed exactly along the coordinate axes, only the shear components can be computed at the two sides of the corner point by either utilising equation (11.15) or equations (11.16). For this case the two equations can not be used concurrently. We therefore propose to use only equation (11.16).

Consider the region in Figure 11.4, where we prescribe *Dirichlet* boundary conditions with zero displacements at nodes 1,2 and 3 and *Neuman* boundary conditions with prescribed tractions **t** along elements 2 and 3.

The discretised integral equations for this case are written as

$$\Delta \mathbf{U}_{1i}^1 \, \mathbf{t}_1^1 + \Delta \mathbf{U}_{2i}^1 \mathbf{t}_2^1 + \Delta \mathbf{U}_{1i}^4 \, \mathbf{t}_1^4 + \Delta \mathbf{U}_{2i}^4 \, \mathbf{t}_2^4 + \left(\Delta \mathbf{T}_{2i}^2 + \Delta \mathbf{T}_{1i}^3 \right) \mathbf{u}_4 =$$
$$- \left(\Delta \mathbf{U}_{1i}^2 \, \mathbf{t}_1^2 + \Delta \mathbf{U}_{2i}^2 \mathbf{t}_2^2 + \Delta \mathbf{U}_{1i}^3 \, \mathbf{t}_1^3 + \Delta \mathbf{U}_{2i}^3 \, \mathbf{t}_2^3 \right) \quad for \quad i = 1,2,3,4 \tag{11.28}$$

or in matrix form as

$$
\begin{bmatrix}
\Delta U^1_{11} & \Delta U^1_{21} & \Delta U^4_{11} & \Delta U^4_{21} & -\left(\Delta T^2_{21}+\Delta T^3_{11}\right) \\
\Delta U^1_{12} & \Delta U^1_{22} & \Delta U^4_{12} & \Delta U^4_{22} & -\left(\Delta T^2_{22}+\Delta T^3_{12}\right) \\
\Delta U^1_{13} & \Delta U^1_{23} & \Delta U^4_{13} & \Delta U^4_{23} & -\left(\Delta T^2_{23}+\Delta T^3_{13}\right) \\
\Delta U^1_{14} & \Delta U^1_{24} & \Delta U^4_{14} & \Delta U^4_{24} & -\left(\Delta T^2_{24}+\Delta T^3_{14}\right)
\end{bmatrix}
\begin{Bmatrix}
t^1_1 \\ t^1_2 \\ t^4_1 \\ t^4_2 \\ u_4
\end{Bmatrix}
=
\begin{Bmatrix}
F_1 \\ F_2 \\ F_3 \\ F_4
\end{Bmatrix}
\qquad (11.29)
$$

where the sub matrices are given by

$$
\Delta \mathbf{U}^e_{ni} = \begin{bmatrix} \Delta U_{xx} & \Delta U_{xy} \\ \Delta U_{yx} & \Delta U_{yy} \end{bmatrix}^e_{ni}
\qquad (11.30)
$$

This system of equations has more unknown than can be determined and additional equations are required.

For the example problems with linear boundary elements these auxiliary equations are given by

$$
\begin{bmatrix}
N_{1,\xi}v^1_x & N_{1,\xi}v^1_y & N_{2,\xi}v^1_x & N_{2,\xi}v^1_y & 0 & 0 & 0 & 0 \\
0 & 0 & 0 & 0 & N_{1,\xi}v^4_x & N_{1,\xi}v^4_y & N_{2,\xi}v^4_x & N_{2,\xi}v^4_y
\end{bmatrix}
\begin{Bmatrix}
t^1_{x1} \\ t^1_{y1} \\ t^1_{x2} \\ t^1_{y2} \\ t^4_{x1} \\ t^4_{y1} \\ t^4_{x2} \\ t^4_{y2}
\end{Bmatrix}
= 0
\qquad (11.31)
$$

These equations can now be implemented into equation 11.29 and written as a single system of equations which can be solved for all unknown.

$$
\begin{bmatrix}
\Delta U_{xx11}^{1} & \Delta U_{xy11}^{1} & \Delta U_{xx21}^{1} & \Delta U_{xy21}^{1} & \cdots \\[4pt]
\Delta U_{yx11}^{1} & \Delta U_{yy11}^{1} & \Delta U_{yx21}^{1} & \Delta U_{yy21}^{1} \\[4pt]
\vdots \\[4pt]
N_{1,\xi}v_{x}^{1} & N_{1,\xi}v_{y}^{1} & N_{2,\xi}v_{x}^{1} & N_{2,\xi}v_{y}^{1} & 0 & 0 & \cdots \\[4pt]
0 & 0 & 0 & 0 & N_{1,\xi}v_{x}^{4} & N_{1,\xi}v_{y}^{4} & \cdots
\end{bmatrix}
\left\{
\begin{array}{c}
t_{x1}^{1} \\ t_{y1}^{1} \\ t_{x2}^{1} \\ t_{y2}^{1} \\ \vdots \\ t_{x1}^{4} \\ t_{y1}^{4} \\ t_{x2}^{4} \\ t_{y2}^{4} \\ \vdots
\end{array}
\right\}
=
\left\{
\begin{array}{c}
F_{1} \\ F_{2} \\ \vdots \\ 0 \\ 0
\end{array}
\right\}
\qquad (11.32)
$$

We can see that when edges and corners are present, then there are changes to the left-hand side of the system of equations, because the system has more unknowns. For 2-D elasticity problems, two unknowns are added for each corner. The auxiliary equations for determining these additional unknowns are added at the end of the matrix.

11.3.1 Region assembly with corners

The procedure for the assembly of regions has to be modified if corners are present. In Figure 11.6 we show three regions with global numbering (enclosed in circles) and the local (region) numbering.

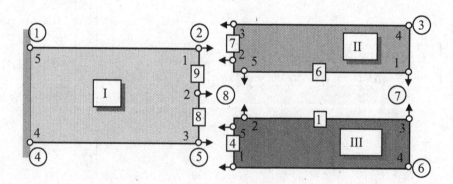

Figure 11.6 Assembly of regions with corners

Following the stiffness matrix assembly procedure outlined in Chapter 10, we obtain the following system of equations for region I:

$$\begin{Bmatrix} t_1 \\ t_2 \\ t_3 \end{Bmatrix}^I = \begin{bmatrix} t_{11} & t_{12} & t_{13} \\ t_{21} & t_{22} & t_{23} \\ t_{31} & t_{32} & t_{33} \end{bmatrix}^I \begin{Bmatrix} u_1 \\ u_2 \\ u_3 \end{Bmatrix}^I \tag{11.33}$$

For region II we have:

$$\begin{Bmatrix} t_1 \\ t_2 \\ t_3 \\ t_5 \end{Bmatrix}^{II} = \begin{Bmatrix} t_{10} \\ t_{20} \\ t_{30} \\ t_{50} \end{Bmatrix}^{II} + \begin{bmatrix} t_{11} & t_{12} & t_{13} \\ t_{21} & t_{22} & t_{23} \\ t_{31} & t_{32} & t_{33} \\ t_{51} & t_{52} & t_{53} \end{bmatrix}^{II} \begin{Bmatrix} u_1 \\ u_2 \\ u_3 \end{Bmatrix}^{II} \tag{11.34}$$

and for region III:

$$\begin{Bmatrix} t_1 \\ t_2 \\ t_3 \\ t_5 \end{Bmatrix}^{III} = \begin{Bmatrix} t_{10} \\ t_{20} \\ t_{30} \\ t_{50} \end{Bmatrix}^{III} + \begin{bmatrix} t_{11} & t_{12} & t_{13} \\ t_{21} & t_{22} & t_{23} \\ t_{31} & t_{32} & t_{33} \\ t_{51} & t_{52} & t_{53} \end{bmatrix}^{III} \begin{Bmatrix} u_1 \\ u_2 \\ u_3 \end{Bmatrix}^{III} \tag{11.35}$$

For assembly of regions and determination of interface displacements, the equations of equilibrium and compatibility are used.

Assuming no tractions are applied at the interface the equations of equilibrium in terms of tractions are:

$$t_1^I + t_3^{II} = 0$$
$$t_2^I + t_2^{II} = 0 \quad \text{and} \quad t_2^I + t_5^{II} = 0$$
$$t_2^{III} + t_5^{II} = 0 \tag{11.36}$$
$$t_3^I + t_1^{III} = 0$$
$$t_3^{III} + t_1^{II} = 0$$

There are 6 x 2 equations and only 4 x 2 unknown displacements at the interface (these are in global numbering u_2, u_8, u_5, u_7). The system is therefore over determined and cannot be solved. A way of overcoming this problem is to work with equivalent nodal forces, as used in the finite element method.

The equivalent nodal point forces are computed using the principle of virtual work. If we apply a virtual displacement $\delta u_x = 1$ to the corner node of region III, as shown in Figure 11.7, then the work done by the tractions must be equal to that done by the equivalent nodal forces.

This can be written as

$$F_x \cdot 1 = \int_S t_x \delta u_x dS \tag{11.37}$$

where the integration is over the surfaces of the two elements connected to the corner.

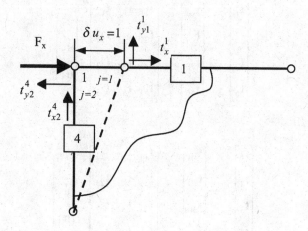

Figure 11.7   Calculation of F_x by principle of virtual work

Substituting the interpolation

$$t_x = \sum_{n=1}^{2(3)} N_n t_{xn}^e; \quad \delta u_x = N_{j(e)} \delta u_x = N_{j(e)} \cdot 1 \tag{11.38}$$

where $j(e)$ is the corner node in local (element) numbering, we obtain

$$F_x \cdot 1 = \sum_{e=1}^{2} \int_{S_e} (N_1 t_{x1}^e + N_2 t_{x2}^e) N_{j(e)} dS_e \tag{11.39}$$

A second equation can be obtained by applying a virtual displacement in the y-direction. The force vector at the corner node can be written as

$$\mathbf{F} = \sum_{e=1}^{2} \left(\int_S N_{j(e)} \mathbf{N} dS_e \right) \mathbf{t}^e \tag{11.40}$$

where for a linear boundary element

$$F = \begin{Bmatrix} F_x \\ F_y \end{Bmatrix}; \quad N = \begin{bmatrix} N_1 & 0 & N_2 & 0 \\ 0 & N_1 & 0 & N_2 \end{bmatrix}; \quad t^e = \begin{Bmatrix} t_{x1} \\ t_{y1} \\ t_{x2} \\ t_{y2} \end{Bmatrix} \qquad (11.41)$$

Equation (11.40) can be simplified to

$$\mathbf{F} = \sum_{e=1}^{2} \mathbf{M}^e \mathbf{t}^e \qquad (11.42)$$

where

$$\mathbf{M}^e = \begin{bmatrix} M_1 & 0 & M_2 & 0 \\ 0 & M_1 & 0 & M_2 \end{bmatrix} \qquad (11.43)$$

with

$$M_n^e = \int_{S_e} N_{j(e)} N_n dS_e \qquad (11.44)$$

The integration over elements can be conveniently carried out using numerical integration (Gauss Quadrature) with three points. For 2-D elasticity we have:

$$M_n^e = \int_{\xi=-1}^{1} N_{j(e)} N_n |J| dS_e = \sum_{m=1}^{3} N_{j(e)} N_n |J| W_m \qquad (11.45)$$

The second and third equations (11.36) can now be replaced by a single equilibrium equation (note that the local node number of point 8 in global numbering is 2 for all regions)

$$\mathbf{F}_2^I + \mathbf{F}_2^{II} + \mathbf{F}_2^{III} = 0 \qquad (11.46)$$

where

$$\mathbf{F}_2^I = \mathbf{M}^9 \mathbf{t}^9 + \mathbf{M}^8 \mathbf{t}^8$$
$$\mathbf{F}_2^{II} = \mathbf{M}^6 \mathbf{t}^6 + \mathbf{M}^7 \mathbf{t}^7 \qquad (11.47)$$
$$\mathbf{F}_2^{III} = \mathbf{M}^1 \mathbf{t}^1 + \mathbf{M}^4 \mathbf{t}^4$$

The notation for tractions is based on element numbering. Traction vectors in local (region) numbering are given by

$$\mathbf{t}^9 = \begin{Bmatrix} \mathbf{t}_1^I \\ \mathbf{t}_2^I \end{Bmatrix} \quad \text{and} \quad \mathbf{t}^8 = \begin{Bmatrix} \mathbf{t}_2^I \\ \mathbf{t}_3^I \end{Bmatrix}$$

$$\mathbf{t}^6 = \begin{Bmatrix} \mathbf{t}_1^{II} \\ \mathbf{t}_5^{II} \end{Bmatrix} \quad \text{and} \quad \mathbf{t}^7 = \begin{Bmatrix} \mathbf{t}_2^{II} \\ \mathbf{t}_3^{II} \end{Bmatrix} \tag{11.48}$$

$$\mathbf{t}^1 = \begin{Bmatrix} \mathbf{t}_2^{III} \\ \mathbf{t}_3^{III} \end{Bmatrix} \quad \text{and} \quad \mathbf{t}^4 = \begin{Bmatrix} \mathbf{t}_1^{III} \\ \mathbf{t}_5^{III} \end{Bmatrix}$$

Substituting equation (11.33) and using (11.42) we get:

$$\mathbf{F}_2^I = \mathbf{M}^9 \begin{bmatrix} t_{11} & t_{12} & t_{13} \\ t_{21} & t_{22} & t_{23} \end{bmatrix}^I \begin{Bmatrix} \mathbf{u}_2 \\ \mathbf{u}_8 \\ \mathbf{u}_5 \end{Bmatrix} + \mathbf{M}^8 \begin{bmatrix} t_{21} & t_{22} & t_{23} \\ t_{31} & t_{32} & t_{33} \end{bmatrix} \begin{Bmatrix} \mathbf{u}_2 \\ \mathbf{u}_8 \\ \mathbf{u}_5 \end{Bmatrix}$$

$$= \mathbf{M}^I \begin{bmatrix} t_{11} & t_{12} & t_{13} \\ t_{21} & t_{22} & t_{23} \\ t_{31} & t_{32} & t_{33} \end{bmatrix}^I \begin{Bmatrix} \mathbf{u}_2 \\ \mathbf{u}_8 \\ \mathbf{u}_5 \end{Bmatrix} \tag{11.49}$$

with :

$$\mathbf{M}^I = \begin{bmatrix} M_1^9 & 0 & M_2^9 + M_1^8 & 0 & M_2^8 & 0 \\ 0 & M_1^9 & 0 & M_2^9 + M_1^8 & 0 & M_2^8 \end{bmatrix} \tag{11.50}$$

It is left to the reader to verify that

$$\mathbf{F}_2^{II} = \mathbf{M}^{II} \begin{Bmatrix} \mathbf{t}_{10} \\ \mathbf{t}_{20} \\ \mathbf{t}_{30} \\ \mathbf{t}_{40} \end{Bmatrix} + \mathbf{M}^{II} \begin{bmatrix} t_{11} & t_{12} & t_{13} \\ t_{21} & t_{22} & t_{23} \\ t_{31} & t_{32} & t_{33} \\ t_{51} & t_{52} & t_{53} \end{bmatrix}^{II} \begin{Bmatrix} \mathbf{u}_7 \\ \mathbf{u}_8 \\ \mathbf{u}_2 \end{Bmatrix}$$

$$\tag{11.51}$$

$$\mathbf{F}_2^{III} = \mathbf{M}^{III} \begin{Bmatrix} \mathbf{t}_{10} \\ \mathbf{t}_{20} \\ \mathbf{t}_{30} \\ \mathbf{t}_{50} \end{Bmatrix} + \mathbf{M}^{III} \begin{bmatrix} t_{11} & t_{12} & t_{13} \\ t_{21} & t_{22} & t_{23} \\ t_{31} & t_{32} & t_{33} \\ t_{51} & t_{52} & t_{53} \end{bmatrix}^{III} \begin{Bmatrix} \mathbf{u}_5 \\ \mathbf{u}_8 \\ \mathbf{u}_7 \end{Bmatrix}$$

where $\mathbf{M}^{II}$ and $\mathbf{M}^{III}$ are 8x 2 matrices assembled with the same method as used to obtain the matrix $\mathbf{M}^{I}$.

Equation (11.46) can now be written as

$$\mathbf{F}_0 + [\mathbf{K}]\begin{Bmatrix} \mathbf{u}_2 \\ \mathbf{u}_8 \\ \mathbf{u}_5 \\ \mathbf{u}_7 \end{Bmatrix} = 0 \tag{11.52}$$

where $[\mathbf{K}]$ is the assembled stiffness matrix and

$$\mathbf{F}_0 = \mathbf{M}^{II}\begin{Bmatrix} \mathbf{t}_{10} \\ \mathbf{t}_{20} \\ \mathbf{t}_{30} \\ \mathbf{t}_{40} \end{Bmatrix}^{II} + \mathbf{M}^{III}\begin{Bmatrix} \mathbf{t}_{10} \\ \mathbf{t}_{20} \\ \mathbf{t}_{30} \\ \mathbf{t}_{50} \end{Bmatrix}^{III} \tag{11.53}$$

For the assembly of regions, the modified equilibrium equations are used for a corner node.

11.4   THREE-DIMENSIONAL ELASTICITY

Turning to 3-D elasticity problems, we have three surfaces at a node inside the domain, where five regions meet, as shown in Figure 11.8: A total of nine traction components exist at that corner (Figure 11.9). Three can be obtained to form the integral equations. Therefore six auxiliary equations are needed.

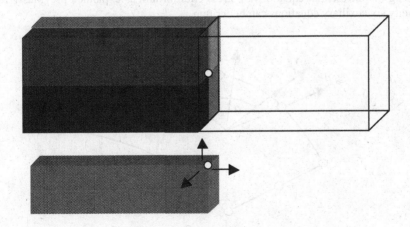

Figure 11.8   Three-dimensional model of cantilever beam with five regions

For obtaining the other auxiliary equations, a local orthogonal coordinate system is created for each boundary element, as shown in Figure 11.19.

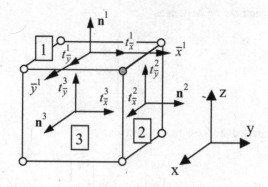

Figure 11.9   Corner node showing local traction components

Direction $\bar{x}$ (specified by vector $\mathbf{v}_1$) is taken to be tangential to the line $\eta = const$, i.e.

$$\mathbf{v}_1 = \frac{1}{J_1}\frac{\partial}{\partial \xi}\mathbf{x} = \frac{1}{J_1}\sum_{n=1}^{4(8)}\frac{\partial N_n}{\partial \xi}\mathbf{x}; \quad J_1 = \sqrt{(\mathbf{v}_{1x})^2 + (\mathbf{v}_{1y})^2 + (\mathbf{v}_{1z})^2} \tag{11.54}$$

Direction $\bar{y}$ is specified by a unit vector $\mathbf{v}_2$ which is assumed to be orthogonal to vectors $\mathbf{v}_1$ and $\mathbf{n}$. This direction is computed by

$$\mathbf{v}_2 = \mathbf{v}_1 \times \mathbf{n} \tag{11.55}$$

Using the differential equations of stress equilibrium, as explained previously for 2-D elasticity six auxiliary equations can be written.

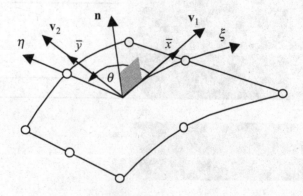

Figure 11.10   Local coordinate system for the definition of tangential traction components

These equations are

$$\frac{\partial t_{\bar{x}j}^e}{\partial \bar{x}^e} = 0 ; \quad \frac{\partial t_{\bar{y}j}^e}{\partial \bar{y}^e} = 0 \quad for \quad e = 1,2,3 \tag{11.56}$$

The equations can be rewritten in terms of intrinsic coordinates ξ, η as

$$\frac{\partial t_{\bar{x}}^e}{\partial \bar{x}^e} = \frac{\partial \xi}{\partial \bar{x}} \frac{\partial t_{\bar{x}j}^e}{\partial \xi} + \frac{\partial \eta}{\partial \bar{x}} \frac{\partial t_{\bar{x}j}^e}{\partial \eta} = 0$$

$$\frac{\partial t_{\bar{y}}^e}{\partial \bar{y}^e} = \frac{\partial \xi}{\partial \bar{y}} \frac{\partial t_{\bar{y}j}^e}{\partial \xi} + \frac{\partial \eta}{\partial \bar{y}} \frac{\partial t_{\bar{y}j}^e}{\partial \eta} = 0 \tag{11.57}$$

where

$$\frac{\partial \xi}{\partial \bar{x}} = \frac{1}{J_1} \quad ; \quad \frac{\partial \eta}{\partial \bar{x}} = \frac{1}{J_1} \frac{cos\theta}{sin\theta} \quad ; \quad \frac{\partial \xi}{\partial \bar{y}} = 0 \quad ; \quad \frac{\partial \eta}{\partial \bar{y}} = \frac{1}{J_2 \, sin\theta} \tag{11.58}$$

Using the interpolation function, the derivatives of the tractions at node j are computed by

$$\frac{\partial t_{\bar{x}j}^e}{\partial \xi} = \sum_{n=1}^{4(8)} \frac{\partial N_n(\xi_j, \eta_j)}{\partial \xi} t_{\bar{x}n}^e \quad ; \quad \frac{\partial t_{\bar{x}j}^e}{\partial \eta} = \sum_{n=1}^{4(8)} \frac{\partial N_n(\xi_j, \eta_j)}{\partial \eta} t_{\bar{x}n}^e$$

$$\frac{\partial t_{\bar{y}j}^e}{\partial \xi} = \sum_{n=1}^{4(8)} \frac{\partial N_n(\xi_j, \eta_j)}{\partial \xi} t_{\bar{y}n}^e \quad ; \quad \frac{\partial t_{\bar{y}j}^e}{\partial \eta} = \sum_{n=1}^{4(8)} \frac{\partial N_n(\xi_j, \eta_j)}{\partial \eta} t_{\bar{y}n}^e \tag{11.59}$$

where the tractions in the the $\bar{x}$ and $\bar{y}$ directions are computed by

$$t_{\bar{x}} = v_{1x} t_x + v_{1y} t_y + v_{1z} t_z$$

$$t_{\bar{y}} = v_{2x} t_x + v_{2y} t_y + v_{2z} t_z \tag{11.60}$$

For linear elements, the auxiliary equations for the example in Figure 11.9 are written as

$$\frac{\partial t_{\bar{x}}^e}{\partial \bar{x}^e} = a_1 t_{\bar{x}1}^e + a_2 t_{\bar{x}2}^e + a_3 t_{\bar{x}3}^e + a_4 t_{\bar{x}4}^e = 0$$

$$\frac{\partial t_{\bar{y}}^e}{\partial \bar{y}^e} = b_1 t_{\bar{y}1}^e + b_2 t_{\bar{y}2}^e + b_3 t_{\bar{y}3}^e + b \, t_{\bar{y}4}^e = 0 \tag{11.61}$$

where

$$a_n = \frac{\partial \xi}{\partial \bar{x}} \frac{\partial N_n}{\partial \xi} + \frac{\partial \eta}{\partial \bar{x}} \frac{\partial N_n}{\partial \eta}; \quad b_n = \frac{\partial \xi}{\partial \bar{y}} \frac{\partial N_n}{\partial \xi} + \frac{\partial \eta}{\partial \bar{y}} \frac{\partial N_n}{\partial \eta} \tag{11.62}$$

The auxiliary equations can be implemented into the system of equations in a similar way as for two-dimensional problems. The system of equation takes the form

$$
\begin{bmatrix}
\Delta \mathbf{U}_1^1 & \cdots & \Delta \mathbf{U}_1^2 & \cdots \\
\vdots & & & \\
[\mathbf{a}]^1 & \cdots & 0 & \\
0 & \cdots & [\mathbf{a}]^2 & \cdots
\end{bmatrix}
\begin{Bmatrix}
\mathbf{t}^1 \\
\vdots \\
\mathbf{t}^2 \\
\vdots
\end{Bmatrix}
=
\begin{Bmatrix}
\mathbf{f} \\
\vdots \\
0 \\
0
\end{Bmatrix}
\tag{11.63}
$$

where

$$
[\mathbf{a}]^e = \begin{bmatrix} \mathbf{a}_1^e & \cdots & \mathbf{a}_N^e \end{bmatrix}
\tag{11.64}
$$

N is the number of element nodes and

$$
\mathbf{a}_n^e = \begin{bmatrix}
a_n v_{1x} & a_n v_{1y} & a_n v_{1z} \\
b_n v_{2x} & b_n v_{2y} & b_n v_{2z}
\end{bmatrix}
\tag{11.65}
$$

11.6 IMPLEMENTATION

The implementation of the corner node logic can be described by the following steps:

1. **Detection of corner nodes**
   Corner nodes are detected by the fact that the outward normals computed at a point are significantly different for the elements that connect to it. This can be checked by taking the dot product of the outward normal of one of the elements with the normal vectors of the other elements. If the dot product is smaller than 1.0, then the point is a corner point. A tolerance is applied in order to avoid the additional computational complexity for slight corners. A tolerance of 0.1 is assigned in the program, but this may be changed. This value means that we declare a point to be a corner if the dot product is smaller than 0.9.

2. **Assembly of system of equations**
   The assembly of the system of equations proceeds exactly as for the problem without corners, except that in the assembly of the left-hand side the $\Delta \mathbf{U}$ terms are not added for corner nodes.

3. **Augmentation of system of equations for a region**
   For corner nodes the system of equations is augmented with additional equations for the additional unknowns which have to be determined. The number of auxiliary equations needed depending on the problem to be solved is shown in Table 11.1. Note that this is reduced by Ndof (number of degrees of freedom) for each edge that has

Neumann boundary conditions applied. Therefore, no auxiliary equations are needed if *Neumann* boundary conditions are applied at one edge for 2-D problems and at two edges for 3-D problems. Auxiliary equations are assembled in a similar way as for the region equations, that is the sub-matrices $[\mathbf{a}]^e$ are put in the location corresponding to the element e.

Table 11.1   Number of additional equations per corner node

	Potential problems		Elasticity problems	
	2-D	*3-D*	*2-D*	*3-D*
Aux. equations	1	2	2	6

4. **Assembly of system of equations for the whole problem**
   The assembly of the system of equations proceeds in the same way as for regions without corners, except that at a node where auxiliary equations have been generated the equlibrium is expressed in nodal point forces instead of tractions. This will modify the assembly procedure as shown in Section 11.3.1.

Once the interface displacements have been obtained, we proceed in the same way as for regions without corners, except that when calculating the unknowns for regions with corners discontinuous tractions are obtained there. However, since these tractions are stored element by element this poses no special problems.

11.5.1   Subroutine for detecting corners

The **SUBROUTINE** Detect_corners finds a discontinuity in the tangent at a point where two elements meet. If the outward normal vectors of all elements which connect to a point differ by a set tolerance (currently 10%), then corner information is generated for this point. The information in Infocorn contains the numbers of the elements that connect to the corner and the local numbers of the connecting node. The number of corners detected is returned in Ncorn.

Note that a node detected by this subroutine does not necessarily have discontinuous interface tractions. This will depend upon any *Neumann* boundary conditions assigned to the elements.

```
SUBROUTINE Detect_corners(maxe,Incie,xP,Ncorn,Infocorn)
!----------------------------------------------------
!      Detects corners and generates corner info
!      in Infocorn which contains a list of elements
!      and local node numbers connected to the corner node
!      Maxcont and Ncorn are
```

```fortran
!     assigned in the calling program
!----------------------------------------------------
INTEGER, INTENT(IN) ::   maxe  ! Number of elements to be checked
INTEGER, INTENT(IN) ::   Incie(:,:)  !  Element incidences
REAL, INTENT(IN)    ::   xP(:,:)     !  Node coordinates
INTEGER, INTENT(OUT)::   Ncorn       !  Number of corners detected
INTEGER, INTENT(OUT)::   Infocorn(:,:) ! Information about corners
INTEGER,ALLOCATABLE ::   Nelcon(:,:),Ncount(:),Nodcon(:,:)
INTEGER,ALLOCATABLE ::   IncieT(:,:)
REAL, ALLOCATABLE :: Vnorm(:,:),Elcor(:,:)
REAL :: Tolerance= 0.1 ! Tolerance for detecting corners
INTEGER :: Cdim,Nodes,Nodel,ldim,nel,n,node
INTEGER :: nel1,n1,ncounts,nc,maxcnt
REAL:: xsi,eta,Jac
Cdim= UBOUND(xP,1)       !  Cartesian dimension
Nodes= UBOUND(xP,2)      !  total number of nodes
Nodel= UBOUND(Incie,2)   !  Nodes/Element
Ldim= Cdim-1
ALLOCATE(Nelcon(Nodes,Maxcont),Ncount(Nodes))
ALLOCATE(Nodcon(Nodes,Maxcont))
ALLOCATE(Vnorm(Cdim,Maxcont),Elcor(Cdim,Nodel))
!   Create Table of element numbers connecting to Nodes
IncieT= Incie
Maxcnt=0
Ncount=0
Element_loop: &
DO nel=1,maxe
Nodes_of_Element_loop: &
DO n=1,nodel
  Node= IncieT(nel,n)
  IF(Node == 0) CYCLE    ! This node has already been considered
Element_loop1: &
  DO nel1=nel+1,maxe
    Nodes_of_Element_loop1: &
    DO n1=1,nodel
      IF(IncieT(nel1,n1) == Node) THEN  ! Node with same no. found
        Ncount(Node)= Ncount(Node)+1
        IF(Ncount(Node) > Maxcont) THEN
          CALL Error_Message('Too many Elements on corner')
          STOP

        END IF
        IncieT(nel1,n1)= 0
        Nelcon(Node,Ncount(node)) = nel1 ! Element number
        Nodcon(Node,Ncount(node))= n1    ! Local node number
      END IF
    END DO &
    Nodes_of_Element_loop1
  END DO &
  Element_loop1
END DO &
```

```
Nodes_of_Element_loop
END DO &
Element_loop
!    Check if outward normals are different
Ncorn= 0
Nodes_Total: &
DO Node=1,Nodes
Ncounts= Ncount(Node)
!    compute outward normals for all elements that connect
Count_loop: &
DO Nc=1,Ncounts
  Nel= Nelcon(Node,Nc)
  Elcor(:,:)= xP(:,Incie(Nel,:))
  CALL Local_coor(xsi,eta,Nodcon(Node,nc),1dim)
  CALL Normal_Jac(vnorm(:,Nc),Jac,xsi,eta,1dim,nodel
                 &,incie(nel,:),Elcor)
 END DO &
 Count_loop
 Count_loop1: &
 DO Nc=2,Ncounts
   IF(1.0 - DOT_PRODUCT(Vnorm(:,1),Vnorm(:,Nc)) > Tolerance)&
                             THEN
     Ncorn= Ncorn+1
     IF(Ncorn > Maxcorn) THEN
       CALL Error_Message('No of corners exceeded')
       STOP
     END IF
     Infocorn(Ncorn,1:Ncounts)= Nelcon(Node,1:Ncounts)
     Infocorn(Ncorn,Ncounts+1:Ncounts*2)= Nodcon(Node,1:Ncounts)
     EXIT
   END IF
 END DO &
 Count_loop1
END DO &
Nodes_Total
END SUBROUTINE Detect_corners
```

11.5.2 Subroutine for computing auxiliary equation coefficients

Subroutine Aux_coeff determines the matrices $[\mathbf{a}]^e$ (equation (11.59)) for assembling the auxiliary equations.

Summarising the above definitions, we have

$$[\mathbf{a}]^e = \begin{bmatrix} \mathbf{a}_1^e & \cdots & \mathbf{a}_N^e \end{bmatrix} \tag{11.66}$$

where for potential problems:

$$\mathbf{a}_n^e = \left[\frac{\partial N_n}{\partial \xi} \right] \tag{11.67}$$

for 2-D elasticity problems

$$a_n^e = \left[\frac{\partial N_n}{\partial \xi} v_x \quad \frac{\partial N_n}{\partial \xi} v_y \right] \tag{11.68}$$

and finally, for 3-D elasticity problems

$$\mathbf{a}_n^e = \begin{bmatrix} a_n v_{1x} & a_n v_{1y} & a_n v_{1z} \\ b_n v_{2x} & b_n v_{2y} & b_n v_{2z} \end{bmatrix} \tag{11.69}$$

with

$$a_n = \frac{\partial \xi}{\partial \overline{x}} \frac{\partial N_n}{\partial \xi} + \frac{\partial \eta}{\partial \overline{x}} \frac{\partial N_n}{\partial \eta}; \quad b_n = \frac{\partial \xi}{\partial \overline{y}} \frac{\partial N_n}{\partial \xi} + \frac{\partial \eta}{\partial \overline{y}} \frac{\partial N_n}{\partial \eta} \tag{11.70}$$

```fortran
SUBROUTINE Aux_coeff(NEL,J,Inci,Elcor,AC)
!-------------------------------------------------
!     Computes auxiliary coefficients for
!     additional equations for corners
!-------------------------------------------------
INTEGER, INTENT (IN) :: Nel        !  Element Number
INTEGER, INTENT (IN) :: J          !  local node number corner
INTEGER, INTENT (IN) :: Inci(:)    !  Element Incidences
REAL, INTENT (IN)    :: Elcor(:,:) !  Element coordinates
REAL, INTENT (OUT)   :: AC(:,:)    !  Auxiliary coeff
REAL, ALLOCATABLE    :: Vxsi(:),Veta(:),DNi(:,:),Ni(:)
REAL :: Jxsi,Jeta, vn(3), CosT, SinT,J1,J2,v2(3)
ALLOCATE (Vxsi(cdim),Veta(cdim),Dni(Nodes,Ldim),Ni(Nodes))
!   Assign xsi,eta coords of node
CALL Local_coor(xsi,eta,j,ldim)
!   Compute derivatives of shape function
CALL Serendip_deriv(DNi,xsi,eta,ldim,nodes,inci)

IF(Ndof == 1) THEN   !   Potential problems
AC= Dni(:,1)
 RETURN
END IF
IF(Cdim == 2) THEN    !   2-D elasticity problems
   Vxsi(1)= Dot_Product(Dni(:,1),Elcor(1,:))
   Vxsi(2)= Dot_Product(Dni(:,1),Elcor(2,:))
   CALL Vector_norm(Vxsi,Jxsi)
   AC(1:Nodel:2)= Dni(:,1)*Vxsi(1)
   AC(2:Nodel+1:2)= Dni(:,1)*Vxsi(2)
   RETURN
END IF
IF(Cdim == 3) THEN
```

```
     Vxsi(1)= Dot_Product(Dni(:,1),Elcor(1,:))
     Vxsi(2)= Dot_Product(Dni(:,1),Elcor(2,:))
     Vxsi(3)= Dot_Product(Dni(:,1),Elcor(3,:))
     CALL Vector_norm(Vxsi,Jxsi)
     Veta(1)= Dot_Product(Dni(:,2),Elcor(1,:))
     Veta(2)= Dot_Product(Dni(:,2),Elcor(2,:))
     Veta(3)= Dot_Product(Dni(:,2),Elcor(3,:))
     CALL Vector_norm(Veta,Jeta)
     Vn= Vector_ex(Vxsi,Veta)
     V2= Vector_ex(Vxsi,vn)
     J1= Jxsi
     CALL Vector_norm(V2,J2)
     CosT= Dot_product(Vxsi,Veta)
     SinT= Dot_product(V2,Veta)
     Dxsidx= 1/J1
     Detadx= 1/J1*Cost/SinT
     Dxsidy= 0.
     Detady= 1/(J2*sinT)
     Element_nodes:&
  DO n=1,nodel
     an= dxsidx*DNi(n,1)+detadx*DNi(n,2)
     bn= dxsidy*DNi(n,1)+detady*DNi(n,2)
     AC((n-1)*3:n*3,1)= an*Vxsi(:)
     AC((n-1)*3:n*3,2)= an*V2(:)
  END DO &
  Element_nodes
  END IF
  RETURN
  END SUBROUTINE Aux_coeff
```

11.6   CONCLUSIONS

In this chapter we have dealt with a problem which is unique to the BEM, and which does not arise in the FEM, namely edges and corners. This is because in the BEM, tractions are used, whereas in the FEM, nodal forces are used. These nodal point forces can be visualised as the resultants of tractions. Whereas for nodal point forces the geometrical shape at the corner is completely irrelevant, tractions which are distributed forces over a boundary surface require special attention. However, this is only necessary when two or more surfaces meet at a point where *Dirichlet* or interface boundary conditions are specified on two sides. The latter may occur in a multi-region analysis if this situation is brought about by a discretisation of the problem into regions for example because material properties change inside the continuum. The tractions are multi-valued at such points, but the integral equations do not supply enough equations to compute them.

The problems of edges and corners has not received a great deal of attention, and in many cases, schemes have been suggested which either introduce additional errors or cause additional effort in the discretisation and solution of equations. We believe that the proposed scheme is fairly straightforward to implement, and because it does not require auxiliary nodes, is also easy to use.

11.6   REFERENCES

1. Bonnet, M. (1995) *Boundary Integral Equation Methods for Solids and Fluids*. Wiley, Chichester.
2. Patterson, C. and Sheikh, M.A. (1984) Interelement continuity in the boundary element method. *Topics in Boundary Element Research* (C.A. Brebbia ed), vol. 1, Springer-Verlag, Berlin.
3. Beer, G. and Watson, J.O. (1995) *Introduction to Finite and Boundary Element Methods for Engineers*. Wiley, Chichester.
4. Gao, X.W. and Davies, T. (2001) *Boundary Element Programming in Mechanics*. Cambridge University Press, London.

12

Body Forces

Oh heavenly bodies!
J. Keppler

12.1 INTRODUCTION

The advantages of the boundary element method over the FEM, that no elements are required inside the domain, also has some disadvantages: loading may only be applied at the boundary but not inside the domain. A number of problems exist where the consideration of applying loading inside the domain is necessary, for example

- where a heat or fluid source has to be considered inside the domain
- where self weight or centrifugal forces exist
- where initial strains are applied inside the domain, for example when material is subjected to swelling.
- where forces are being generated inside the domain, for example ones created by acceleration of mass as they occur in dynamics.

In addition, as we will see later, for the analysis of domains exhibiting non-linear material behaviour, for which we cannot find fundamental solutions, the problem can be considered as one where initial stresses are generated inside the domain.

In this chapter we will discuss methods which allow us to consider such loads, known in general terms as body forces. Here we will distinguish between those which are constant, such as, self weight, and those which vary inside the domain. We will find that we can deal with constant body forces in a fairly straightforward way, since the volume integrals which occur can be transformed into surface integrals. In the case where they are not constant, however, the only way to deal with the volume integration is by providing additional volume discretisation.

We will start this chapter by revisiting the Betti theorem as derived for the integral equations, but now considering the additional effect of body forces.

12.2 GRAVITY

First we deal with gravity forces, for example, those generated by the self weight. If the material is homogeneous then these forces are constant inside the domain. We expand Betti's theorem used in Chapter 4 to derive the integral equations by taking into consideration the effect of body forces.

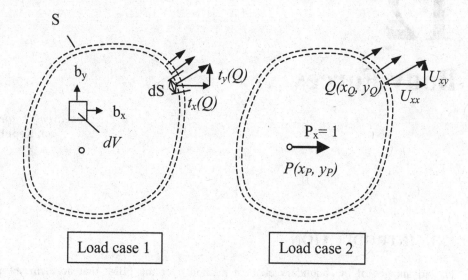

Figure 12.1   Application of Betti's theorem including the effect of body forces

As shown in Figure 12.1 for 2-D problems, the forces of load case one consist of boundary tractions **t** (components t_x and t_y) and of body forces **b** (components b_x, b_y) which are defined as forces per unit volume.

The work done by the loads of load case one times the displacements of load case two W_{12} is computed by

$$W_{12} = \int_S (t_x(Q) U_{xx}(P,Q) + t_y(Q) U_{xy}(P,Q)) dS(Q) + \int_V (b_x U_{xx} + b_y U_{xy}) dV \qquad (12.1)$$

and the work W_{21} is

$$W_{21} = \int_S \left[u_x(Q) T_{xx}(P,Q) + u_y(Q) T_{xy}(P,Q) \right] dS + 1\, u_x(P) \qquad (12.2)$$

The integral equations including the body force effect can be written as:

$$\mathbf{u}(P) = \int_S \mathbf{t}^T(Q) \mathbf{U}(P,Q)\, dS - \int_S \mathbf{u}^T(Q) \mathbf{T}(P,Q)\, dS + \int_V \mathbf{b}^T(Q) \mathbf{U}(P,Q)\, dV \qquad (12.3)$$

where the last integral in equation (12.3) is a volume integral. It can be shown[1] that for body forces which are constant over volume V, this integral can be transformed into a surface integral

$$\int_V \mathbf{b}^T(Q)\mathbf{U}(P,Q)dV = \int_S \mathbf{G}dS \qquad (12.4)$$

where for 2-D and 3-D problems

$$\mathbf{G} = \begin{bmatrix} G_x \\ G_y \end{bmatrix}; \quad \mathbf{G} = \begin{bmatrix} G_x \\ G_y \\ G_z \end{bmatrix} \qquad (12.5)$$

For 3-D problems the coefficients of $\mathbf{G}$ may be computed from[1]

$$G_i = \frac{1}{8\pi\,\mathrm{G}}\left(b_i\cos\theta - \frac{1}{2(1-v)}n_i\cos\psi \right) \quad i=x,y,z \qquad (12.6)$$

where $\mathbf{G}$ is the shear modulus and

$$\cos\theta = \mathbf{n}\bullet\frac{1}{r}\mathbf{r}; \quad \cos\psi = \mathbf{b}\bullet\frac{1}{r}\mathbf{r} \qquad (12.7)$$

Vectors $\mathbf{n}$ and $\mathbf{r}$ and scalar r have already been defined in Chapter 4. For plane strain problems we have[1]:

$$G_i = \frac{1}{8\pi G}\left(2\ln\frac{1}{r}-1\right)\left(b_i\cos\theta - \frac{1}{2(1-v)}n_i\cos\psi\right) \quad i=x,y \qquad (12.8)$$

The discretised form of equation (10.3) can be written as

$$\mathbf{cu}(P_i)+\sum_{e=1}^{E}\sum_{n=1}^{N}\Delta\mathbf{T}_{ni}^e\,\mathbf{u}_n^e = \sum_{e=1}^{E}\sum_{n=1}^{N}\Delta\mathbf{U}_{ni}^e\,\mathbf{t}_n^e + \sum_{e=1}^{E}\Delta\mathbf{G}^e \qquad (12.9)$$

where E is the number of elements, N the number of element nodes and

$$\Delta\mathbf{G}^e = \int_{S_e}\mathbf{G}(P,Q)\,dS(Q) \qquad (12.10)$$

The integral may be evaluated using numerical integration.
For 2-D problems we can write

$$\Delta\mathbf{G}^e \approx \sum_{m=1}^{M}\mathbf{G}(P,Q(\xi_m))J(\xi_m)W_m \qquad (12.11)$$

and for 3-D problems

$$\Delta \mathbf{G}^e \approx \sum_{k=1}^{K} \sum_{m=1}^{M} \mathbf{G}(P, Q(\xi_m, \eta_k)) J(\xi_m, \eta_k) W_k W_m \tag{12.12}$$

where J is the Jacobian and W_k, W_m are the Gaussian weights introduced in Section 3.9.1. For the three-dimensional case no singularity occurs as P approaches Q, and therefore the minimum integration order (K, M) with which we are able to accurately compute the surface area of the element can be used. The analysis of problems with constant body forces proceeds the same way as before except that an additional right-hand side term is assembled as shown in equation (12.9).

12.2.1 Postprocessing

In postprocessing the effect of body forces has to be included. For the calculation of displacements at point P_a, equation (8.44) is modified to

$$\mathbf{u}(P_a) = \sum_{e=1}^{E} \sum_{n=1}^{N} \Delta \mathbf{U}_n^e \, \mathbf{t}_n^e - \sum_{e=1}^{E} \sum_{n=1}^{N} \Delta \mathbf{T}_n^e \, \mathbf{u}_n^e + \sum_{e=1}^{E} \Delta \mathbf{G}^e \tag{12.13}$$

whereas for computation of stresses we have

$$\boldsymbol{\sigma}(P_a) = \sum_{e=1}^{E} \sum_{n=1}^{N} \Delta \mathbf{S}_n^e \, \mathbf{t}_n^e - \sum_{e=1}^{E} \sum_{n=1}^{N} \Delta \mathbf{R}_n^e \, \mathbf{u}_n^e + \sum_{e=1}^{E} \Delta \hat{\mathbf{S}}^e \tag{12.14}$$

where

$$\Delta \hat{\mathbf{S}}^e = \int_{S_e} \hat{\mathbf{S}}(P_a, Q) dS(Q) \tag{12.15}$$

Matrix $\mathbf{S}$ obtained by differentiation of equations (12.6) or (12.8) is given by

$$\hat{\mathbf{S}} = \begin{Bmatrix} S_{xx} \\ S_{yy} \\ S_{xy} \end{Bmatrix} \text{ for } 2-D \text{ and } \hat{\mathbf{S}} = \begin{Bmatrix} S_{xx} \\ S_{yy} \\ S_{zz} \\ S_{xy} \\ S_{yz} \\ S_{xz} \end{Bmatrix} \text{ for } 3-D \tag{12.16}$$

For 3-D problems we have[1]

$$\hat{S}_{ij} = \frac{1}{8\pi r}\left\{\begin{array}{l} cos\theta(b_i r_j + b_j r_i) + \dfrac{1}{1-v}v\delta_{ij}(cos\theta\,cos\psi - cos\phi) \\ -\dfrac{1}{2}\big[cos\psi(n_i r_j + n_j r_i) + (1-2v)(b_i n_j + b_j n_i)\big] \end{array}\right\} \quad i,j = x,y,z \qquad (12.17)$$

where $cos\theta$ and $cos\psi$ have been defined previously in equation (12.7), r_x, r_y, r_z is as defined in (8.42), δ_{ij} is the Kronecker delta defined in equation (2.2) and

$$cos\phi = \mathbf{b}\bullet\mathbf{n} \qquad (12.18)$$

For plane strain problems we have[1]

$$\hat{S}_{ij} = \frac{1}{8\pi}\left\{\begin{array}{l} 2cos\theta(b_i r_j + b_j r_i) \\ +\dfrac{1}{1-v}\left[v\delta_{ij}\left(\begin{array}{l}2cos\theta\,cos\psi \\ +(1-2ln\dfrac{1}{r})cos\phi\end{array}\right) - cos\psi(n_i r_j + n_j r_i) \right. \\ \left. +\dfrac{1-2v}{2}\left(1-2ln\dfrac{1}{r}\right)(b_i n_j + b_j n_i)\right] \end{array}\right\} \quad i,j = x,y \qquad (12.19)$$

Two subroutines, Grav_dis and Grav_stress, which compute matrices $\mathbf{G}$ and $\hat{\mathbf{S}}$, needed for the gravity load case, are added to the library Elasticity_lib. The subroutines can be used to compute the element contributions for assembly of the right-hand side and for the internal stress computation.

```
SUBROUTINE Grav_dis(GK,dxr,r,Vnor,b,G,ny)
!------------------------------------------------
!    FUNDAMENTAL SOLUTION FOR Displacements
!    Gravity Loads(Kelvin solution)
!------------------------------------------------
IMPLICIT NONE
REAL              :: GK(:)            !  Fundamental solution
REAL,INTENT(IN)   :: dxr(:)           !  rx/r etc.
REAL,INTENT(IN)   :: r                !
REAL,INTENT(IN)   :: Vnor(:)          !  normal vector
REAL,INTENT(IN)   :: b (:)            !  gravity force vector
REAL,INTENT(IN)   :: G                !  Shear modulus
REAL,INTENT(IN)   :: ny               !  Poisson's ratio
INTEGER           :: Cdim             !  Cartesian dimension
REAL              :: c1,c2,costh,Cospsi !  Temps
```

```fortran
C1= 1.0/(8*Pi*G)
C2=1.0/(2.0*(1.0-ny))
Costh=  DOT_PRODUCT(Vnor ,DXR); Cospsi= DOT_PRODUCT(b,DXR)
 IF(Cdim == 2) THEN
 C1= C1*(2.0*LOG(1.0/r)-1.0)
 GK= C1*(b*costh - C2*Vnor*cospsi)
ELSE
 GK= C1*(b*costh - C2*Vnor*cospsi)
END IF
RETURN
END


SUBROUTINE Grav_stress(SK,dxr,r,Vnor,b,G,ny)
!-------------------------------------------------
!    FUNDAMENTAL SOLUTION FOR Stresses
!    Gravity Loads(Kelvin solution)
!-------------------------------------------------
IMPLICIT NONE
REAL             :: SK(:)          !    Kernel
REAL,INTENT(IN) :: dxr(:)         !    rx/r etc.
REAL,INTENT(IN) :: r              !
REAL,INTENT(IN) :: Vnor(:)        !    normal vector
REAL,INTENT(IN) :: b (:)          !    body force vector
REAL,INTENT(IN) :: G              !    Shear modulus
REAL,INTENT(IN) :: ny             !    Poisson's ratio
INTEGER          :: Cdim          !    Cartesian dimension
INTEGER          :: II(6),JJ(6) !  Order of stress components
REAL             :: c,c1,c2,c3,c4,costh,Cospsi,Cosphi   !   Temps
C2=1.0/(1.0-ny)
C3= 1-2.0*ny ; C5= (1-2.0*ny)/2.0
Costh=  DOT PRODUCT(Vnor ,DXR)
Cospsi= DOT PRODUCT(b,DXR)
Cosphi= DOT PRODUCT(b,Vnor)
IF(Cdim == 2) THEN      !   Two-dimensional solution
 C1= 1.0/(8*Pi)
 C4= 1.0 - 2.0*LOG(1.0/r)
 II(1:3)= (/1,2,1)
 JJ(1:3)= (/1,2,2)
 Stress_components: &
 DO N=1,3
  I= II(N) ; J= JJ(N)
  IF(I == J) THEN
   C= ny*(2.0*costh*cospsi + C4*cosphi)
  ELSE
   C= 0.0
  END IF
  SK(N)= 2.0*COSth*(b(i)*dxr(j)+ b(j)*dxr(i))&
       + C2*(C - cospsi*(Vnor(i)*dxr(j)+ Vnor(j)*dxr(i)) &
       + C5*c4*(b(i)*vno(j)+b(j)*vnor(i)  ; SK(N)= SK(n)*C1
 END DO
 Stress_components
```

```
   SK= C1*SK
  ELSE                        !    Three-dimensional solution
     II= (/1,2,3,1,2,3)
   JJ= (/1,2,3,2,3,1)
     C1= 1.0/(8*Pi*r)
   Stress_components1: &
    DO N=1,6
     I= II(N) ; J= JJ(N)
     C=0.
     IF(I == J) THEN
      C= c2*ny*(costh*cospsi-cosphi)
     ELSE
       C= 0.0
     END IF
     SK(N)= 2.0*costh*(b(I)dxr(J)+ b(J)dxr(I))+ C &
         -0.5 *(cospsi*(Vnor(I)*dxr(J)+ Vnor(J)*dxr(I))&
                  + C3*(b(I)*Vnor(J)  + b(J)*Vnor(I)))
   END DO &
   Stress_components1
   SK= C1*SK
  END IF
  RETURN
  END
```

12.3 INITIAL STRAINS

There are often problems where strains are generated inside a domain that are not associated with loading by forces. Examples are thermal strains generated by a temperature increase and strains which are due to swelling of soil due to water ingress. Invariably these strains will not be constant over the whole domain, as was the case with gravity forces. Therefore, it will no longer be possible to transform the volume integrals which arise into surface integrals.

In this section we will introduce the concept of volume cells for the evaluation of volume integrals. We find that these cells are identical to finite elements. We start with the Betti theorem, but instead of a body force we consider initial strains applied inside the domain. It is obvious that if these strains exist, additional work will be carried out. Referring to Figure 12.2, the work done by the displacements/strains of load case one times the forces/stresses of load case two is given by:

$$W_{12} = \int_{S} (u_x(Q)T_{xx}(P,Q) + u_y(Q)T_{xy}(P,Q))dS(Q)$$

$$+ \iint \varepsilon_{x0}dx\,\Sigma_{xx}(P,\overline{Q})dy + \varepsilon_{y0}dy\,\Sigma_{yx}(P,\overline{Q})dx$$

(12.20)

where $\Sigma_{xx}(P,\overline{Q})$ and $\Sigma_{yx}(P,\overline{Q})$ are the fundamental solutions for stresses at $\overline{Q}$ due to a unit x-force at P. Here we assume that only volumetric initial strains are present, even though shear strains could easily be included. The work done by the displacements of load case two, times the forces/stresses of load case one, is the same as for the case where no initial strains are applied. After placing the unit load in the y-direction, we obtain for 2-D as well as for 3-D problems

$$\mathbf{u}(P)=\int_S \mathbf{t}^T(Q)\mathbf{U}(P,Q)\,dS-\int_S \mathbf{u}^T(Q)\mathbf{T}(P,Q)\,dS+\int_V \mathbf{\Sigma}(P,\overline{Q})\mathbf{\varepsilon}_0(\overline{Q})\,dV \qquad (12.21)$$

where the last integral is a volume integral.

For 3-D problems, matrix $\mathbf{\Sigma}$ is given by

$$\mathbf{\Sigma}=\begin{bmatrix} \Sigma_{xx} & \Sigma_{xy} & \Sigma_{xz} \\ \Sigma_{yx} & \Sigma_{yy} & \Sigma_{yz} \\ \Sigma_{zx} & \Sigma_{zy} & \Sigma_{zz} \end{bmatrix} \qquad (12.22)$$

where[2]

$$\Sigma_{kk}=\frac{C_2}{r^n}\left(C_3 r_k+C_4 r_k^3\right); \quad \Sigma_{kj}=\frac{C_2}{r^n}\left(-C_3 r_k+C_4 r_k^2 r_j\right)$$
$$\text{for} \quad k,j=x,y,z \qquad (12.23)$$

The values for the constants are given in Table 12.1.

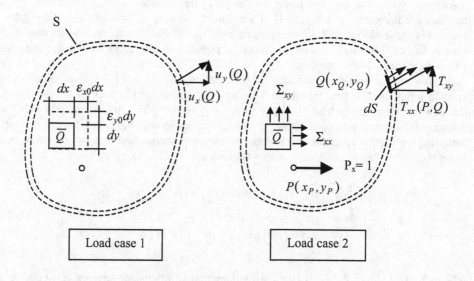

Figure 12.2   Application of the Betti theorem including the effect of initial strains

Table 12.1   Constants for fundamental solution initial strains

	Plane strain	Plane stress	3-D
n	1	1	2
C_2	$1/4\pi(1-\nu)$	$(1+\nu)/4\pi$	$1/8\pi(1-\nu)$
C_3	$1-2\nu$	$(1-\nu)/(1+\nu)$	$1-2\nu$
C_4	2	2	3

A FUNCTION for computing Matrix Σ is written and added to the Elasticity_lib. FUNCTION SigmaK returns an array of dimension 2 x 2 or 3 x 3 with fundamental solutions for normal stresses.

```
FUNCTION SigmaK(dxr,r,E,ny,Cdim)
!-------------------------------------------------
!   FUNDAMENTAL SOLUTION FOR Normal Stresses
!   isotropic material (Kelvin solution)
!-------------------------------------------------
REAL,INTENT(IN)    :: dxr(:)              ! rx/r etc.
REAL,INTENT(IN)    :: r                   ! r
REAL,INTENT(IN)    :: E                   ! Young's modulus
REAL,INTENT(IN)    :: ny                  ! Poisson's ratio
INTEGER,INTENT(IN)                        :: Cdim            !
                 Cartesian dimension
REAL               :: SigmaK(Cdim,Cdim) ! Returns array CdimxCdim
INTEGER            :: n,i,j
REAL               :: G,c,c2,c3,c4        ! Temps
G= E/(2.0*(1+ny))
SELECT CASE (Cdim)
 CASE (2)         !       Plane strain solution
   n= 1
   c2= 1.0/(4.0*Pi*(1.0-ny))
   c3= 1.0-2.0*ny
   c4= 2.0
 CASE(3)          !       Three-dimensional solution
   n= 2
   c2= 1.0/(8.0*Pi*(1.0-ny))
   c3= 1.0-2.0*ny
   c4= 3.0
 CASE DEFAULT
END SELECT
Direction_Pi: &
DO i=1,Cdim
  Direction_Sigma: &
  DO j=1,Cdim
   IF(i == j) THEN
     SigmaK(i,i)= c2/r**n*(c3*dxr(i)+c4*dxr(i)**3)
   ELSE
     SigmaK(j,i)= c2/r**n*(-c3*dxr(j)+c4*dxr(j)*3*dxr(i))
   END IF
```

```
END DO &
Direction_Sigma
END DO &
Direction_Pi
RETURN
END FUNCTION SigmaK
```

The discretised form of equation (10.3) can be written as

$$\mathbf{c}u(P_i) + \sum_{e=1}^{E}\sum_{n=1}^{N}\Delta\mathbf{T}_{ni}^{e}\,\mathbf{u}_{n}^{e} = \sum_{e=1}^{E}\sum_{n=1}^{N}\Delta\mathbf{U}_{ni}^{e}\,\mathbf{t}_{n}^{e} + \int_{V}\mathbf{\Sigma}\,\mathbf{\varepsilon}_0\,dV \qquad (12.24)$$

or in matrix form for the case where only *Neumann* boundary conditions are specified

$$[\mathbf{T}]\{\mathbf{u}\}=\{\mathbf{F}\}+\{\mathbf{F}\}_{\varepsilon} \qquad (12.25)$$

We propose to evaluate this integral numerically with the Gauss Quadrature method that has already been used for evaluating boundary integrals.

To apply this method, however, the volume where initial strains are specified needs to be discretised, i.e. subdivided into elements. We use two-dimensional elements for the discretisation of 2-D problems and three-dimensional elements for 3-D problems.

12.3.1 Volume cells

The cell elements used for the evaluation of volume integrals are identical to the isoparametric elements used in the finite element method. In Figure 12.3 we show a linear cell element used for 2-D problems with local node numbering and intrinsic coordinates. For 3-D problems we use the cell element in Figure 12.4.

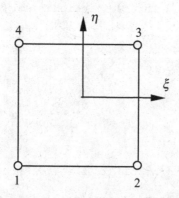

Figure 12.3   Cell element for 2-D problems

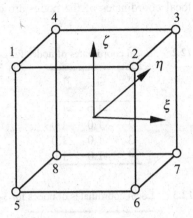

Figure 12.4   Isoparametric cell element with linear shape function

The interpolation of the geometry and the strain is given by

$$\mathbf{x} = \sum_{n=1}^{4(8)} N_n \mathbf{x}_n^e \quad \text{and} \quad \boldsymbol{\varepsilon}_0 = \sum_{n=1}^{4(8)} N_n \boldsymbol{\varepsilon}_{0n}^e \qquad (12.26)$$

where for 2-D problems

$$\mathbf{x} = \begin{Bmatrix} x \\ y \end{Bmatrix} \quad \text{and} \quad \boldsymbol{\varepsilon}_0 = \begin{Bmatrix} \varepsilon_{x0} \\ \varepsilon_{yx0} \end{Bmatrix} \qquad (12.27)$$

For 3-D problems we have

$$\mathbf{x} = \begin{Bmatrix} x \\ y \\ z \end{Bmatrix} \quad \text{and} \quad \boldsymbol{\varepsilon}_0 = \begin{Bmatrix} \varepsilon_{x0} \\ \varepsilon_{y0} \\ \varepsilon_{z0} \end{Bmatrix} \qquad (12.28)$$

$\mathbf{x}_n^e$ and $\boldsymbol{\varepsilon}_{0n}^e$ are the values of coordinates and initial strains at the nth node of element e.
The shape functions are defined as

$$N_n = \frac{1}{4}(1 + \xi_n \xi)(1 + \eta_n \eta) \qquad (12.29)$$

for 2-D problems, and

$$N_n = \frac{1}{8}(1 + \xi_n \xi)(1 + \eta_n \eta)(1 + \zeta_n \zeta) \qquad (12.30)$$

for 3-D problems, where local coordinates of the nodes are defined in Tables 12.2 and 12.3.

Table 12.2   Local coordinates of nodes for 2-D cells

n	ξ_n	η_n
1	-1.0	-1.0
2	1.0	-1.0
3	1.0	1.0
4	-1.0	1.0

Table 12.3   Local coordinates of nodes for 3-D cells

n	ξ_n	η_n	ζ_n
1	-1.0	-1.0	+1.0
2	1.0	-1.0	1.0
3	1.0	1.0	1.0
4	-1.0	1.0	1.0
5	-1.0	-1.0	-1.0
6	1.0	-1.0	-1.0
7	1.0	1.0	-1.0
8	-1.0	1.0	-1.0

12.3.2 Numerical evaluation of volume integrals

For the purpose of explaining the treatment of initial strains, we assume that the initial strains are constant within the domain discretised by volume cells, although it is obvious that an initial strain which varies over the domain can also be easily considered, as will be shown in the next section.

The volume integral can be replaced by a sum over elements

$$\{F\}_\varepsilon = \int_V \Sigma \, \varepsilon dV = \sum_{c=1}^{N_c} \Delta\Sigma^c \varepsilon_0 \tag{12.31}$$

where c is the cell number and N_c is the number of volume cells.

For two-dimensional problems we define

$$\Delta\Sigma_n^c = \int_{S_c} \Sigma \, dS_c = \int_{-1}^{1}\int_{-1}^{1} \Sigma \, J \, d\xi \, d\eta \tag{12.32}$$

For 2-D elements the Jacobian J is given by

$$J = \frac{\partial x}{\partial \xi}\frac{\partial y}{\partial \eta} - \frac{\partial y}{\partial \xi}\frac{\partial x}{\partial \eta} \tag{12.33}$$

whereas for 3-D problems we have

$$J = det \begin{vmatrix} \dfrac{\partial x}{\partial \xi} & \dfrac{\partial y}{\partial \xi} & \dfrac{\partial z}{\partial \xi} \\[2mm] \dfrac{\partial x}{\partial \eta} & \dfrac{\partial y}{\partial \eta} & \dfrac{\partial z}{\partial \eta} \\[2mm] \dfrac{\partial x}{\partial \zeta} & \dfrac{\partial y}{\partial \zeta} & \dfrac{\partial z}{\partial \zeta} \end{vmatrix} \tag{12.34}$$

where

$$\frac{\partial x}{\partial \xi} = \sum_{n=1}^{8} \frac{\partial N_n}{\partial \xi} x_n, \text{ etc.} \tag{12.35}$$

The integral in equation (12.32) may be evaluated numerically as

$$\Delta \Sigma^e \approx \sum_{m=1}^{M} \sum_{k=1}^{K} \Sigma(P, \overline{Q}(\xi_m, \eta_k)) J(\xi_m, \eta_k) W_m W_k \tag{12.36}$$

where M and K are the number of Gauss points in the ξ and η directions. For 3-D problems we have

$$\Delta \Sigma_n^c = \int_{S_c} \Sigma \, dS_c = \int_{-1}^{1} \int_{-1}^{1} \int_{-1}^{1} \Sigma \, J \, d\xi \, d\eta \, d\zeta \tag{12.37}$$

The numerical evaluation is given by

$$\Delta \Sigma^e \approx \sum_{l=1}^{L} \sum_{m=1}^{M} \sum_{k=1}^{K} \Sigma(P, \overline{Q}(\xi_m, \eta_k, \zeta_l)) |J(\xi_m, \eta_k, \zeta_l)| W_m W_k W_l \tag{12.38}$$

where L, M, K are the number of Gauss points in the ξ, η and ζ directions.

12.3.3 Postprocessing

In postprocessing the effect of the initial strains has to be included. For the calculation of displacements, equation (8.44) is modified to

$$\mathbf{u}(P_a) = \sum_{e=1}^{E} \sum_{n=1}^{N} \Delta \mathbf{U}_n^e \, \mathbf{t}_n^e - \sum_{e=1}^{E} \sum_{n=1}^{N} \Delta \mathbf{T}_n^e \, \mathbf{u}_n^e + \sum_{c=1}^{N_c} \Delta \Sigma^e \boldsymbol{\varepsilon}_0 \tag{12.39}$$

For the computation of stresses at point P_a we can write

$$\sigma(P_a) = \sum_{e=1}^{E}\sum_{n=1}^{N}\Delta\mathbf{S}_n^e\,\mathbf{t}_n^e - \sum_{e=1}^{E}\sum_{n=1}^{N}\Delta\mathbf{R}_n^e\,\mathbf{u}_n^e + \sum_{c=1}^{N_c}\Delta\hat{\boldsymbol{\Sigma}}^c\boldsymbol{\varepsilon}_0 \qquad (12.40)$$

The matrices $\Delta\hat{\boldsymbol{\Sigma}}^e$ are computed by taking derivatives of the displacements and by applying Hooke's law

$$\Delta\hat{\boldsymbol{\Sigma}}^c = \int_{S_c}\hat{\mathbf{E}}(P_a,Q)\mathbf{D}dS(Q) \qquad (12.41)$$

where $\mathbf{D}$ is the elasticity matrix and $\hat{\mathbf{E}}$ is obtained by differentiation of (12.39)

$$\hat{\mathbf{E}} = \begin{bmatrix} \hat{E}_{xxx} & \hat{E}_{yxx} \\ \hat{E}_{xyy} & \hat{E}_{yyy} \\ \hat{E}_{xxy} & \hat{E}_{yxy} \end{bmatrix}^{T} \quad \text{for} \quad 2-\text{D} \quad \text{and} \quad \hat{\mathbf{E}} = \begin{bmatrix} \hat{E}_{xxx} & \hat{E}_{yxx} & \hat{E}_{zxx} \\ \hat{E}_{xyy} & \hat{E}_{yyy} & \hat{E}_{zyy} \\ \hat{E}_{xzz} & \hat{E}_{yzz} & \hat{E}_{zzz} \\ \hat{E}_{xxy} & \hat{E}_{yxy} & \hat{E}_{zxy} \\ \hat{E}_{xyz} & \hat{E}_{yyz} & \hat{E}_{zyz} \\ \hat{E}_{xxz} & \hat{E}_{yxz} & \hat{E}_{zxzz} \end{bmatrix}^{T} \quad \text{for} \quad 3-\text{D} \quad (12.42)$$

The expression for $\hat{\mathbf{E}}$ is given by[3]

$$\hat{E}_{ijk} = \frac{C_2}{r^n}\left[\begin{matrix} C_3\left(\delta_{ik}\delta_{kj} + \delta_{jk}\delta_{ki} + \delta_{ij} + n\delta_{ij}r_k^2\right) + \\ n\nu\left(\delta_{ij}r_jr_k + \delta_{jk}r_kr_j + \delta_{ik}r_kr_j + \delta_{jk}r_ir_k\right) + n\left(r_ir_j - C_6r_ir_jr_k^2\right) \end{matrix}\right] \qquad (12.43)$$

where x,y,z may be substituted for $i,j,k,$), r_x, r_y, r_z is as defined in equation (8.42), δ_{ij} is the Kronecker Delta as defined in equation (2.2) and the constants are given in Table 12.4.

Table 12.4   Constants for fundamental solution $\hat{\mathbf{E}}$

	Plane strain	Plane stress	3-D
n	2	2	3
C_2	$1/4\pi(1-\nu)$	$(1+\nu)/4\pi$	$1/8\pi(1-\nu)$
C_3	$1-2\nu$	$(1-\nu)/(1+\nu)$	$1-2\nu$
C_6	4	4	5

Note that functions $\hat{\mathbf{E}}$ are *hypersingular,* and this has to be considered in the selection of the number of integration points for the numerical evaluation. If an interior point coincides exactly with the node point of a cell, then a limiting value of the integral has to be taken. We will discuss this further in Chapter 13 (non-linear problems).

We now add a FUNCTION EK for computation of $\hat{\mathbf{E}}$ to the Elasticity_lib. The function returns an array of dimension 3 x 2 for 2-D problems and 6 x 3 for 3-D problems.

```
FUNCTION EK(dxr,r,E,ny,Cdim,Nstres)
!-------------------------------------------------
!    Fundamental solution derivative
!    for the computation of interior stresses
!    effect of volumetric initial strains
!    isotropic material (Kelvin solution)
!-------------------------------------------------
REAL,INTENT(IN)      :: dxr(:)        !   rx/r etc.
REAL,INTENT(IN)      :: r             ! distance to P
REAL,INTENT(IN)      :: E             ! Young's modulus
REAL,INTENT(IN)      :: ny            ! Poisson's ratio
INTEGER,INTENT(IN) :: Cdim            ! Cartesian dimension
INTEGER, INTENT (IN):: Nstres         ! No. of stress components
REAL                 :: EK(Nstres,Cdim) ! returns array of dim Cdim
iNTEGRE              :: i,j,k
REAL                 :: G,c,c2,c3,c6,n,dij,dik,djk  !   Temps
INTEGER :: II(6), JJ(6) !  sequence of stresses in pseudo-vector
G= E/(2.0*(1+ny))
SELECT CASE (Cdim)
  CASE (2)              !       Plane strain solution
   n= 2
   c2= 1.0/(4.0*Pi*(1.0-ny))
   c3= 1.0-2.0*ny
   c6= 4.0
   II(1:3)= (/1,2,1/)
   JJ(1:3)= (/1,2,2/)
  CASE(3)               !       Three-dimensional solution
   n= 3
   c2= 1.0/(8.0*Pi*(1.0-ny))
   c3= 1.0-2.0*ny
   c6= 5.0
   II= (/1,2,3,1,2,3/)
   JJ= (/1,2,3,2,3,1/)
 CASE DEFAULT
END SELECT
Direction_Eps: &
DO k=1,Cdim
 Direction_Sigma: &
DO n=1,Nstres
   i= II(n)
   j= JJ(n)
   dij= 0.0
```

```
  IF(i==j) dij=1.0
 dik= 0.0
 IF(i==k) dik=1.0
 djk= 0.0
 IF(j==k) djk= 1.0
 C= C3*(dik*djk+djk*dik-dij+n*dij*dxr(k)**2)+ &
  n*ny*(dik*dxr(j)*dxr(k)+djk*dxr(k)*dxr(i)+ &
  dik*dxr(k)*dxr(j)+djk*dxr(i)*dxr(k))+ &
  n*(dxr(i)*dxr(j)-C6*dxr(i)*dxr(j)*dxr(k)**2)
  EK(n,i)= C2/r**n*C
 END DO &
 Direction_Sigma
 END DO &
 Direction_Eps
 RETURN
 END FUNCTION EK
```

12.4 INITIAL STRESSES

The last type of body forces considered here are initial stresses. These can be thought of occurring due to non-linear material response. As will be seen later in the chapter on plasticity, these correspond to the plastic stresses generated when the stresses at a point reach the elastic limit. The capability to deal with initial stresses, therefore, will be important for the application of the BEM to non-linear material response, in particular plasticity. The consideration of initial stresses follows a similar line as the consideration of initial strains.

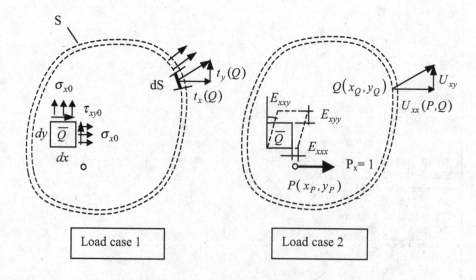

Figure 12.5   Application of Betti 's theorem, of including the effect of initial strains

We again start with the theorem of Betti, but instead of initial strains we consider stresses σ_0 applied inside the domain. If they exist, it is obvious that additional work will be done.

Referring to Figure 12.5, the work done by the tractions/stresses of load case one, times the displacement/strains of load case two, is given by:

$$W_{12} = \int_S (t_x(Q)U_{xx}(P,Q) + t_y(Q)U_{xy}(P,Q))dS(Q)$$

$$+ \iint \sigma_{x0}dyE_{xxx}(P,\overline{Q})dx + \sigma_{y0}dxE_{xyy}(P,\overline{Q})dy + \tau_{xy0}dxE_{xxy}(P,\overline{Q})dy$$

(12.44)

where $E_{xxx}, E_{xyy}, E_{xxy}$ are the fundamental solutions for strains $\varepsilon_x, \varepsilon_y$ and γ_{xy} at point $\overline{Q}$ due to a unit x-force at P. The work done by the forces/stresses of load case two times the displacements of load case one is the same as for the case without initial stresses. After placing the unit load in the y-direction and in the z-direction we obtain

$$\mathbf{u}(P) = \int_S \mathbf{t}^T(Q)\mathbf{U}(P,Q)dS - \int_S \mathbf{u}^T(Q)\mathbf{T}(P,Q)dS + \int_V \mathbf{E}(P,\overline{Q})\boldsymbol{\sigma}_0(\overline{Q})dV \quad (12.45)$$

or assuming *Neumann* boundary conditions everywhere

$$[\mathbf{T}]\{\mathbf{u}\} = \{\mathbf{F}\} + \{\mathbf{F}\}_\sigma \quad (12.46)$$

For 2-D problems matrix $\mathbf{E}$ is given by

$$\mathbf{E} = \begin{bmatrix} E_{xxx} & E_{yxx} \\ E_{xyy} & E_{yyy} \\ E_{xxy} & E_{yxy} \end{bmatrix}^T \quad (12.47)$$

For 3-D problems we have

$$\mathbf{E} = \begin{bmatrix} E_{xxx} & E_{yxx} & E_{zxx} \\ E_{xyy} & E_{yyy} & E_{zyy} \\ E_{xzz} & E_{yzz} & E_{zzz} \\ E_{xxy} & E_{yxy} & E_{zxy} \\ E_{xyz} & E_{yyz} & E_{zyz} \\ E_{xxz} & E_{yxz} & E_{zxzz} \end{bmatrix}^T \quad (12.48)$$

where[2]

$$E_{ijk} = -\frac{C}{r^n}\left[C_3\left(r_k\delta_{ij} + r_j\delta_{ik}\right) - r_j\delta_{jk} + C_4 r_i r_j r_k\right]$$ (12.49)

where x,y,z may be substituted for i,j,k, r_x, r_y, r_z is as defined in equation (8.42) and the values for the constants are given in Table 12.5.

Table 12.5   Constants for fundamental solution

	Plane strain	Plane stress	3-D
n	1	1	2
C	$1/8\pi G(1-v)$	$(1+v)/8\pi\, G$	$1/16\, G\, \pi(1-v)$
C_3	$1-2\,v$	$(1-v)/(1+v)$	$1-2\,v$
C_4	2	2	3

A FUNCTION EpsK is added to the Elasticity_lib which returns matrix **E** for 2-D and 3-D elasticity problems.

```
FUNCTION EpsK(dxr,r,E,ny,Cdim,Nstres)
!------------------------------------------------
!    FUNDAMENTAL SOLUTION FOR Strains
!    isotropic material (Kelvin solution)
!------------------------------------------------
REAL,INTENT(IN)     :: dxr(:)            ! rx/r etc.
REAL,INTENT(IN)     :: r                 ! distance to P
REAL,INTENT(IN)     :: E                 ! Young's modulus
REAL,INTENT(IN)     :: ny                ! Poisson's ratio
INTEGER,INTENT(IN) :: Cdim               ! Cartesian dimension
INTEGER, INTENT(IN) :: Nstres            ! No.of strain components
REAL                :: EpsK(Nstres,Cdim)! Function returns array
INTEGER             :: n,i,j,k
REAL                :: G,c,c3,c4,dij,dik,djk  !    Temps
INTEGER :: II(6), JJ(6) ! sequence of stresses in pseudo-vector
G= E/(2.0*(1+ny))
SELECT CASE (Cdim)
  CASE (2)        !      Plane strain solution
     n= 1
     c= 1.0/(8.0*Pi*G*(1.0-ny))
     c3= 1.0-2.0*ny
     c4= 2.0
     II(1:3)= (/1,2,1/)
     JJ(1:3)= (/1,2,2/)
  CASE(3)         !      Three-dimensional solution
     n= 2
     c= 1.0/(16.0*Pi*G*(1.0-ny))
     c3= 1.0-2.0*ny
     c4= 3.0
     II= (/1,2,3,1,2,3/)
```

```
      JJ= (/1,2,3,2,3,1/)
   CASE DEFAULT
 END SELECT
 Direction_Pi: &
 DO k=1,Cdim
  Component_Eps: &
  DO n=1,Cdimi= II(n)
    j= JJ(n)
    dij= 0.0
    IF(i==j) dij=1.0
    dik= 0.0
 .  IF(i==k) dik=1.0
    djk= 0.0
    IF(j==k) djk=1.0
    EpsK(n,k)= C/r**n*(C3*(dxr(k)*dij+dxr(j)*dik) &
               -dxr(j)*djk+C4*dxr(i)*dxr(j)*dxr(k))
  END DO &
  Component_Eps
 END DO &
 Direction_Pi
 RETURN
 END FUNCTION EpsK
```

12.4.1 Numerical evaluation of volume integrals

We assume that the variation of initial stresses over the domain is specified locally over each volume cell.

For 2-D elasticity problems, the stresses at a point with intrinsic coordinates ξ, η inside the cell c are given by

$$\sigma_0 = \sum_{n=1}^{4} N_n(\xi,\eta)\sigma_{0n}^c \tag{12.50}$$

where σ_{0n}^c are the initial stresses at node n of cell c.

For 3-D problems we have

$$\sigma_0 = \sum_{n=1}^{8} N_n(\xi,\eta,\zeta)\sigma_{0n}^e \tag{12.51}$$

The volume integral can be replaced by a sum of integrals over volume cells

$$\{\mathbf{F}\}_\sigma = \int_V \mathbf{E}\sigma_0 dV = \sum_{c=1}^{N_c} \sum_{n=1}^{4(8)} \Delta\mathbf{E}_n^c\sigma_{0n} \tag{12.52}$$

where N_c is the number of volume cells.

For two-dimensional problems we define

$$\Delta E_n^c = \int_{V_c} \mathbf{E} N_n \, dV_c = \int_{-1}^{1}\int_{-1}^{1} \mathbf{E} N_n J \, d\xi \, d\eta \tag{12.53}$$

This integral may be evaluated numerically as

$$\Delta E_n^c \approx \sum_{m=1}^{M}\sum_{k=1}^{K} \mathbf{E}(P,\overline{Q}(\xi_m,\eta_k)) N_n \, J(\xi_m,\eta_k) \, W_m W_k \tag{12.54}$$

where M and K are the number of Gauss points in the ξ and η directions.

For 3-D problems we have

$$\Delta E_n^c = \int_{V_c} \mathbf{E} N_n \, dV_c = \int_{-1}^{1}\int_{-1}^{1}\int_{-1}^{1} \mathbf{E} N_n \, J \, d\xi \, d\eta \, d\zeta \tag{12.55}$$

The numerical evaluation is given by

$$\Delta \mathbf{E}_n^c \approx \sum_{l=1}^{L}\sum_{m=1}^{M}\sum_{k=1}^{K} \mathbf{E}(P,\overline{Q}(\xi_m,\eta_k,\zeta_l)) N_n \, J(\xi_m,\eta_k,\zeta_l) \, W_m W_k W_l \tag{12.56}$$

where L,M,K are the number of Gauss points in the ξ, η and ζ directions.

The kernel to be integrated is of the same order as kernel T, and therefore the number of integration points needed is the same as for performing the integrations for the coefficient matrices as explained in Chapter 6.

12.4.2 Postprocessing

In postprocessing the effect of initial stresses has to be included. For the calculation of displacements,   equation (8.34) is expanded to

$$\mathbf{u}(P_a) = \sum_{e=1}^{E}\sum_{n=1}^{N} \Delta \mathbf{U}_n^e \, \mathbf{t}_n^e - \sum_{e=1}^{E}\sum_{n=1}^{N} \Delta \mathbf{T}_n^e \, \mathbf{u}_n^e + \sum_{c=1}^{N_c}\sum_{n=1}^{N} \Delta \mathbf{E}_n^c \, \boldsymbol{\sigma}_{0n}^c \tag{12.57}$$

whereas for computation of stresses we have

$$\boldsymbol{\sigma}(P_a) = \sum_{e=1}^{E}\sum_{n=1}^{N} \Delta \mathbf{S}_n^e \, \mathbf{t}_n^e - \sum_{e=1}^{E}\sum_{n=1}^{N} \Delta \mathbf{R}_n^e \, \mathbf{u}_n^e + \sum_{c=1}^{N_c}\sum_{n=1}^{N} \Delta \hat{\mathbf{E}}_n^c \, \boldsymbol{\sigma}_{0n}^c \tag{12.58}$$

where

$$\Delta \hat{\mathbf{E}}_n^c = \int_{S_c} \hat{\mathbf{E}}(P_a,Q) N_n \, dS(Q) \tag{12.59}$$

Matrix $\hat{\mathbf{E}}$ is essentially the same as that defined in equation (12.38), except that it has to be expanded so that initial shear stresses can also be considered.

For 2-D problems the matrix is defined as

$$
\hat{\mathbf{E}} = \begin{bmatrix} E_{xxxx} & E_{xxyy} & E_{xxxy} \\ E_{yyxx} & E_{yyyy} & E_{yyxy} \\ E_{xyxx} & E_{xyyy} & E_{xyxy} \end{bmatrix}
\tag{12.60}
$$

whereas for 3-D problems we have

$$
\hat{\mathbf{E}} = \begin{bmatrix} E_{xxxx} & E_{xxyy} & E_{xxzz} & E_{xxxy} & E_{xxyz} & E_{xxxz} \\ E_{yyxx} & \cdots & & & & \vdots \\ E_{zzxx} & & & & & \\ E_{xyxx} & & & & & \\ E_{yzxx} & & & & & \\ E_{xzxx} & \cdots & & & & E_{xzxz} \end{bmatrix}
\tag{12.61}
$$

The expression for $\mathbf{E}$ is given by[2]

$$
\hat{E}_{ijkl} = \frac{C_2}{r^n} \left[\begin{array}{l} C_3\left(\delta_{ik}\delta_{kj} + \delta_{jk}\delta_{ki} + \delta_{ij}\delta_{kl} + n\delta_{ij}r_k\, r_l\right) + \\ nv\left(\delta_{ij}r_j r_k + \delta_{jk}r_k r_j + \delta_{ik}r_k r_j + \delta_{jk}r_i r_k\right) + n\left(r_i r_j - C_6 r_i r_j r_k r_l\right) \end{array} \right]
\tag{12.62}
$$

where x,y,z may be substituted for $i,j,k,$), r_x, r_y, r_z is as defined in equation (8.42), δ_{ij} has been defined previously and the constants are given in Table 12.6.

Table 12.6   Constants for fundamental solution

	Plane strain	Plane stress	3-D
n	2	2	3
C_2	$1/4\pi(1-v)$	$(1+v)/4\pi$	$1/8\pi(1-v)$
C_3	$1-2v$	$(1-v)/(1+v)$	$1-2v$
C_6	4	4	5

As with the solution for initial strains, the kernel is hypersingular and appropriate intergration orders have to be selected, depending on the proximity to the point where the stresses are evaluated. If the interior point is one of the nodes of a volume cell, then a limiting value has to be taken. This will be discussed in more detail in the chapter on plasticity.

FUNCTION ExK is added to Elasticity_lib. This is similar to FUNCTION EK, except that it is expanded to cater for shear stresses as well.

```fortran
FUNCTION ExK(dxr,r,E,ny,Cdim,Nstres)
!-------------------------------------------------
!    Fundamental solution derivative
!    for the computation of interior stresses
!    effect of initial stresses
!    isotropic material (Kelvin solution)
!-------------------------------------------------
REAL,INTENT(IN)      :: dxr(:)           ! rx/r etc.
REAL,INTENT(IN)      :: r                ! distance to P
REAL,INTENT(IN)      :: E                ! Young's modulus
REAL,INTENT(IN)      :: ny               ! Poisson's ratio
INTEGER,INTENT(IN)   :: Cdim             ! Cartesian dimension
INTEGER, INTENT (IN) :: Nstres           ! No of stress
                                         !        components
REAL                 :: ExK(Nstres,Nstres) ! returns array
INTEGER              :: i,j,k
REAL                 :: G,c,c2,c3,c6,n,dij,dik,djk,dkl    !   Temps
INTEGER :: II(6), JJ(6)  !  sequence of stresses in pseudo-vector
                         G= E/(2.0*(1+ny))
SELECT CASE (Cdim)
  CASE (2)                ! Plane strain solution
   n= 2
   c2= 1.0/(4.0*Pi*(1.0-ny))
   c3= 1.0-2.0*ny
   c6= 4.0
   II(1:3)= (/1,2,1/)
   JJ(1:3)= (/1,2,2/)
  CASE(3)                             !  Three-dimensional solution
   n= 3
   c2= 1.0/(8.0*Pi*(1.0-ny))
   c3= 1.0-2.0*ny
   c6= 5.0
   II= (/1,2,3,1,2,3/)
   JJ= (/1,2,3,2,3,1/)
 CASE DEFAULT
END SELECT
Direction_Eps: &
DO m=1,Nstres
 k= II(m)
 l= JJ(m)
 Direction_Sigma: &
 DO n=1,Nstres
   i= II(n)
   j= JJ(n)
   dij= 0.0
   IF(i==j) dij=1.0
```

```
    dik= 0.0
    IF(i==k) dik=1.0
    djk= 0.0
    IF(j==k) djk= 1.0
    dkl= 0.0
    IF(k==l) dkl= 1.0
    djl= 0.0
    IF(j==l) djl=1.0
    C= C3*(dik*djk+djk*dik-dij*dkl+n*dij*dxr(k)**2)+ &
       n*ny*(dik*dxr(j)*dxr(k)+djk*dxr(k)*dxr(i)+ &
          dik*dxr(k)*dxr(j)+djl*dxr(i)*dxr(k))+ &
       n*(dxr(i)*dxr(j)-C6*dxr(i)*dxr(j)*dxr(k)* dxr(l))
      ExK(n,m)= C2/r**n*C
  END DO &
  Direction_Sigma
END DO &
Direction_Eps
RETURN
END FUNCTION ExK
```

12.5 IMPLEMENTATION

The implementation of body forces will not be discussed in more detail here. In the case of gravity, an additional contribution to the right-hand side has to be computed by the application of numerical integration (equations (12.11) and (12.12)). In the case of initial strains, additional volume discretisation has to be specified in the area where these strains occur in order to perform the integration. In postprocessing the effect of body forces has to be taken into account.

12.6 EXAMPLE

An example in geomechanics is presented to demonstrate the application of the methods just explained. It is the case of a circular tunnel in an infinite soil mass, where parts of the ground is subject to swelling. This is an application in soil mechanics where minerals such as gypsum present in the soil mass cause an increase in volume when subjected to moisture.

In Figure 12.6 we show a diagram of the problem to be analysed. The 5 m thick layer of soil subjected to swelling is 10m above the tunnel with a diameter of 20m. The question to be answered from the analysis was what effect the swelling had on the tunnel, that is what displacements were generated. In a practical application, the effect on the stresses in the tunnel lining would also be important.

Figure 12.7 shows the mesh used to analyse this problem. The tunnel surface is discretised into 16 linear boundary elements and the swelling zone into eight linear cells. The material properties for the soil are assumed to be E = 1000.0 and ν = 0.0. The swelling zone is assumed to be subjected to an initial strain of 10% in the vertical direction. One of the results of the analysis namely the displaced shape of the tunnel is shown in Figure 12.7. Note that the displacements are magnified about 10 times.

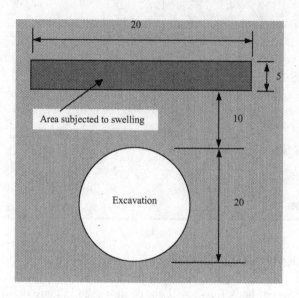

Figure 12.6   Problem specification

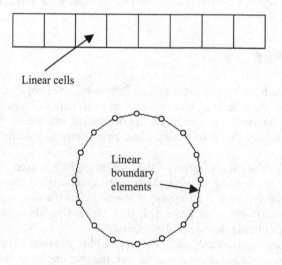

Figure 12.7   Mesh used

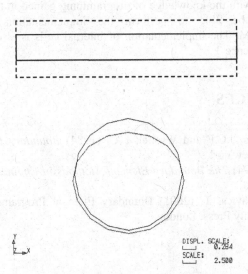

Figure 12.8   Displaced shape due to swelling (displacements magnified)

12.7 CONCLUSIONS

In this chapter we have dealt with the treatment of various types of effects which occur inside the domain where no elements exist. We have loosely called these effects body forces, even though some of these, for example the initial strains were not forces at all. With the capability to deal with these effects the range of application of the BEM has been expanded, but we have also laid the foundations for the next chapter that deals with plasticity. Indeed, in the solution of material non-linear problems we can visualise the problem of the redistribution of stresses as body forces which are generated once plasticity occurs.

The effect of body forces is that an additional right-hand side is generated in the system of equations. If the body forces are constant then this term is computed as a surface integral and no additional discretisation is necessary. Otherwise a volume mesh has to be created in order to enable a volume integration to be carried out. Those critical of the BEM might suggest that the main attraction, surface instead of volume discretisation, of the method would be lost. However, this is not the case. Volume cells are only needed where internal loading occurs and there are no additional degrees of freedom associated with the nodes of a volume mesh. This is demonstrated by the example shown where volume elements were only provided in the swelling zone. An equivalent finite element discretisation would have to cover all the domain, including the one where no swelling occurs. Also the number of unknowns at the nodes of the boundary elements was not increased by the fact that volume cells were needed to compute the right-hand side of the system of equations for this case.

The implementation of body forces has not been discussed in the level of detail of the other chapters here. The reason for this is that the implementation is fairly straightforward, and with the knowledge of programming gained in the previous chapters the reader should be able to perform the implementation into the program General_purpose_BEM. The implementation of internal cells is also discussed in much greater detail in Reference 3.

12.8   REFERENCES

1. Brebbia, A., Telles, J.C.F and Wrobel, L.C. (1984) *Boundary Element Techniques*. Springer-Verlag, NewYork.
2. Banerjee, P.K. (1994) *The Boundary Element Methods in Engineering*. McGraw-Hill, London.
3. Gao, X.W. and Davies, T. (2001) Boundary Element Programming in Mechanics. Cambridge University Press, London.

13

Non-linear Problems

13.1   INTRODUCTION

So far we have discussed problems where there is a linear relationship between applied loading and displacement, or between flow and temperature/potential. The system of equations

$$[\mathbf{T}]\{\mathbf{u}\}=\{\mathbf{F}\} \tag{13.1}$$

corresponds to a linear analysis, if $\{\mathbf{u}\}$ is a linear function of $\{\mathbf{F}\}$.

The linearity of equation (13.1) is only guaranteed if certain assumptions are made when deriving the system of equations. These assumptions are:

1. The relationships between flux and temperature/potential or stresses and strains are linear.

2. Matrix $[\mathbf{T}]$ is not affected by changes in geometry of the boundary due to displacements.

3. Boundary conditions do not change during loading.

Indeed, we have implicitly relied on these assumptions to be true in all our previous derivations of the theory. An example where the first assumption is violated is elasto- or viscoplastic material behaviour (this is generally referred to as material non-linear behaviour). The second one is violated if the displacements significantly change the shape of the boundary and this has an influence on the solution (large displacement problems).

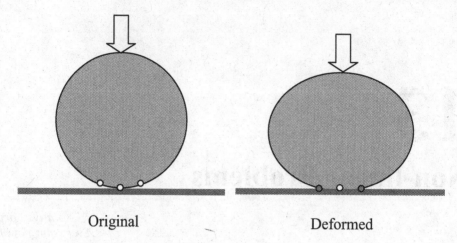

Original   Deformed

Figure 13.1   Example of non-linear analysis: contact problem

Finally, the third no longer holds true for contact problems, where either the boundary or the interface conditions between regions change during loading, thereby affecting the assembly of $[\mathbf{T}]$. This is illustrated by the example of an elastic sphere on a rigid surface, shown in Figure 13.1. After deformation two nodes, indicated by dark circles may change from *Neumann* to *Dirichlet* boundary condition because they are in contact with the plane the sphere is resting on.

If one of the assumptions for linearity are not satisfied, then the relationship between $\{\mathbf{u}\}$ and $\{\mathbf{F}\}$ will become non-linear. In a non-linear analysis matrix, $[\mathbf{T}]$ itself becomes a function of unknown $\{\mathbf{u}\}$. It is therefore not possible to solve the system of equations directly.

In this chapter we shall discuss solution methods for non-linear problems starting with the general solution process. We will then discuss two different types of non-linear behaviour, plasticity and contact problems. We shall see that solution methods for these types of problems are very similar to the ones employed by the finite element method. We will also find that the BEM is well suited to deal with contact problems because boundary tractions are used as primary unknowns.

13.2   GENERAL SOLUTION PROCEDURE

The method proposed is to first find a solution, assuming that the conditions for linearity are satisfied by solving

$$[\mathbf{T}]_0\{\mathbf{u}\}_0 - \{\mathbf{F}\}_0 = 0 \qquad\qquad (13.2)$$

where $[\mathbf{T}]_0$ is the 'linear' coefficient matrix. With solution vector $\{\mathbf{u}\}_0$ a check is then made to see whether all linearity assumptions have been satisfied, for example, we may check if the internal stresses (computed by postprocessing) violate any yield condition, or if boundary conditions have changed because of deformations. If any one of these linearity conditions have not been satisfied, this means that matrix $[\mathbf{T}]$ has changed during loading, i.e. instead of equation (12.2) we have

$$[\mathbf{T}]_1\{\mathbf{u}\}_0 - \{\mathbf{F}\}_0 = \{\mathbf{R}\}_1 \tag{13.3}$$

Here $[\mathbf{T}]_1$ is the changed matrix, also referred to as the 'tangent' matrix, and $\{\mathbf{R}\}_1$ is a residual vector. For equation (13.2) to be satisfied, the solution vector has to be modified.

We compute the first correction to $\{\mathbf{u}\}$ as

$$[\mathbf{T}]_1\Delta\{\mathbf{u}\}_1 = \{\mathbf{R}\}_1 \tag{13.4}$$

and proceed with these corrections until residual vector $\{\mathbf{R}\}$ approaches zero.

Final displacements are obtained by summing all corrections:

$$\{\mathbf{u}\} = \{\mathbf{u}\}_0 + \Delta\{\mathbf{u}\}_1 + \cdots + \Delta\{\mathbf{u}\}_N \tag{13.5}$$

where N is the number of iterations to achieve convergence. The solution is assumed to have converged if the norm of the current residual vector is much smaller than the first residual vector, i.e. when

$$\frac{\|\mathbf{R}_N\|}{\|\mathbf{R}_1\|} \leq Tol \tag{13.6}$$

where Tol is a specified tolerance.

Alternative to the system of equations (13.4), we may use the 'linear' matrix throughout the iteration, that is, equation (13.4) is modified to

$$[\mathbf{T}]_0\Delta\{\mathbf{u}\}_1 = \{\mathbf{R}\}_1 \tag{13.7}$$

This will obviously result in slower convergence, but will save us computing a new left hand side.

13.3   PLASTICITY

There are two principal ways in which non-linear material behaviour may be considered: elastoplasticity and viscoplasticity[1]. Regardless of the method used, the solution is achieved by calculating initial strains or stresses. Using the procedures outlined in Chapter 12, residuals $\{\mathbf{R}\}$ may then be computed directly from these initial stresses.

13.3.1 Elastoplasticity

In the theory of elastoplasticity, we define a yield function $F(\boldsymbol{\sigma}, C_1, C_2....) = 0$ which specifies a limiting state of stress $\boldsymbol{\sigma}$ (C_1, C_2, etc., are plastic material parameters). Stress states can only be such that F is negative (elastic states) or zero (plastic states). Positive values of F are not allowed.

A popular yield function for soil and rock material is the Mohr-Coulomb[2] condition which can be expressed as a surface in principal stress space by

$$F(\sigma) = \frac{\sigma_1 + \sigma_3}{2}\sin\phi - \frac{\sigma_1 - \sigma_3}{2} - c\cos\phi = 0 \qquad (13.8)$$

where σ_1 and σ_3 are maximum and minimum principal stresses, c is cohesion and ϕ the angle of friction. The yield function is plotted as a surface in the principal stress space in Figure 13.2. After the solution of equation (13.2), it may be that stresses that were in an elastic state at increment n-1 ($F < 0$) change to an inadmissible state ($F > 0$) at increment n (Figure 13.2).

Therefore, this stress state has to be corrected back to the yield surface. To do this we have to isolate the plastic and elastic components.

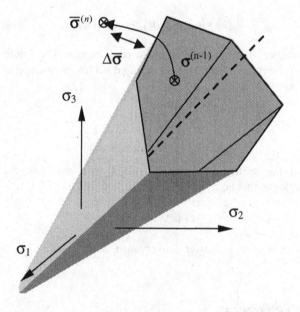

Figure 13.2 Mohr-Coulomb yield surface showing elastic and inadmissible stress states

If the state of stress is such that $F(\sigma) < 0$, then the theory of elasticity governs the relationship between stress and strain, i.e. (see Chapter 4)

$$\sigma_e = \mathbf{D}\varepsilon_e \qquad (13.9)$$

For stress states where $F(\sigma) = 0$, elastic strains ε_e as well as plastic strains ε_p are present, that is, the total strain ε contains both components, but equation (13.9) can only be applied to the elastic component.

The stress-strain law can now be written incrementally as

$$d\sigma = \mathbf{D}d\varepsilon_e = \mathbf{D}(d\varepsilon - d\varepsilon_p) = (\mathbf{D} - \mathbf{D}_p)d\varepsilon = \mathbf{D}_{ep}d\varepsilon \qquad (13.10)$$

where $\mathbf{D}_{ep}$ is the elastoplastic matrix and $d\boldsymbol{\varepsilon}$ is the total strain increment.

Since equation (13.10) only applies when stresses are on the yield surface, we have to determine the intersection where the stress path crosses the yield surface.

This point may be computed using the condition that

$$F(\boldsymbol{\sigma}^{(n-1)} + r\Delta\boldsymbol{\sigma}) = 0 \tag{13.11}$$

where $\Delta\boldsymbol{\sigma} = \bar{\boldsymbol{\sigma}}^{(n)} - \boldsymbol{\sigma}^{(n-1)}$ and r is a scale factor. If we assume that F is a linear function of $\boldsymbol{\sigma}$, then we can write

$$F(\boldsymbol{\sigma}^{(n-1)}) + r\left[F(\bar{\boldsymbol{\sigma}}^{(n)}) - F(\boldsymbol{\sigma}^{(n-1)})\right] = 0 \tag{13.12}$$

from which scale factor r can be computed as

$$r = \frac{F(\boldsymbol{\sigma}^{(n-1)})}{(F(\bar{\boldsymbol{\sigma}}^{(n)}) - F(\boldsymbol{\sigma}^{(n-1)})} \tag{13.13}$$

The proportion of total stress increment $\Delta\boldsymbol{\sigma}$ which lies outside the yield surface is now computed as

$$\Delta\bar{\boldsymbol{\sigma}} = (1 - r)\Delta\boldsymbol{\sigma} \tag{13.14}$$

This stress was wrongly assumed to be elastic and a correction has to be made. The strain which has incorrectly been assumed to be elastic can be computed by

$$\Delta\boldsymbol{\varepsilon}_P = \mathbf{D}^{-1}\Delta\bar{\boldsymbol{\sigma}} \tag{13.15}$$

Finally, a stress state which satisfies the yield condition can be computed as

$$\boldsymbol{\sigma}^{(n)} = \boldsymbol{\sigma}^{(n-1)} + r\Delta\boldsymbol{\sigma} + \Delta\bar{\boldsymbol{\sigma}}_P \tag{13.16}$$

where

$$\Delta\bar{\boldsymbol{\sigma}}_P = \mathbf{D}_{ep}\mathbf{D}^{-1}(1 - r)\Delta\bar{\boldsymbol{\sigma}} \tag{13.17}$$

The stress increment to be used for computation of residual $\{\mathbf{R}\}$ is the difference between the stress state that violates the yield condition ($\bar{\boldsymbol{\sigma}}^{(n)}$) and that found on the yield surface ($\boldsymbol{\sigma}^{(n)}$):

$$\Delta\boldsymbol{\sigma}_P = \bar{\boldsymbol{\sigma}}^{(n)} - \boldsymbol{\sigma}^{(n)} \tag{13.18}$$

Equation (13.17) is only an approximation, since $\mathbf{D}_{ep}$ is valid only for infinitesimally small increments. Therefore, if the corrected stresses are checked again, the yield condition may not be satisfied exactly. Various 'return algorithms' may be applied to ensure that stresses always satisfy the condition $F \leq 0$ [3].

Apart from the case of the stress state going from an elastic to a plastic state discussed here, other cases may occur during iterations and this will affect the computation of value r. Table 13.1 shows how the value of r is computed depending on the values of yield function F at iteration $n-1$ and n.

Table 13.1   Values of r for various cases

Case	$n-1$	n	r
1	$F < 0$	$F < 0$	1.0
2	$F < 0$	$F > 0$	Eq. (13.13)
3	$F = 0$	$F > 0$	0
4	$F = 0$	$F < 0$	1.0

13.3.2 Viscoplasticity

The concept of viscoplasticity allows $F(\boldsymbol{\sigma})$ to be greater than zero [4]. A positive yield function simply means that the stress state has a higher plastic potential.

The stresses are then allowed to creep back to a lower plastic potential (Figure 13.3) and eventually to the yield surface. This takes into consideration the fact that the material requires time to 'react' to changes in stress, and also allows the consideration of creep behaviour.

The strain rate at which 'creeping' takes place is assumed to be proportional to the plastic potential. That is

$$\frac{\partial}{\partial t}\boldsymbol{\varepsilon}_P = \dot{\boldsymbol{\varepsilon}}_P = \frac{1}{\eta}\Phi(F)\frac{\partial Q}{\partial \boldsymbol{\sigma}} \tag{13.19}$$

where

$$\Phi(F)=0 \quad ; \quad for \ \ F<0$$
$$\Phi(F)=F \quad ; \quad for \ \ F>0 \tag{13.20}$$

In the above equations η is a material parameter describing its time dependent behaviour (viscosity) and Q is a plastic potential. For associated plasticity we have $Q \equiv F$. The material parameter η can be determined by controlled creep tests on specimens of the material but is not easily obtained. In rock mechanics applications for example the viscous behaviour of the rock mass and not of an individual rock specimen is important. If η is not known one can still do a visco-plastic analysis with a fictitious time step. This would then be similar to an elasto-plastic analysis where the iteration steps are replaced by fictitious time steps.

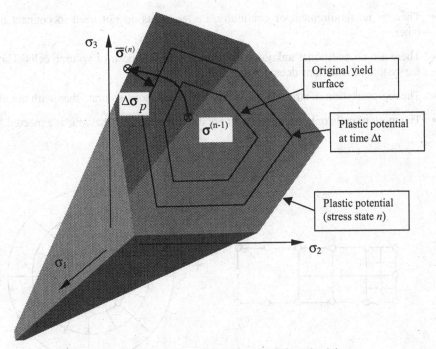

$$\sigma_3$$

$$\overline{\sigma}^{(n)}$$

$$\Delta\sigma_p$$

$$\sigma^{(n-1)}$$

Original yield
surface

Plastic potential
at time Δt

Plastic potential
(stress state n)

$$\sigma_2$$

$$\sigma_1$$

Figure 13.3   Explanation of the concept of viscoplasticity

At each time step and a viscoplastic strain increment is computed at each time step by:

$$\Delta\boldsymbol{\varepsilon}_P = \dot{\boldsymbol{\varepsilon}}_P \Delta t \tag{13.21}$$

where Δt is a time increment. The stress increment for the computation of the residual $\{\mathbf{R}\}$ is

$$\Delta\boldsymbol{\sigma}_P = \mathbf{D}\Delta\boldsymbol{\varepsilon}_P \tag{13.22}$$

The time increment Δt cannot be chosen freely, but has to satisfy certain stability conditions to prevent oscillations[4].

13.3.3 Method of solution

For the solution of problems in plasticity we have to amend the discretisation of the problem. In addition to surface elements we require the specification of volume cells for the integration of initial stresses. These volume cells have been discussed in the previous chapter. Figure 13.4 shows examples of discretisations for a cantilever beam and a circular hole in an infinite domain. The discretisations actually almost look like finite element meshes, and it could be argued that one might as well use finite elements for this problem.

However, there are subtle differences:

- There is no requirement of continuity, i.e. elements do not need to connect to each other.

- There are no additional unknown associated with the mesh of volume cells. Therefore the system of equations does not increase in size.

- The representation of stress is still not affected and more accurate than with the FEM.

- The mesh of cells only needs to cover zones where plastic behaviour is expected.

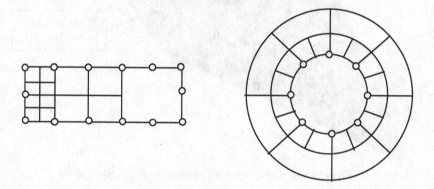

Figure 13.4   Volume cells for the example of a cantilever beam and a circular hole

An elasto- and viscoplastic analysis proceeds as follows:

1. A linear elastic analysis is performed. Results $\{u\}_0$ are obtained.

2. Stresses $\sigma^{(0)}$ are computed at all nodes of the cells using the procedures outlined in Chapter 8. If a cell node lies exactly on the boundary then the computation of stresses is carried out with the 'traction recovery method' using the boundary interpolation of the displacements.

3. The yield condition F is checked at all nodes of cells. If $F > 0$ then plastic stress increments $\Delta\sigma^c_{Pn}$ are computed.

4. For each cell ΔE^c_n is computed using numerical integration by applying equations (12.55) or (12.56). The cell contributions are added and a right-hand side vector $\{F\}_\sigma$ obtained, this is equivalent to residual force vector $\{R\}$. The norm of residual forces $\|R\|$ is determined.

5. The system of equations is solved with this new right-hand side and displacement increments $\Delta\{u\}_n$ obtained. Total displacements are computed as $\{u\}_n = \{u\}_{n-1} + \Delta\{u\}_n$.

6. Stresses $\boldsymbol{\sigma}^{(n)}$ are computed again at all the nodes of the cells. This time, however, the effect of the initial stresses has to be considered, i.e. equation (12.58) must be used.

7. Steps 3, 4, 5, 6 are repeated until the norm of residual forces $\|\mathbf{R}\|$ is sufficiently small.

13.3.4 Evaluation of singular integrals $\Delta \mathrm{E}_n^c$

The evaluation of integrals $\Delta \mathrm{E}_n^c$ has been discussed in section 12.4.1 However, the special case where one of the cell nodes coincides with a collocation point, P_i, has not been considered. This occurs when an internal cell is adjacent to a boundary element. As a collocation point is approached, the kernel $\mathbf{E}$ tends to infinity with $o(1/r)$ for 2-D problems and $o(1/r^2)$ for 3-D problems.

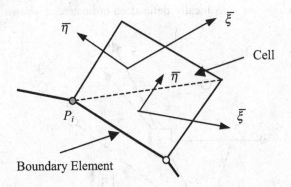

Figure 13.5   Case where cell point is a collocation point

To evaluate the volume integral for the case where one of the cell nodes coincides with a collocation point P_i we subdivide a cell into two subelements, as shown in Figure 13.5. For 2-D problems the subdivision is carried out in exactly the same way as for the evaluation of the boundary integrals for 3-D problems, i.e. the square domain is mapped into triangular domains where the apex of the triangle is located at Pi (see Section 6.3.4).

Equation (12.52) is rewritten as

$$\Delta \mathrm{E}_{ni}^e = \sum_{s=1}^{2} \int_{-1}^{1}\int_{-1}^{1} \mathbf{E}(P_i, \overline{Q}(\xi, \eta)) N_n J(\xi, \eta) \frac{\partial(\xi, \eta)}{\partial(\overline{\xi}, \overline{\eta})} d\overline{\xi} d\overline{\eta} \approx$$

$$\sum_{s=1}^{2} \sum_{m=1}^{M} \sum_{k=1}^{K} \mathbf{E}(P_i, \overline{Q}(\overline{\xi}_m, \overline{\eta}_k)) N_n(\overline{\xi}_m, \overline{\eta}_k) J(\overline{\xi}_m, \overline{\eta}_k) \overline{J}(\overline{\xi}_m, \overline{\eta}_k) W_m W_k$$

(13.23)

The computation of the Jacobian $\bar{J}$ of the transformation from subelement coordinates $\bar{\xi},\bar{\eta}$ to intrinsic coordinates ξ,η is explained in Section 6.3.4. Since the Jacobian of this transformation tends to zero with $o(r)$ as point P_i is approached, the singularity is cancelled out.

For three-dimensional problems, the Gauss formula is given by:

$$\int_{-1}^{1}\int_{-1}^{1}\int_{-1}^{1} \mathbf{E}(P_i,Q(\xi,\eta))N_n|J|\,d\xi d\eta d\zeta \approx$$

$$\sum_{l=1}^{L}\sum_{m=1}^{M}\sum_{k=1}^{K} \mathbf{E}(P_i,Q(\xi_l,\eta_m,\zeta_k))N_n(\xi_l,\eta_m,\zeta_k)J(\xi_l,\eta_m,\zeta_k)W_l W_m W_k$$

(13.24)

If one of the nodes of the element is a collocation point, a subdivision, analogous to the 2-D case, into tetrahedra with locally defined co-ordinates, as shown in Figure 13.6, is used.

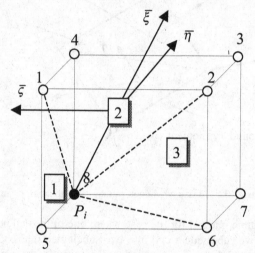

Figure 13.6   Subdivision method for computing singular volume integrals (3-D problems)

Using this subdivision, the integral is expressed as

$$\sum_{s=1}^{3}\int_{-1}^{1}\int_{-1}^{1}\int_{-1}^{1} \mathbf{E}(P_i,Q(\xi,\eta,\varsigma))N_n J(\xi,\eta,\varsigma)\frac{\partial(\xi,\eta,\varsigma)}{\partial(\bar{\xi},\bar{\eta},\bar{\varsigma})}\,d\bar{\xi} d\bar{\eta} d\bar{\varsigma} \approx$$

$$\sum_{s=1}^{3}\sum_{l=1}^{L}\sum_{m=1}^{M}\sum_{k=1}^{K} \mathbf{E}(P_i,Q(\bar{\xi}_l,\bar{\eta}_m,\bar{\varsigma}_k))N_n(\bar{\xi}_l,\bar{\eta}_m,\bar{\varsigma}_k)J(\bar{\xi}_l,\bar{\eta}_m,\bar{\varsigma}_k)\bar{J}W_l W_m W_k$$

(13.25)

where $\bar{J}$ is the Jacobian of the transformation from ξ,η,ς to $\bar{\xi},\bar{\eta},\bar{\varsigma}$ coordinates.

This transformation is given by

$$\xi = \sum_{n=1}^{5} \bar{N}_n(\bar{\xi},\bar{\eta},\bar{\varsigma})\xi_{l(n)}; \quad \eta = \sum_{n=1}^{5} \bar{N}_n(\bar{\xi},\bar{\eta},\bar{\varsigma})\eta_{l(n)}$$

$$\varsigma = \sum_{n=1}^{5} \bar{N}_n(\bar{\xi},\bar{\eta},\bar{\varsigma})\varsigma_{l(n)}$$

(13.26)

where $l(n)$ is the node number of the nth node (i.e. $l(1)=1$, $l(2)=2$, $l(3)=3$, $l(4)=4$ and $l(5)=8$ for sub-cell 2 in Figure 13.6) and the shape functions are defined as

$$\bar{N}_1 = \frac{1}{8}(1+\bar{\xi})(1-\bar{\eta})(1-\bar{\varsigma}); \quad \bar{N}_2 = \frac{1}{8}(1+\bar{\xi})(1+\bar{\eta})(1-\bar{\varsigma})$$

$$\bar{N}_3 = \frac{1}{8}(1+\bar{\xi})(1-\bar{\eta})(1+\bar{\varsigma}); \quad \bar{N}_4 = \frac{1}{8}(1-\bar{\xi})(1+\bar{\eta})(1+\bar{\varsigma})$$

(13.27)

$$\bar{N}_5 = \frac{1}{8}(1-\bar{\xi})$$

The Jacobian is defined as

$$|J| = det \begin{vmatrix} \dfrac{\partial \xi}{\partial \bar{\xi}} & \dfrac{\partial \eta}{\partial \bar{\xi}} & \dfrac{\partial \varsigma}{\partial \bar{\xi}} \\ \dfrac{\partial \xi}{\partial \bar{\eta}} & \dfrac{\partial \eta}{\partial \bar{\eta}} & \dfrac{\partial \varsigma}{\partial \bar{\eta}} \\ \dfrac{\partial \xi}{\partial \bar{\varsigma}} & \dfrac{\partial \eta}{\partial \bar{\varsigma}} & \dfrac{\partial \varsigma}{\partial \bar{\varsigma}} \end{vmatrix}$$

(13.28)

where

$$\frac{\partial \xi}{\partial \bar{\xi}} = \sum_{n=1}^{5} \frac{\partial \bar{N}_n}{\partial \bar{\xi}}(\bar{\xi},\bar{\eta},\bar{\varsigma})\xi_{l(n)}, \quad \text{etc.}$$

(13.29)

The Jacobian tends to zero with $o(r^2)$, thereby cancelling out the singularity.

13.3.5 Computation of internal stresses

For computation of the internal stress results, equation (12.58) is used and the integrals ΔE_n^c can be evaluated using Gauss Quadrature. However, if the point on which we determine stresses P_a coincides with the nodes of cells, then the integrand tends to infinity with $o(r^2)$ for 2-D problems, and $o(r^3)$ and cannot be evaluated numerically.

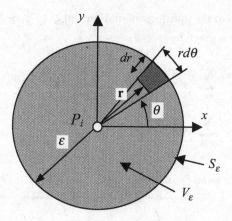

Figure 13.7   Polar coordinates for integration over a circular volume of exclusion

To explain the evaluation of these integrals we rewrite equation (12.58):

$$\boldsymbol{\sigma}(P_a) = \sum_{e=1}^{E}\sum_{n=1}^{N}\Delta\mathbf{S}_n^e\,\mathbf{t}_n^e - \sum_{e=1}^{E}\sum_{n=1}^{N}\Delta\mathbf{R}_n^e\,\mathbf{u}_n^e + \hat{\boldsymbol{\sigma}}(P_a) \qquad (13.30)$$

Here we assume that point P_a coincides with a boundary point, and

$$\hat{\boldsymbol{\sigma}}(P_a) = \int_V \hat{\mathbf{E}}(P_a,Q)\Delta\boldsymbol{\sigma}_P(Q)dV(Q) \qquad (13.31)$$

where V is the volume of integration and $\Delta\boldsymbol{\sigma}_P$ the plastic stress increment. We rewrite this integral[5]

$$\int_V \hat{\mathbf{E}}(P_a,Q)\Delta\boldsymbol{\sigma}_P(Q)dV(Q) = \int_V \hat{\mathbf{E}}(P_a,Q)(\Delta\boldsymbol{\sigma}_P(Q) - \Delta\boldsymbol{\sigma}_P(P_a))dV(Q)$$

$$+ \left[\int_V \hat{\mathbf{E}}(P_a,Q)dV(Q)\right]\Delta\boldsymbol{\sigma}_P(P_a) \qquad (13.32)$$

Now the singularity of the first integral is reduced, and it can be integrated numerically, as explained previously. The second integral can be evaluated analytically using polar coordinates.

Referring to Figure 13.7, we have for 2-D problems

$$\int_{V_\varepsilon}\hat{\mathbf{E}}dV = \int_0^{2\pi}\lim_{\varepsilon\to0}\int_0^\varepsilon \frac{1}{r^2}\hat{\mathbf{E}}\,r\,d\theta\,dr \qquad (13.33)$$

where $\hat{\mathbf{E}}$ is the part of the kernel which is not a function of r (i.e. term within square parentheses in equation (12.62)).

After some algebra, the volume integral can be replaced with a surface integral

$$\int_{V_\varepsilon} \hat{\mathbf{E}}dV = \int_{S_\varepsilon} \hat{\mathbf{E}}(\mathbf{r} \bullet \mathbf{n})ln(r)dS \qquad (13.34)$$

where $\mathbf{r}$ is the vector specified in Figure 13.7 and $\mathbf{n}$ the outward normal to surface V_ε.

In the numerical implementation we divide the cells into 'degenerate' subcells as shown in Figure 13.8. The first part of the integral in equation (13.32) is carried out over volume V^s of the subcell, whereas the second integral is evaluated over surface S^s of the subcell.

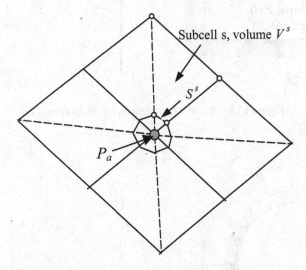

Figure 13.8   Numerical treatment of integration computation of stress at P_a

The implementation in 3-D follows the same procedure as for 2-D. Indeed, the result of the integration in equation (13.34) turns out to be exactly the same as in 2-D. This method for determining stresses cannot be used for points exactly on the boundary. We have already presented an alternative method for computing the stress tensor on the boundary itself, using variations of the displacements and tractions over boundary elements, in Chapter 8. All that is required here is to modify this procedure by taking into consideration the effect of the initial stresses.

13.3.6 Example

To illustrate the application of the method we present the analysis of a circular excavation in a viscoplastic pre-stressed infinite soil mass. For the soil, a Mohr-Coulomb yield condition was assumed.

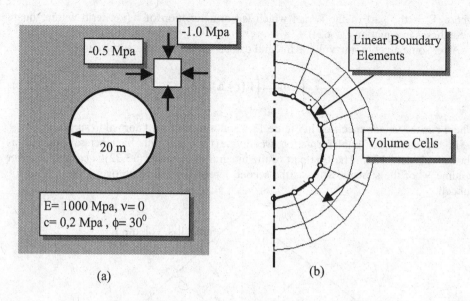

Figure 13.9   Example problem and discretisation used

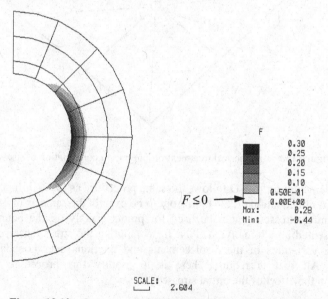

Figure 13.10   Contours of yield function F, time increment 3

The problem dimensions and the material properties are depicted in Figure 13.9 (a). The soil mass is assumed be in a stress state of 1 Mpa compression in the vertical direction and 0.5 Mpa in the horizontal. The method of solution is the same as explained in Chapter 9, except that the computation of an additional right-hand side due to plasticity using internal cells has been added.

The solution proceeds in increments where the yield condition is checked at all the nodes of cells after each time step. If no changes are observed in the stresses then a stationary situation has occurred and the analysis is stopped.

The mesh consist of eight linear boundary elements and 24 linear volume cells (Figure 13.9 (b)). The system of equations to be solved has only 16 unknown. Symmetry about the y-axis is assumed.

Some results of the non-linear analysis are presented. Figure 13.10 shows a plot of the yield function after time increment 3. The dark zones can be interpreted as material which is undergoing plastic straining, darker areas indicating increasing plastic potential (F). The distribution of maximum compressive stress after three time increments is shown in Figure 13.11. It can be seen that due to plasticity taking place near the surface of the excavation the tangential stresses which in an elastic analysis increase as the boundary is approached decrease instead.

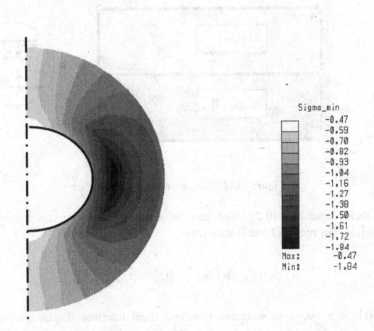

Figure 13.11 Contours of maximum compressive stress after time increment 3

13.4   CONTACT PROBLEMS

The second type of non-linear problem we will discuss here is where boundary conditions change during loading. An example of a contact problem is where interface conditions between regions change. A practical application of this is delamination /slip and crack propagation. For these type of problems we have a condition, similar to the yield condition discussed previously, which determines when the continuity conditions for an interface no longer apply. In the case of crack propagation, for example, we may have a condition based on tensile strength of the material which determines when nodes separate.

For problems with joints we may have a criterion based on the angle of friction and cohesion which determines when slip occurs.

In our discussion here we will concentrate on relatively simple problems: ones where contact initially exists and where it is lost due to some conditions being violated. We will see that the theory we will develop can be applied to delamination and joint problems.

13.4.1 Method of analysis

We start with the multi-region method developed in Chapter 10. Consider the beam in Figure 13.12 consisting of two regions.

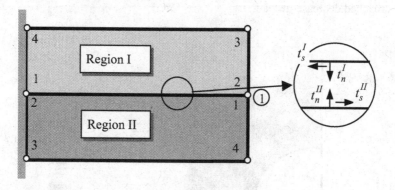

Figure 13.12 Cantilever beam with interface

We recall equation 10.25, that can be used to compute the tractions at the interface $\{\mathbf{t}\}_c$. For regions I and II we have

$$\{\mathbf{t}\}_c^I = \{\mathbf{t}\}_{c0}^I + \mathbf{K}^I\{\mathbf{u}\}_c^I ; \quad \{\mathbf{t}\}_c^{II} = \{\mathbf{t}\}_{c0}^{II} + \mathbf{K}^{II}\{\mathbf{u}\}_c^{II} \tag{13.35}$$

where $\{\mathbf{t}\}_{c0}$ is a vector of tractions assuming fixed interface displacements, $\mathbf{K}$ is the stiffness matrix of the interface nodes and $\{\mathbf{u}\}_c$ is a vector containing displacements at the interface nodes. Since the beam is fixed on the left, our problem has only two interface unknowns.

For contact problems it is convenient to work with components in a direction normal to interface t_n and in a direction tangential to interface t_s, instead of global components t_x and t_y. Also, to separate delamination and slip it is required that the local components be used for the displacements as well. The relationship between global and local components is given by:

$$\bar{\mathbf{t}} = \mathbf{T}_g^T\mathbf{t}; \quad \mathbf{u} = \mathbf{T}_g\bar{\mathbf{u}} \tag{13.36}$$

where $\mathbf{T}_g$ is the transformation matrix, as discussed in Chapter 3, and

$$\mathbf{\bar{t}} = \begin{Bmatrix} t_n \\ t_s \end{Bmatrix} \quad ; \quad \mathbf{t} = \begin{Bmatrix} t_x \\ t_y \end{Bmatrix} ; \quad \mathbf{\bar{u}} = \begin{Bmatrix} u_n \\ u_s \end{Bmatrix} \quad ; \quad \mathbf{u} = \begin{Bmatrix} u_x \\ u_y \end{Bmatrix} \tag{13.37}$$

In terms of local components, equations (13.35) are rewritten as

$$\mathbf{\bar{t}}_c^I = \mathbf{T}_g \mathbf{t}_{c0}^I + \mathbf{\bar{K}}^I \mathbf{u}_c^I \quad ; \quad \mathbf{\bar{t}}_c^{II} = \mathbf{T}_g \mathbf{t}_{c0}^{II} + \mathbf{\bar{K}}^{II} \mathbf{u}_c^{II} \tag{13.38}$$

where $\mathbf{\bar{K}}^N$ is the transformed stiffness matrix, i.e.

$$\mathbf{\bar{K}}^N = \mathbf{T}_g^T \mathbf{K}^N \mathbf{T}_g \tag{13.39}$$

The conditions at the interface normally stipulate that the equations of equilibrium and compatibility have to be satisfied, i.e.

$$\mathbf{\bar{t}}^I = \mathbf{\bar{t}}^{II} \quad ; \quad \mathbf{\bar{u}}^I = \mathbf{\bar{u}}^{II} \tag{13.40}$$

We may now define conditions for compatibility to exist. For example, the condition

$$t_n \leq T \tag{13.41}$$

stipulates that the traction normal to the interface has to be smaller than or, at most, equal to, the tensile strength of the material. If t_n has reached T then delamination occurs, that is, the compatibility condition is no longer applied to that point in the n direction.

Analogous to plasticity, the yield function can be written as

$$F_1(t_n) = t_n - T \tag{13.42}$$

Another condition may be that the shear traction is limited by

$$|t_s| \leq c + t_n \tan\varphi \tag{13.43}$$

where c is the cohesion and φ the angle of friction. If $|t_s| = c + t_n \tan\phi$ slip occurs, that is, the compatibility condition is no longer applied to that point in the s direction.

The corresponding yield function is written as:

$$F_2(t_s, t_n) = |t_s| - c - t_n \tan\varphi \tag{13.44}$$

The consequence is that when either F_1 or F_2 is zero the assembly is changed: instead of adding all stiffness coefficients we assemble the corresponding stiffness coefficients for region I and II into different locations in $\mathbf{K}$.

Consider, for example, the problem in Figure 13.12. The equations for compatibility at node 1 are (since only one node is involved we have left out the subscript denoting the local (region) node number):

$$u_{n1} = u_n^I = u_n^{II} \quad \text{and} \quad u_{s1} = u_s^I = u_s^{II} \tag{13.45}$$

With the vector of interface unknown only involving the ones at node 1

$$\mathbf{u}_c = \begin{Bmatrix} u_{n1} \\ u_{s1} \end{Bmatrix} \tag{13.46}$$

For the example with only one interface node, we may write for the region stiffness matrices:

$$\mathbf{K}^I = \begin{bmatrix} t_{nn}^I & t_{ns}^I \\ t_{sn}^I & t_{ss22}^I \end{bmatrix} \quad \text{and} \quad \mathbf{K}^{II} = \begin{bmatrix} t_{nn}^{II} & t_{ns}^{II} \\ t_{sn}^{II} & t_{ss}^{II} \end{bmatrix} \tag{13.47}$$

and the following assembled interface stiffness matrix is obtained:

$$\mathbf{K} = \begin{bmatrix} t_{nn}^I + t_{nn}^{II} & t_{ns}^I + t_{ns}^{II} \\ t_{sn}^I + t_{sn}^{II} & t_{ss}^I + t_{ss}^{II} \end{bmatrix} \tag{13.48}$$

If $F_1 = 0$ then the normal displacement and – as a consequence – also the shear displacement of region I are independent of region II. The vector of interface unknown is expanded to

$$\mathbf{u}_c = \begin{Bmatrix} u_n^I \\ u_s^I \\ u_n^{II} \\ u_s^{II} \end{Bmatrix} \tag{13.49}$$

and the stiffness matrix is defined as

$$\mathbf{K} = \begin{bmatrix} t_{nn}^I & t_{ns}^I & 0 & 0 \\ t_{sn}^I & t_{ss}^I & 0 & 0 \\ 0 & 0 & t_{nn}^{II} & t_{ns}^{II} \\ 0 & 0 & t_{sn}^{II} & t_{ss}^{II} \end{bmatrix} \tag{13.50}$$

If $F_2 = 0$ then slip occurs and compatibility does not apply to the shear displacement. The vector of interface unknown is given by

$$\mathbf{u}_c = \begin{Bmatrix} u_{n1} \\ u_s^I \\ u_s^{II} \end{Bmatrix}$$

(13.51)

In the stiffness matrix only the terms associated with the normal components are added:

$$\mathbf{K} = \begin{bmatrix} t_{nn}^I + t_{nn}^{II} & t_{ns}^I & t_{ns}^{II} \\ t_{sn}^I & t_{ss}^I & 0 \\ t_{sn}^{II} & 0 & t_{ss11}^{II} \end{bmatrix}$$

(13.52)

13.4.2 Solution procedure

Only in exceptional cases will a point be reached when the yield functions are exactly 0. As with plasticity, we will have the condition that if the yield function is checked with traction $\bar{\mathbf{t}}$ we find that either $F_1(\bar{\mathbf{t}}) > 0$ or $F_2(\bar{\mathbf{t}}) > 0$. In the first case this means that the material has been stressed beyond the tensile strength, in the second that the friction law has been violated.

In the first instance the excessive stress, i.e. the one which caused the yield condition to be violated, is computed by

$$\Delta t_p = \bar{t}_n - T$$

(13.53)

whereas in the second case,

$$\Delta t_p = |\bar{t}_s| - c - \bar{t}_n \tan\varphi$$

(13.54)

We now propose the following solution procedure:

1. The system is solved in the normal way using the interface compatibility and equilibrium conditions (equations (13.40)).

2. The yield conditions F_1 and F_2 are computed at each interface node. If both are zero then the analysis is finished.

3. If one of the yield conditions is greater than zero residual tractions are computed according to equations (13.53) or (13.54).

4. The interface matrix **K** is re-assembled taking into consideration the relaxed continuity conditions for interface points which are separating or slipping.

5. The system is solved with the residual tractions applied as loading in the opposite direction.

6. Points 2 to 5 are repeated until the yield conditions are satisfied at all interface nodes.

The extension of the method to three dimensions is straightforward. In 3-D we have two instead of one shear traction ($\bar{t}_{s1}$ and $\bar{t}_{s1}$), and when we check the yield condition we have to work with a resultant shear traction. This is given by

$$|\bar{t}_s| = \sqrt{(\bar{t}_{s1})^2 + (\bar{t}_{s2})^2} \tag{13.55}$$

13.4.3 Example application

As an example of application we present an analysis of the delamination of a cantilever beam. The beam consists of two finite regions described by quadratic boundary elements, as shown in Figure 13.13. At the interface the tensile strength of the material was assumed to be zero. Shear loading is applied to the bottom half of the beam.

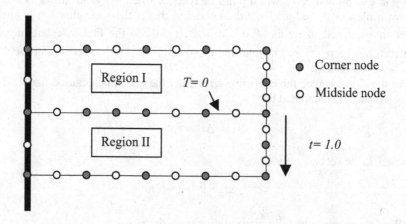

Figure 13.13 Mesh used for cantilever analysis

Figure 13.14 shows the distribution of normal stress at the interface after the linear analysis. It can be clearly seen that the yield condition for tension is violated in the interface on the right hand side of the beam. As a consequence the interface conditions are changed and the interface nodes are disconnected in the areas where tensile stresses were encountered. Figure 13.15 shows that after iteration step 1, delamination starts as a consequence.

Further examples of the application of the method to the modelling of faulted rock in mining and tunnelling applications as well as reservoir engineering can be found in References 7 and 8. The method can also be applied to the simulation of dynamic crack propagation[9] in the simulation of borehole blasting.

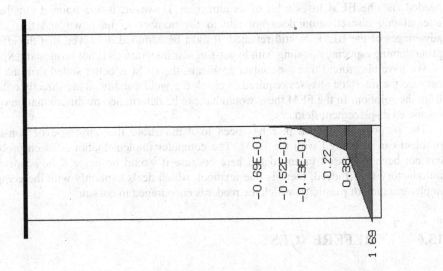

Figure 13.14   Distribution of normal stress at the interface

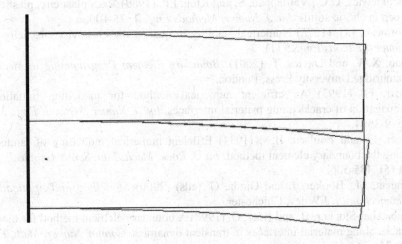

Figure 13.15   Displaced shape of beam after first iteration

13.5 CONCLUSIONS

In this chapter we attempted to show that the treatment of non-linear problems is almost as straightforward as with the FEM. Only two types of non-linear problems have been discussed: plasticity and contact problems. In the first, additional volume discretisation is needed and the BEM loses a bit of its attraction. However, it was pointed out that the internal cell discretisation does not add to the number of unknown and that all the advantages of the BEM are still retained. It must be admitted, however, that the effort in programming especially dealing with hyper-singular integrations is not unsubstantial.

We have also found that for contact problems, the BEM is better suited than the FEM because the interface stresses required to check the yield conditions are directly obtained from the solution. In the FEM these would have to be determined by differentiation of the computed displacement field.

The purpose of this chapter has been to demonstrate that any type of non-linear problem can be solved with the BEM. The computer implementation of such problems has not been discussed in any detail here because it would be beyond the scope of an introductory text. Indeed, there is one textbook which deals explicitly with the computer implementation of plasticity, which the reader is encouraged to consult[5].

13.6 REFERENCES

1. Hill, R. (1950) *The Mathematical Theory of Plasticity.* Oxford University Press.
2. Pande, G., Beer, G. and Williams, J. (1990) *Numerical Methods in Rock Mechanics.* Wiley, Chichester.
3. Zienkiewicz, O.C., Valliappan, S. and King, I.P. (1969) Viscoplasticity, plasticity and creep in elastic solids. *Int. J. Numer. Methods Eng.*, **1**, 75-100.
4. Cormeau, I.C. (1975) Numerical stability in quasi-static elastoviscoplasticity. *Int. J. Numer. Methods Eng.*, **9** (1).
5. Gao, X.W. and Davies, T. (2001). *Boundary Element Programming in Mechanics.* Cambridge University Press, London.
6. Beer, G. (1993) An efficient numerical method for modelling initiation and propagation of cracks along material interfaces. *Int. J. Numer. Methods Eng.*, **36** (21), 3579-3594.
7. Beer, G. and Poulsen, B.A. (1994) Efficient numerical modelling of faulted rock using the boundary element method. *Int. J. Rock. Mech. Min. Sci. & Geomech. Abstr.*, **31** (5), 485-506.
8. Zaman, M., Booker, J. and Gioda, G. (eds) (2000) *Modelling and Applications in Geomechanics.* J.Wiley, Chichester.
9. Tabatabai-Stocker, B. and Beer, G. (1998) A boundary element method for modelling cracks along material interfaces in transient dynamics. *Comun. Numer. Meth. Engng.*, **14**, 355-365.

14

Coupled Boundary Element/ Finite Element Analysis

Marriage à la mode
O. C. Zienkiewicz

14.1   INTRODUCTION

In the introduction we compared the Boundary Element Method (BEM) with its main 'competitor' the Finite Element Method (FEM). Although in the specific example the impression was given that a BEM analysis would be superior to the FEM, this was not meant to imply that this is always the case. In a famous paper written more than two decades ago[1], O.C. Zienkiewicz pointed out that benefits could be gained by combining the two methods of analysis, thereby gaining the 'best of both worlds'. This was at a time when BEM protagonists claimed that the BEM could do everything better and there was almost no collaboration between the two groups. Zienkiewicz, in his inimitable style, chose the title "Marriage à la mode" which shows a double meaning: marriage a la mode means a marriage of convenience, not love, but also there is a double meaning with the word mode (displacement mode = shape function).

There are several reasons why one would want to consider the combination of the two methods:

- Some problems, for example those involving highly inhomogeneous material, still require significant effort to solve with the BEM.

- For some problems no fundamental solutions of the governing differential equations can be found and in certain cases the solutions are extremely complex.

- Users familiar and happy with a FEM package may want to upgrade the program capabilities by including, for example, an efficient modelling of an infinite domain. Many well known commercial packages have capabilities for the specification of a user defined element stiffness, so the implementation would be fairly straightforward.

As will be seen here, the coupling of the FEM and BEM is not very difficult. Indeed, we have already set the foundation for this in Chapter 10, where we explained how the 'stiffness matrix' of a region can be computed. In essence, for coupling we have to find a way of harmonising the differences between the two methods. The main difference is that nodal tractions are used, as primary unknown in the BEM, whereas nodal point forces are used in the FEM. The 'stiffness matrix' we obtain for the BE region turns out to be unsymmetrical and this may cause some problems because symmetric solvers are usually employed in the FEM.

14.2   COUPLING THEORY

There are basically two approaches to coupling the boundary and finite element methods. In the first approach, the BE region is treated as a large finite element and its stiffness is computed and assembled into the global stiffness matrix. In the second approach, finite elements are treated as equivalent BE regions and their 'stiffness matrix' is determined and assembled, as explained in Chapter 10 for multiple regions.

The choice of coupling method depends mainly on the software available for the implementation, i.e. if boundary element capabilities are to be added to a finite element program or finite element capabilities to a boundary element one.

In the following we will discuss the theoretical basis and implementation of each approach. The coupling theory is discussed for problems in elasticity. However, as demonstrated throughout this book, potential problems can be considered in an analogous way.

14.2.1   Coupling to finite elements[2]

The FEM leads to a system of simultaneous equations which relate displacements at all the nodes to *nodal forces*. In the BEM, on the other hand, a relationship between nodal displacements and *nodal tractions* is established.

Consider the cantilever beam in Figure 14.1 consisting of one BE region connected to two finite elements. We refer to the assembly of two finite elements as a *finite element region*. Following the procedures in Chapter 10, we can obtain for the BE region a relationship between tractions $\{t\}_c$ and displacements $\{u\}_c$ at the interface (equation (10.22)):

$$\{t\}_c = \{t\}_{c0} + \mathbf{K}_{BE}\{u\}_c \tag{14.1}$$

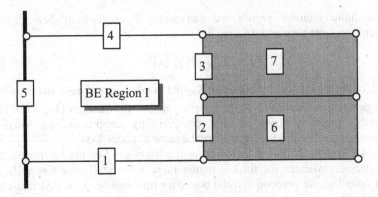

Figure 14.1   Cantilever beam: discretisation into finite and boundary elements

In the above, $\{t\}_{c0}$ is a vector containing tractions, if the interface is fixed, and $\mathbf{K}_{BE}$ is the pseudo 'stiffness matrix' of the BE region.

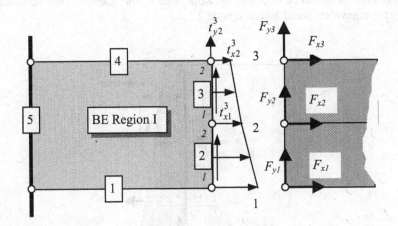

Figure 14.2   Interface between finite element and boundary regions showing interface forces

For the example problem with a smooth interface we have:

$$\{t\}_c = \begin{Bmatrix} t_{x1} \\ t_{y1} \\ t_{x2} \\ t_{y2} \\ t_{x3} \\ t_{y3} \end{Bmatrix}; \quad \{u\}_c = \begin{Bmatrix} u_{x1} \\ u_{y1} \\ u_{x2} \\ u_{y2} \\ u_{x3} \\ u_{y3} \end{Bmatrix} \tag{14.2}$$

For the finite element region we can write a relationship between interface displacements and interface nodal forces as

$$\{\mathbf{F}\}_c = \{\mathbf{F}\}_{c0} + \mathbf{K}_{FE}\{\mathbf{u}\}_c \tag{14.3}$$

where $\{\mathbf{F}\}_{c0}$ is an initial force vector and $\mathbf{K}_{FE}$ the condensed stiffness matrix of the finite element region which only involves the interface nodes. In equations (14.1) and (14.3) we have already implicitly assumed that compatibility conditions are satisfied (i.e. displacements of the BE and FE regions are the same at nodes 1-3).

Figure 14.2 shows the forces that exist at the interface. For the BE region these are boundary stresses, whereas for the FE region these are nodal point forces. In the first method of coupling, we propose that the boundary tractions be converted into equivalent nodal point forces.

The method for computing equivalent nodal point forces is exactly the same as the one shown in Chapter 11 for setting up the equilibrium equations, for the case where we have internal edges and corners in a multi-region analysis.

To compute the x-component of the equivalent nodal point force at node 2, for example, we apply a unit virtual displacement in the x-direction at that point (Figure 14.3). For equilibrium to be satisfied, the work done by the tractions must be equal to that done by the equivalent nodal forces at node 2.

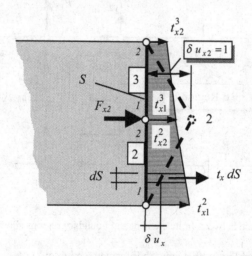

Figure 14.3   Calculation of F_{x2} by principle of virtual work

The total work done is computed by

$$F_{x2} \cdot 1 = \int_S t_x \delta u_x dS \tag{14.4}$$

Substituting the interpolation for tractions and displacements

$$t_x = \sum_{n=1}^{2} N_n t_{xn}^e; \quad \delta u_x(\xi) = N_1(\xi) \cdot 1 \text{ for } e = 1$$

$$\delta u_x(\xi) = N_2(\xi) \cdot 1 \text{ for } e = 2 \tag{14.5}$$

we obtain

$$F_{x2} \cdot 1 = \int_{S_2} (N_1 t_{x1}^2 + N_2 t_{x2}^2) N_2 dS_2 + \int_{S_3} (N_1 t_{x1}^3 + N_2 t_{x2}^3) N_1 dS_3 \qquad (14.6)$$

A second equation can be written for F_{y2} by applying a unit virtual displacement in the y-direction.

Based on this approach a general equation can be derived for computing the equivalent nodal point force at a point k

$$\mathbf{F}_k = \sum_e \sum_{n=1}^{N} \mathbf{M}_{jn}^e t_n^e \qquad (14.7)$$

where the outer sum is over all boundary elements that connect to point k, the inner sum is over all nodes of the element and j is the local (element) node number of node k.

For 2-D problems we have

$$\mathbf{M}_{jn}^e = \begin{bmatrix} M_{jn} & 0 \\ 0 & M_{jn} \end{bmatrix} \qquad (14.8)$$

with

$$M_{jn}^e = \int_{S_e} N_j N_n dS_e \qquad (14.9)$$

The integration over elements can be conveniently carried out using numerical integration (Gauss Quadrature) with three points. For 2-D problems we have:

$$M_{jn} = \int_{\xi=-1}^{1} N_j N_n J dS_e \approx \sum_{m=1}^{3} N_j N_n J W_m \qquad (14.10)$$

whereas for 3-D problems

$$M_{jn} = \int_{\eta=-1}^{1} \int_{\xi=-1}^{1} N_j N_n J dS_e \approx \sum_{k=1}^{3} \sum_{m=1}^{3} N_j N_n J W_m W_k \qquad (14.11)$$

Equation (14.1) can now be expressed in terms of equivalent nodal point forces by pre-multiplying with $\mathbf{M}$

$$\{\mathbf{F}\}_c = \mathbf{M}\{\mathbf{t}\}_c = \mathbf{M}\{\mathbf{t}\}_{c0} + \mathbf{M}\mathbf{K}_{BE}\{\mathbf{u}\}_c \qquad (14.12)$$

where $\mathbf{M}$ is assembled from element contributions $\mathbf{M}_{jn}^e$.

For the example in Figure 14.1 this matrix is given by

$$\mathbf{M} = \begin{bmatrix} \mathbf{M}_{11}^2 & \mathbf{M}_{12}^2 & 0 \\ \mathbf{M}_{21}^2 & \mathbf{M}_{22}^2 + \mathbf{M}_{11}^3 & \mathbf{M}_{12}^3 \\ 0 & \mathbf{M}_{21}^3 & \mathbf{M}_{22}^3 \end{bmatrix} \tag{14.13}$$

Matrix $\mathbf{MK}_{BE}$ is now a 'true' stiffness matrix in the finite element sense, i.e. one that relates nodal point displacements to nodal point forces. However, since $\mathbf{K}_{BE}$ is not symmetric, this matrix is also unsymmetrical.

Although there is no problem in dealing with unsymmetrical matrices, and they do occasionally occur in FEM analysis, some solvers used for finite elements are specialised in dealing with symmetric system of equations and, in some cases, it may be convenient if all stiffness matrices are symmetric. One way of getting a symmetric matrix is to use the principle of minimum potential energy to derive the equilibrium equations at the interface[3].

For simplicity, we only consider the forces/tractions due to interface displacements, so we can compute the total potential energy at the interface as

$$\Pi = \{\mathbf{u}\}_c^T \mathbf{K}_{FE} \{\mathbf{u}\}_c + \{\mathbf{u}\}_c^T \mathbf{MK}_{BE} \{\mathbf{u}\}_c \tag{14.14}$$

where the first expression on the left-hand side is the work done by the FE region and the second is that done by the BE region. Taking the minimum of potential energy, we obtain

$$\frac{\partial \Pi}{\partial \{\mathbf{u}\}_c} = \frac{1}{2} \left(\mathbf{K}_{FE}^T + \mathbf{K}_{FE} \right) \{\mathbf{u}\}_c + \frac{1}{2} \left[\left(\mathbf{MK}_{BE} \right)^T + \mathbf{MK}_{BE} \right] \{\mathbf{u}\}_c = 0 \tag{14.15}$$

The operation in the square parentheses means that a symmetric stiffness matrix for the BE region can be obtained by adding the transpose and by halving the result. This way of stating the equilibrium condition is commonly used in the FEM. However, its application here is not quite correct, since in the derivation of $\mathbf{K}_{BE}$ an interpolation of both displacement and traction has been assumed at the interface. It has been shown however that this error is only significant in some exceptional cases and acceptable results can be obtained for most applications.

Having obtained a 'true' stiffness matrix for the BE region, the further steps in the computation of coupled problems are fairly straightforward. The boundary element region is treated as a super (finite) element, and its stiffness is assembled in the usual way to obtain the system equations.

In the implementation we distinguish between fully coupled and partially coupled analyses. In a fully coupled analysis all nodes of the boundary element region are connected to the finite element region and no loading is assumed at the interface. An example of this type of analysis is the problem of an excavation in an infinite domain solved by a coupled discretisation, as shown in Figure 14.4 (a). In this case, the infinite boundary element region can be considered as a large finite element which accurately represents the effect of the infinite domain. This is a good example of gaining the 'best of both worlds' because the alternative to the coupled mesh shown would be either to extend the finite element region a large distance away and to apply artificial boundary conditions there or to use infinite finite elements. Both methods require more mesh generation and computational effort and result in a loss of accuracy. The reasons for opting for a coupled analysis rather than a pure boundary analysis in this case may be that sequential excavation or the installation of rock bolts and/or thin concrete lining has to be considered. In a fully coupled analysis only, the stiffness matrix of the BE region needs to be determined and pre-multiplied with $\mathbf{M}$ in order to change it to a true stiffness matrix that can be assembled.

An example of a partially coupled analysis is shown in Figure 14.4 (b). Here we consider the additional effect of a ground surface and another (existing) excavation. In a partially coupled analysis we first solve the problem with the interface nodes fixed and obtain an interface traction vector $\{\mathbf{t}\}_{c0}$. Then we compute the pseudo-stiffness matrix of the region. Before we assemble our finite element system, both $\{\mathbf{t}\}_{c0}$, and $\mathbf{K}_{BE}$ have to be pre-multiplied with $\mathbf{M}$, yielding a nodal point force vector as well as a stiffness matrix.

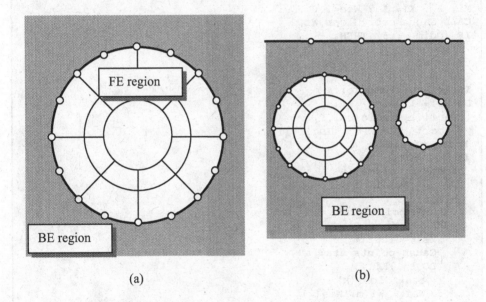

(a)   (b)

Figure 14.4 Fully and partially coupled discretisations

The only additional programming required for the implementation of a coupled analysis capability is the assembly of transformation matrix $\mathbf{M}$ and the pre-multiplication of the stiffness matrix $\mathbf{K}_{BE}$ and, in the case of a partially coupled analysis, the traction vector $\mathbf{t}_{BE}$ with this matrix. If required, a 'symmetrisation' procedure may be applied, as explained above.

We develop a function **Mtrans** which returns matrix $\mathbf{M}$ of dimension Ndofsc x Ndofsc where Ndofsc is the number of interface degrees of freedom. The input parameters of this function are the number of interface elements, the number of interface nodes, the incidence vector for each element and the coordinates of interface nodes.

```
FUNCTION MTrans(Nelc,Ndofsc,xPc,Incic)
!----------------------------------------
!   Function returns the assembled matrix M
!   for the conversion of a pseudo stiffnes matrix
!   into a true stiffness matrix
!----------------------------------------
INTEGER, INTENT (IN):: Nelc          ! No. of interface elements
INTEGER, INTENT (IN):: Ndofsc        ! No. of interface nodes
REAL, INTENT (IN)    :: xPc(:,:)      ! Coords of interface nodes
INTEGER, INTENT (IN):: Incic(:,:)    ! Incidences of interface elem
REAL     ::  Mtrans(Ndofsc,Ndofsc)   ! Function returns array
REAL     ::  MMjn(Ndof,Ndof)
REAL     ::  Glcor(2),Wi(2),Wie(2),Ni(Nodel),Elcor(Cdim,Nodel)
REAL     ::  xsi,eta,Jac,Weit,Mjn
INTEGRE ::  Inci(nodel)
Mtrans= 0.
ldim= Cdim - 1
Mi= 2 ; Ki= 1 ; Wie=1.0
CALL Gauss_coor(Glcor,Wi,2)          ! 2x2 integration
IF (Cdim == 3) THEN
 Ki=2
 Wie= Wi
END IF
Interface_elements: &
DO Nel= 1,Nelc
  Inci(:)= Incic(nel,:)
  Elcor(:,:)= xPc(:,Inci(:))
 Nodes_of_elem1: &
 DO j=1,nodel
  Nodes_of_elem2: &
  DO n=1,nodel
    Mjn= 0.
   Gauss_points_xsi: &
   DO m=1,Mi
     xsi= Glcor(m)
     Gauss_points_eta: &
     DO k=1,Ki
       eta= Glcor(k)
       Weit= Wi(m)*Wie(k)
       CALL Serendip_func(Ni,xsi,eta,ldim,nodel,Inci)
       Jac= Jacobian(xsi,eta,zeta,ldim,nodel,Inci,Elcor
       Mjn= Mjn + Ni(j)*Ni(n)*Jac*Weit
     END DO &
     Gauss_points_eta
   END DO &
   Gauss_points_xsi
    MMjn= 0.
    DO nd=1,ndof
      MMjn(nd,nd)= Mjn
    END DO
    nrow= (Inci(j)-1)*Ndof+1
    ncol= (Inci(n)-1)*Ndof+1
    Mtrans(nrow:,ncol:)= MMjn
```

```
   END DO &
   Nodes_of_elem2
  END DO &
 Nodes_of_elem1
 END DO &
 Interface_elements
 RETURN
 END FUNCTION Mtrans
```

14.2.2 Coupling to boundary elements

The coupling of finite elements to boundary elements follows the same steps as for the multi-region method discussed in Chapter 10. We may consider an assembly of finite elements as a boundary element region. Using standard FEM procedures, we obtain the following system of equations for the finite element assembly

$$\{\mathbf{F}\}_c = \{\mathbf{F}\}_{co} + \mathbf{K}_{FE}\{\mathbf{u}\}_c \tag{14.16}$$

where the notation has been defined at the beginning of this chapter. The equations which we get for each region in the BEM are

$$\{\mathbf{t}\}_c^I = \{\mathbf{t}\}_{c0}^I + \mathbf{K}_{BE}^I\{\mathbf{u}\}_c^I \tag{14.17}$$

where the roman superscript denotes the region number.

For coupling the finite element region all that is required is to convert equation (14.16) into a form such as that in equation (14.17). This is simply the inverse relationship to equation (14.12), i.e.

$$\{\mathbf{t}\}_c = \mathbf{M}^{-1}\{\mathbf{F}\}_c = \mathbf{M}^{-1}\{\mathbf{F}\}_{c0} + \mathbf{M}^{-1}\mathbf{K}_{FE}\{\mathbf{u}\}_c \tag{14.18}$$

The inverse of $\mathbf{M}$ has to be determined but since $\mathbf{M}$ is a sparsely populated and diagonally dominant matrix, this does not pose any problems.

After having obtained the pseudo-stiffness matrix of the finite element region $\mathbf{M}^{-1}\mathbf{K}_{FE}$ and, for partially coupled problems, the equivalent traction vector $\mathbf{M}^{-1}\{\mathbf{F}\}_{c0}$, we proceed in the same way as for multi-region problems.

14.3 EXAMPLES

The example presented here is that of a circular excavation in an infinite domain. This problem is similar to that shown to illustrate the application of plasticity in Chapter 13. Here we present an elastic analysis and compare it with the theoretical solution.

The problem geometry, material properties and initial stress field assumed are shown in Figure 14.5 (a). The discretisation into quadratic finite and boundary elements is shown in Figure 14.5 (b). One plane of symmetry is assumed.

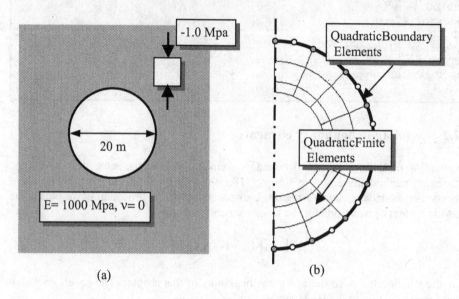

(a)   (b)

Figure 14.5   Example problem specification and coupled mesh used for analysis

Some results of the analysis are shown here. Figure 14.6 shows the displaced shape and Figure 14.7 the distribution of the maximum compressive stress in the finite element region.

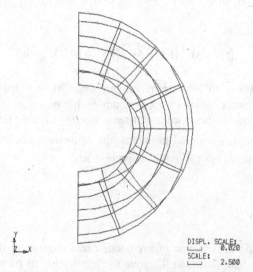

Figure 14.6   Displaced shape after excavation

Note that due to the use of locally defined shape functions the contours are not as smooth as the ones shown in Figure 13.11,which were obtained using fundamental solutions.

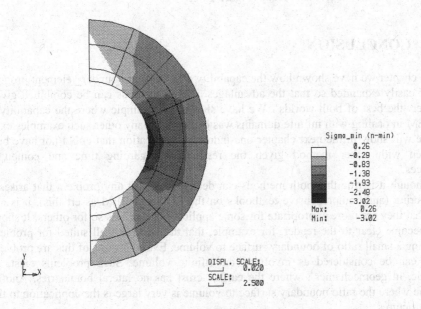

Figure 14.7   Contours of maximum compressive stress

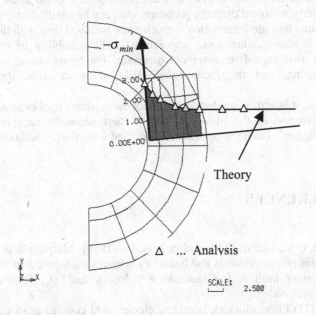

Figure 14.8   Distribution of maximum compressive stress, comparison with theory

The distribution of maximum compressive stress along a nearly horizontal line through the Gauss points of finite elements and inside the boundary element region shows good agreement with the theoretical results (Figure 14.8).

14.4   CONCLUSION

In this chapter we have shown how the capability of a finite or boundary element program can be easily expanded so that the advantages of both methods can be combined giving the user 'the best of both worlds'. We have shown one example where the capability of the BEM in dealing with infinite domains was exploited. Many other such examples exist, and we will show in the next chapter one industrial application that could not have been analysed with either method given the restrictions regarding time and computing resources.

Although it is true that both methods can deal with almost any problem that arises in engineering (and comprehensive textbooks on the FEM and BEM assert this), it is also clear that they are more appropriate for some applications and less so for others. It should have become clear to the reader, for example, that the BEM is well suited for problems involving a small ratio of boundary surface to volume. Extreme cases of this are problems which can be considered as involving an infinite volume. Such problems exist, for example, in geomechanics[4], where the earth's crust has no lateral boundaries. Another extreme where the ratio boundary surface to volume is very large is the application to thin shell structures.

Another aspect is the importance that is given to surface stresses. As we have seen in Chapter 8, stresses at the surface are computed more accurately with the BEM than with the FEM. We have shown that problems where 'body forces' occur in the domain, as for example, plasticity, transient dynamic problems, etc., can be handled with the BEM, but it has to be admitted that implementation is much more involved than with the FEM. A final aspect which is also gaining more importance is the suitability of the methods for implementation with regards to computer hardware. The future seems to lie in massive parallel processing, and the multi-region BEM seems to lend itself to parallel programming.

Given all this, it is surprising that there are not more commercial programs which make use of a combination of the FEM and BEM. There seems to be currently only one commercial program[5] which incorporates a combined use of both methods for the benefit of users.

14.5   REFERENCES

1. Zienkiewicz, O.C., Kelly, D.W. and Bettess, P. (1979) Marriage a la mode- the best of both worlds (finite elements and boundary integrals). Chapter 5 of *Energy Methods in Finite Element Analysis* (R. Glowinski, E Y. Rodin, and O.C. Zienkiewicz, eds.) pp. 82-107, Wiley, London.
2. Beer, G. (1977) Finite element, boundary element and coupled analysis of unbounded problems in elastostatics. *Int. J. Numer. Methods Eng.*, **11**, 355-376.

3   Beer, G. (1998) Marriage a la mode (finite and boundary elements) revisited. *Computational Mechanics New Trends and Applications* (E. Onate, and S.R. Idelsohn, eds), IACM, Barcelona.

4   Beer, G., Golser, H., Jedlitschka, G. and Zacher, P. (1999) Coupled finite element/boundary element analysis in rock mechanics - industrial applications. *Rock Mechanics for Industry* (B. Amadei, R. L. Kranz, G. A. Scott and P. Smeallie, eds). Balkema, Rotterdam pp 133-140.

5   Beer, G. and Sigl, O. (1998) BEFE helps design new CERN facility in Geneva. BENCHmark, 17-19.

15

Industrial Applications

15.1 INTRODUCTION

So far in this book we have developed software which can be applied to compute test examples. The purpose of this was to enable the reader to become familiar with the method, ascertain its accuracy and get a feel for the range of problems that can be solved. The emphasis in software development has been on implementation, which was concise and clear and could be well understood. As pointed out in the introduction to programming, this is not necessarily the most efficient code in terms of storage and computer resources.

If one wants to tackle real engineering problems one is inevitably faced with the need to develop efficient code because of the complexity of the tasks that need to be handled and the large number of unknown that have to be solved. The programs developed here would be unsuitable for such a task.

Aspects of the software that need to be improved are:

- Greater efficiency in the computation of coefficient matrices by rearranging DO loops, so that calculations which are independent of the DO loop variable are taken outside the loop.

- Greater efficiency in data and memory management so that data are only stored in RAM when they are needed, use of hard disk storage in order to achieve this (see, for example Reference 1.

- Greater efficiency in the solution of equations by using iterative solvers such as conjugate gradient methods[2].

- Adaptation to run on computers with several processors running in parallel[3].

In this chapter we attempt to show applications of the boundary element and coupled methods which have been compiled from a number of jobs that have been carried out over more than two decades using BEFE[4], a combined finite element/boundary element program. The purpose of the chapter is twofold. Firstly, an attempt is made to demonstrate the applications for which the BEM may have a particular advantage over the FEM. These applications include:

- Problems involving stress concentrations at the boundary, such as they occur in mechanical engineering.

- Problems consisting of infinite or semi-infinite domains, such as those occurring in geotechnical engineering.

- Problems involving slip and separation at material interfaces, such as they appear in mechanical and geotechnical engineering.

- Contact and crack propagation problems.

The second purpose of this chapter is to show how the very complex problems that invariably arise in industrial applications can be simplified so that the analysis can be performed in a reasonably short time.

It is very rarely the case that a problem can be modelled exactly as it is. In most cases we have to decrease its complexity. For example, if we were to model a jet fighter we would not be able to consider each one of its million parts. The process of *modelling* a given complex structure requires a lot of engineering ingenuity and experience. When we simplify a complex problem we must ensure that the important influences are retained while we can neglect other less important ones. For example, in a structural problem some parts of the structure may not contribute significantly to its load carrying capacity, but are there because of design considerations.

One very significant modelling decision is if a 3-D analysis has to be carried out. Obviously this would result in much greater analysis effort. As an example in geotechnical engineering, consider a tunnel which is very long compared to its diameter. If we are only interested in the displacements and stresses at a cross-section far away from the tunnel face, then a plane strain analysis would obviously suffice. Another way to simplify a problem is the introduction of planes of symmetry. As we have seen in some of the examples in Chapter 9, this results in considerable savings. Obviously if the prototype to be analysed is symmetric there is no loss in modelling accuracy. In some cases, however, symmetry planes can be assumed without significant loss in accuracy even if the prototype itself is not exactly symmetric.

In the following we present background information on each application, in some cases together with a story associated with it. We will start with the description of the problem and how it was simplified. We show the boundary element mesh generated and the results obtained. Comments are made on the quality of the results. The problem areas are divided into mechanical, geotechnical and civil engineering.

15.2 MECHANICAL ENGINEERING

15.2.1 A cracked extrusion press causes concern

A small company in Austria manufactures rolled thin tubes by extrusion. The extrusion press that was in use was 35 years old and made of cast iron (see Figure 15.1). During a routine inspection cracks were detected on the surface of the cast iron casing, as indicated. The company was in the process of ordering a new press, however delivery was expected to take more than six months. There was some concern that something dramatic might happen during the extrusion process, with the press suddenly breaking, meaning not only a danger to lives but also the possibility of losing the press. With full order books the latter was a very serious economic threat.

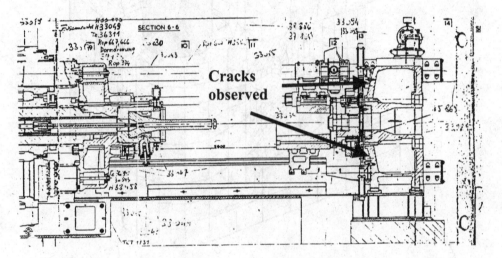

Figure 15.1 35 year old drawing of extrusion press with location of cracks indicated

The aim of the analysis was therefore to determine:

- If the existing cracks would propagate.
- If this propagation would lead to a sudden collapse of the structure.

The geometry of the part to be analysed was fairly complicated, and had to be reconstructed from the original plans. For the purpose of the analysis, it was assumed that there were two planes of symmetry, as shown in Figure 15.2, although this was not strictly true.

The cylindrical bar restraining the casing was not explicitly modelled but instead appropriate *Dirichlet* boundary conditions were applied. Each time a tube is extruded the casing is loaded with a force of 3700 tons (37 MN), as shown by the arrows. Although this load is actually applied dynamically, it was assumed to be static for the purpose of the analysis.

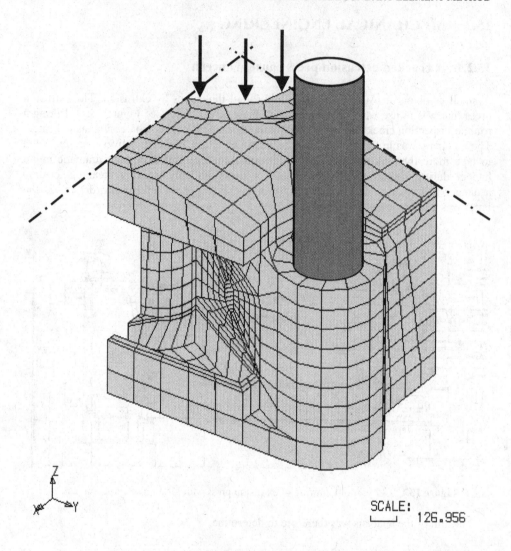

SCALE: 126.956

Figure 15.2   Boundary element model showing axes of symmetry and holding bar

The drawing in Figure 15.2 actually looks like a finite element mesh, but if viewed from the symmetry planes (Figure 15.3) one can notice that, in contrast to a FEM discretisation, there are no elements inside the material. The mesh consists of a total of 1437 linear boundary elements and has 4520 degrees of freedom.

There were two reasons why a boundary element analysis was chosen for this problem. Firstly, the generation of the mesh was found to be easier, since no internal elements and connection between surfaces had to be considered. Secondly, the task was to determine surface stresses and then to investigate crack propagation. As outlined previously, the BEM is well suited for this type of analysis.

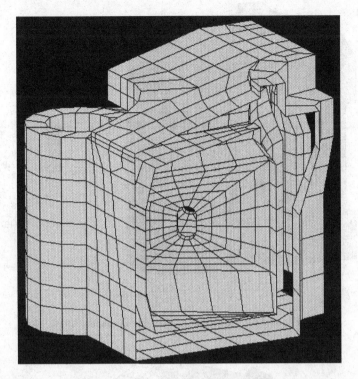

Figure 15.3   Boundary element mesh viewed from one of the symmetry planes

Initially, an analysis with only one region was carried out without considering the presence of cracks. This was done in order to check that the analysis was able to predict crack initiation. The criteria chosen for this was the maximum tensile strength of the material, taking into consideration the dynamic nature of the loading and the number of cycles that the press had so far sustained (approx. 2 million cycles). This analysis was also carried out to see if the BE model was adequate and to enable the client to get confidence in the BEM analysis proposed. The contours of maximum stress obtained from the single region analysis, shown in Figure 15.4, clearly indicate a stress concentration at the locations where cracks were observed, of a magnitude which would cause crack initiation there after a number of cycles.

After this verification of the model, a multi-region analysis was carried out. For this each of the flanges where the crack was observed was divided into two regions. For simplicity, it was assumed that the crack path is known *a priori* and is in the diagonal direction, as observed. Along this assumed crack path an interface was assumed between regions and the interface was allowed to slip and separate. It was found that in the worst case (lowest parameters assumed for the material), the crack would tend to propagate to the corners of the flange (Figure 15.5). However, even with the crack propagated that far, the model predicts that there would be no dramatic failure of the casing. Instead, the deformations would become so large that the press would become inoperable.

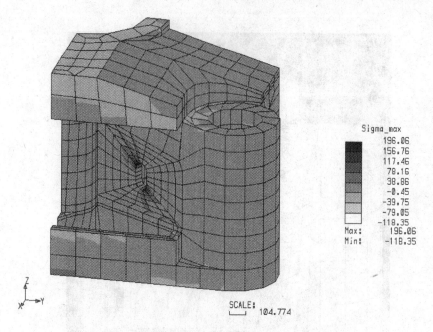

Figure 15.4   Contours of maximum principal stress

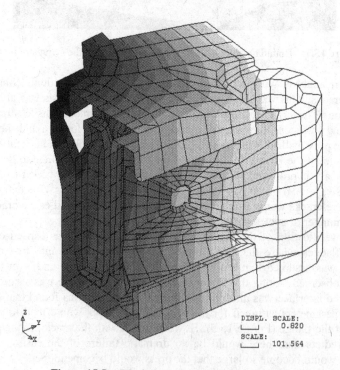

Figure 15.5   Displacements showing crack opening

After six months the new press arrived and was installed. The old press gave service without any major problems prior to replacement.

The advantages of the BEM over a FEM model may be summarised as:

- The fact that there are no elements inside and no connection was required between elements on the boundary the mesh generation was simplified. The number of unknowns and elements was also reduced.

- The stress concentrations were computed more accurately, because they are not obtained using an extrapolation from inside the domain.

- The method was well suited to model crack propagation.

15.3   GEOTECHNICAL ENGINEERING

15.3.1   Instability of slope threatens village

A village in Austria was experiencing severe problems from rock falls from a nearby hill. Huge boulders descended on the village, some houses were destroyed and people had to be evacuated. An underground mine was situated inside the hill and after the events it was claimed that the mine was responsible for the instability. A number of geological experts made a number of claims, none of which could be substantiated. It was decided to perform a numerical analysis so as to verify or reject some of the claims. The aim of the analysis was to determine how much subsidence was actually caused by the mine, and whether this had a destabilising effect.

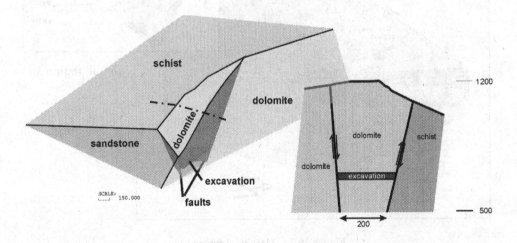

Figure 15.6   Schematic of the hill showing different rock types, interfaces and mining excavation

The hill consists mainly of three different material types (dolomite, schist and sandstone). Figure 15.6 shows a schematic view of the hill and the location of the mine. The material properties assumed for the different rock types are presented in Table 15.1. At the interfaces between the different materials, joints were assumed with the properties listed in Table 15.2.

Table 15.1   Elastic rock mass properties

	Dolomite	Schist	Sandstone
Young's Modulus [MPa]	80.000	25.000	50.000
Poisson ratio	0.3	0.3	0.3

Table 15.2   Joint parameters

	Joint plane 1	Joint plane 2	Joint plane 3
Friction angle [°]	35.0	25.0	25.0
Cohesion [MPa]	0.0	0.0	0.0

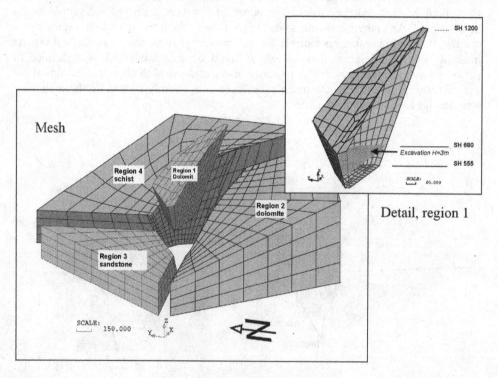

Figure 15.7   Mesh used for analysis

The hill was modelled by five BE regions, four of which are shown in Figure 15.7. The fifth region was assumed to model the semi-infinite domain surrounding the hillside. The mine is situated within region 1. The excavation of the mine was simulated in the following way: an initial stress field prior to excavation was computed due to the effect of self weight. The excavation was then simulated by applying *Neumann* boundary conditions equal in magnitude, but opposite in sign, to the *in situ* boundary stresses.

One of the results of the analysis was the subsidence predicted due to the mining operation. Figure 15.8 shows contours of settlement after the complete excavation of the mine. This picture represents a combination of displacement results obtained on the surface of boundary element region 1 and of internal results computed by postprocessing and plotted on 'dummy' planes inside the region.

The maximum amount of settlement predicted on the surface was 6 mm on the surface and 20 mm immediately above the excavation. The conclusion of the analysis was that the mine could not be the cause of the instability.

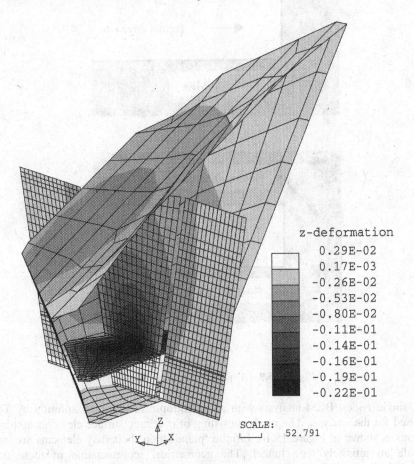

z-deformation

0.29E-02
0.17E-03
-0.26E-02
-0.53E-02
-0.80E-02
-0.11E-01
-0.14E-01
-0.16E-01
-0.19E-01
-0.22E-01

SCALE: 52.791

Figure 15.8   Contours of settlement

15.3.2 Analysis of tunnel advance in anisotropic rock

In many cases, the rock mass through which a tunnel is driven has a significant anisotropy, that is, the deformation moduli are direction dependent. Experience shows that in this case an important influence on displacements is the direction in which the tunnel is advanced. Two types of orthogonal anisotropy were considered for this example and shown in Figures 15.9 (a) and (b), one where the weak direction is dipping towards the advancing tunnel, the other where it dips against it.

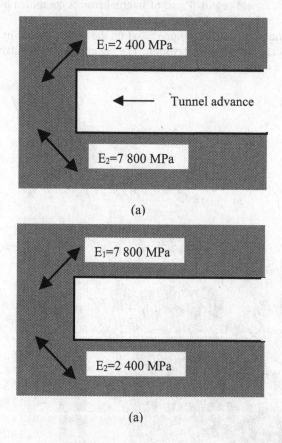

(a)

(a)

Figure 15.9   Types of anisotropy considered

A single region BEM analysis with an anisotropic fundamental solution by Tonon[1] was used for the analysis. The mesh consisting of quadratic surface elements and infinite elements is shown in Figure 15.10. Infinite 'plane strain' boundary elements are used to simulate an infinitely long tunnel. The geometrical representation of these infinite elements is discussed in Chapter 3.

For the variation of the unknown displacement a constant value is assumed in the infinite direction, whereas a quadratic variation was assumed in the finite direction. This element is well suited to simulate the plane strain conditions that exist further away form the tunnel face.

Figure 15.9 shows the effect of the anisotropy on tunnel wall displacement. The figure plots the ratio of wall displacement v at a distance x from the tunnel face (Diameter D) to the displacement very far away (v_f), i.e. under plane strain conditions. One can see that significantly smaller displacements occur near the tunnel face, if the tunnel is driven in the same direction as the dip of the anisotropy.

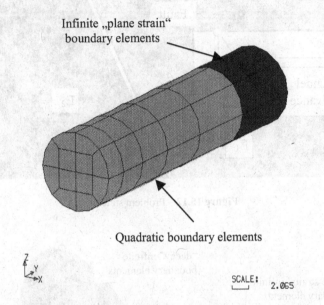

Infinite „plane strain" boundary elements

Quadratic boundary elements

SCALE: 2.065

Figure 15.10   Boundary element mesh

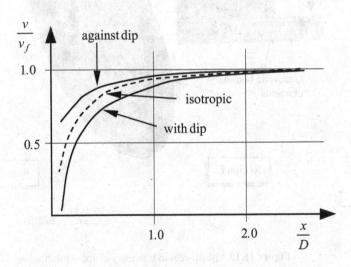

Figure 15.11   Effect of schistosity on tunnel wall displacement

15.3.3 Tunnel approaching fault

Tunnel engineers in Austria have a theory that with the use of careful displacement monitoring, one could predict when a fault was approached during tunnel advance.

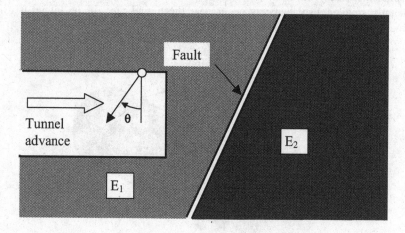

Figure 15.12   Problem statement

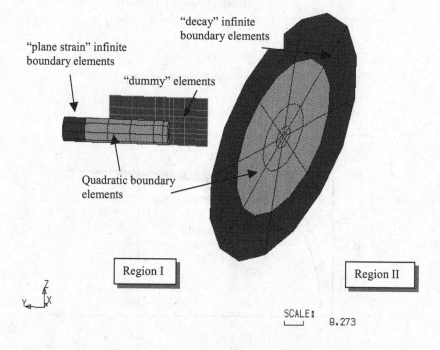

Figure 15.13   Multi-region boundary element mesh

As one cannot see through the rock mass and geological survey data are sometimes incomplete, it is very helpful for the tunnel engineer on site to have knowledge of such conditions ahead of time A numerical analysis was performed to verify this theory[7]. Figure 15.12 shows a schematic cross-section of the problem, as well as the definition of a positive angle in the inclination of the computed displacement vector. In the analysis it was assumed that the region to the right hand side of the fault was stiffer, i.e. had a higher modulus of elasticity. The boundary element mesh for this problem is shown in Figure 15.13.

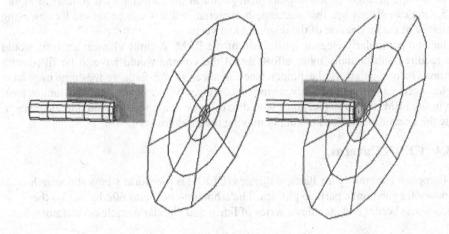

Figure 15.14 Change of maximum shear stress as tunnel advances fault

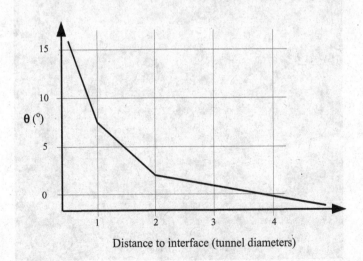

Figure 15.15 Change in angle of displacement vector with distance from tunnel face

Different types of infinite boundary elements are used at the interface between regions and where the tunnel is assumed to extend in infinite direction. In the first case, 'decay' type infinite elements were employed, whereas for the second, 'plane strain' infinite elements, as discussed in the previous sections, were applied. For 'decay' elements, a function of 1/r for the variation of displacements in the infinite direction was assumed, whereas for the finite direction a quadratic variation was assumed. For the display of stresses inside the rock mass, 'dummy elements' were used. The tunnel advance was modelled simply by moving the boundary element mesh that models the tunnel surface forward towards the fault.

The change in angle of the displacement vector at the tunnel wall is plotted in Figure 15.15. One can clearly see that a change in material stiffness can be sensed by observing the change in the inclination of the displacement vector.

This is a particularly elegant application of the BEM. A finite element analysis would have required substantially more effort, as all the volume would have to be filled with elements. In particular, as the tunnel face advances to the fault, re-meshing may have become necessary because highly distorted elements would occur near the tunnel face. Also in the FEM, the tunnel advance would have to be modelled by removing elements to create the excavation, instead of simply moving the mesh forward.

15.3.4 CERN Caverns

The European Laboratory for Particle Physics (CERN) is the world's largest research laboratory for subatomic particle physics. The laboratory occupies 602 ha. across the Franco-Swiss border and includes a series of linear and circular particle accelerators.

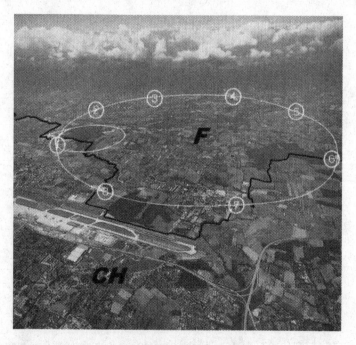

Figure 15.16 Photo showing location of the CERN particle accelerator

The main Large Electron Positron (LEP) accelerator has a circumference of 26.7 km
and a series of underground structures situated at eight access and detector points (Figure
15.16). The LEP accelerator has been operating since 1989, but in 2000 it was due to be
shut down to be replaced by the Large Hadron Collider (LHC) in 2005. This will use all
existing LEP structures, but will also require new surface and underground works. Two
new detectors will be installed in two separated cavern systems, called Point 1 and 5.

Here we will present the three-dimensional analysis of the new caverns of Point 5[6,7]
(Figure 15.17). This is an interesting application because point 1 of the LHC was analysed
using the finite element method and a picture of the results appeared on the cover of the
book *Programming the Finite Element Method*[8]. According to a report published on that
study[9] the mesh had approx 300 000 degrees of freedom, and a supercomputer was
required to solve the problem.

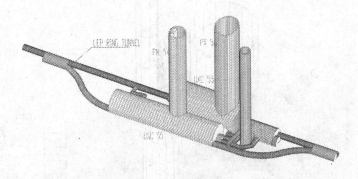

Figure 15.17 Cavern system at Point 5: existing and new structures

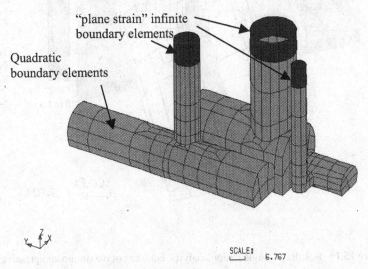

Figure 15.18 Boundary element mesh, single region analysis

Initially, an elastic analysis was carried out with the single region BE mesh shown in Figure 15.18. The aim of the analysis was to ascertain the range of validity of 2-D analyses carried out with a distinct element code. The overburden above the crown is about 75 m. In the analysis, the ground surface was assumed to be far away enough so that its influence on the cavern was neglected. In order to reduce the number of unknown further 'plane strain' infinite elements were used, as indicated in Figure 15.18. The mesh has a total of 4278 unknowns, and took 10 minutes to analyse on a PC. The results of the analysis are shown in Figure 15.19. Here the maximum compressive stress is plotted on two planes inside the rock mass. Looking at the horizontal result plane, it can be seen that at a cross-section between the vertical shafts, nearly plane strain conditions are obtained, warranting a 2-D analysis there.

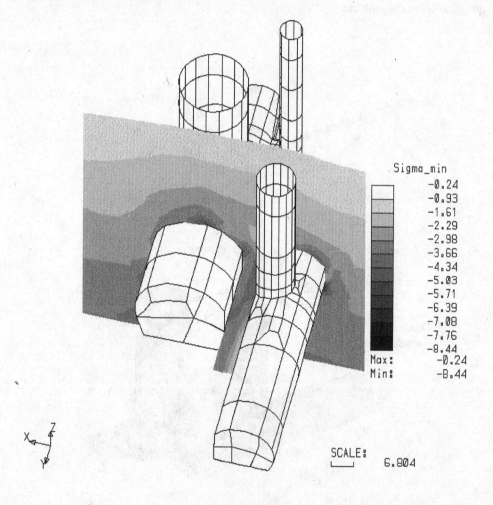

Figure 15.19 Results of single region analysis: contours of maximum compressive stress

Geologists found that a portion of the soil above the cavern could swell significantly if subjected to moisture. Therefore, an analysis had to be carried out to determine the effect of swelling on the final concrete lining. Obviously, this cannot be simplified as a 2-D problem, because the concrete lining acts as a 3-D shell structure. For this analysis a coupled finite element/boundary element analysis, as explained in Chapter 14, was performed with the thin concrete shell and the swelling zone modelled by finite elements. The main reason for the choice of this method was that due to time limitations the job had to be completed quickly, and only standard PCs were available for performing the analysis.

The coupled mesh of cavern USC 55 is shown in Figure 15.20. The mesh has a total of 7575 degrees of freedom, and the run took 45 minutes on a standard PC. Most of the computing time was for computation of the stiffness matrix of the boundary element region.

The displacements of the concrete lining due to swelling were determined from the analysis. These are shown in Figure 15.21. From these displacements the internal forces in the shell (bending moment and normal force) could be determined and used for designing the reinforcement.

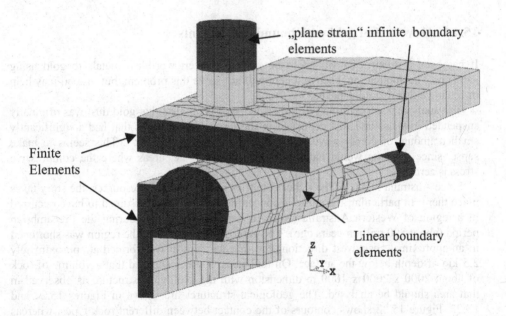

Figure 15.20 Coupled boundary element / finite element mesh of USC55 cavern

The analysis shown here demonstrates that with limited resources available (time and computer) boundary element and coupled analysis offer an efficient alternative to the FEM. Note that the coupled analysis presented here could have been performed with a multi-region analysis using internal cells, where one region would be used to represent the infinite rock mass and the others the concrete lining.

The volume cells would model the effect of the swelling zone, as explained in detail in Chapter 12.

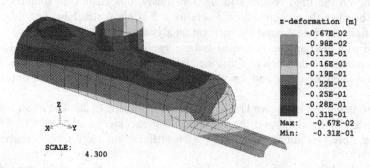

Figure 15.21 Displacements of the concrete shell due to swelling

15.4   GEOLOGICAL ENGINEERING

15.4.1 How to find gold with boundary elements

It has been a dream of mankind to be able to convert worthless metals to gold using alchemy. The application discussed here will not solve this problem, but instead may help in finding existing gold deposits.

The analysis was performed to test a theory of geologists that gold dust was originally suspended in water and was deposited in the ground in locations that had a significantly smaller amount of compressive stress than the surrounding rock[10]. This seems to make sense, since deposits would naturally occur in voids, i.e. areas where the compressive stress is zero.

Since Australia is one of the richer countries in terms of gold resources, the story takes place there. In particular, the analysis concentrates on what is presumed to have occurred in a region of Western Australia (where a deposit was found) during the Precambrian period (about 800 million years ago). The geologists assume that the region was shortened in an approximate east/west direction, and that the deposit was formed at approximately 2.5 km of depth below the surface. On this basis, it was suggested that a volume of rock of about 2000 x 2000 x 1000 m dimension with the geological structure as observed in that area should be analysed. The geological structures are shown in Figures 15.22 and 15.23. Figure 15.22 shows contours of the contact between different rock types, whereas Figure 15.23 shows contours of two faults (Lucky and Golden).

It was assumed that the block was subjected to 1000 m of overburden (which was subsequently eroded) and to tectonic stresses which were estimated from the presumed shortening of the region. For the analysis, a multi-region boundary element method was used with special contact/joint algorithms implemented on the interfaces between regions.

Figure 15.24 shows a view of the four regions considered. Figure 15.25 shows the block analysed with stress boundary conditions applied. In this figure the deformation of the blocks and the movements on the Golden and Lucky faults can be seen.

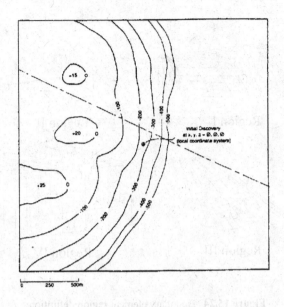

Figure 15.22 Contours of contact between different rock types

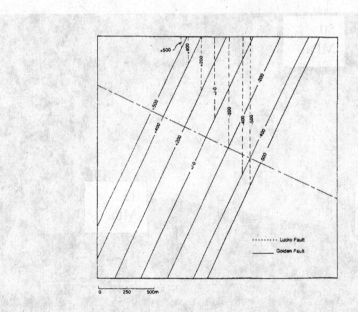

Figure 15.23 Contours of Lucky and Golden faults

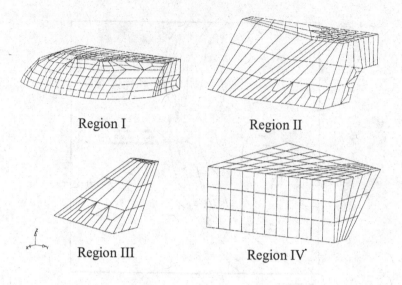

Region I   Region II

Region III   Region IV

Figure 15.24 Boundary element region definition

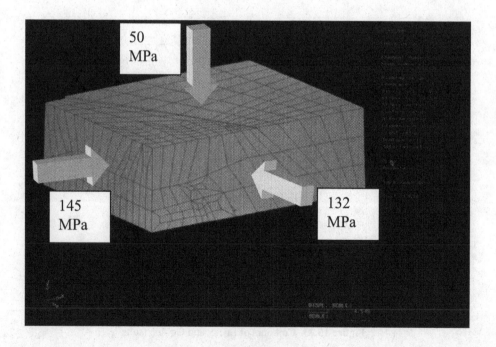

Figure 15.25 Block analysed and stress boundary conditions

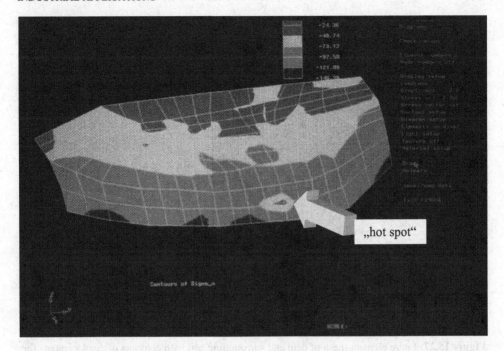

Figure 15.26   Contours of maximum compressive principal stress (light grey = low stress)

The results of the analysis can be seen in Figure 15.26 as contours of the maximum (compressive) principal stress on the contact between regions I and II. One can clearly see an anomaly of the compressive stress, and this is near the location where the gold deposit was assumed to be. So the boundary element method was successfully applied to find gold deposits.

15.5   CIVIL ENGINEERING

15.5.1   Arch dam

There are basically two different types which can be built: gravity or arch dams. The design of the second one needs the most effort in terms of stability analysis and parameter variation. While for gravity dams only a plane analysis is required arch dam analysis is always a full 3D problem, where the stress state and deformation behaviour is controlled by a range of parameters such as:

- stiffness of foundation rock and dam concrete
- valley shape (bottom width, abutment slopes)
- dam shape (vertical and horizontal curvatures)
- base cracking due to upstream tensile stresses
- intersection angle of dam into abutments, etc.

For determining the most optimal shape, a number of analyses may need to be carried out and it is a requirement that these be efficient[7]. We can see in Figure 15.27 that for an analysis with the FEM a large number of elements is used to model the soil/rock mass surrounding the dam.

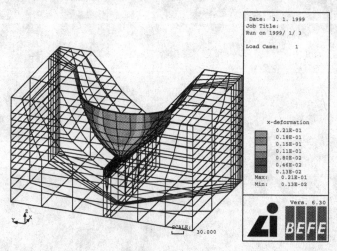

Figure 15.27 Finite element mesh of dam and surrounding soil with contours of displacements for load case reservoir full

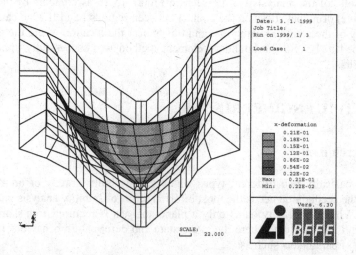

Figure 15.28 Coupled boundary element/finite element mesh of dam and valley floor with contours of displacements for load case reservoir full

A compromise is a coupled analysis where the dam is discretised into finite elements and the valley floor modelled by boundary elements. This coupled mesh is shown in

Figure 15.28. Although mesh generation and run-times of the coupled mesh are significantly smaller than for the FEM mesh, the results of both analysis are comparable. One could go one step further in this by discretising both the dam and the valley floor into boundary elements and performing a multi-region analysis.

It is common procedure to allow some small amount of crack opening at the foundations of the dam. This can be considered by implementing the contact algorithm, as explained in Chapter 13.

15.6   CONCLUSIONS

In this chapter we have attempted to show using some practical applications, that the method is not only of academic interest but can be used to solve real life problems. We have purposely concentrated on applications where the BEM has been shown to have a distinct advantage over the FEM in terms of effort to generate the mesh and in terms of computing resources.

However, we do not make the claim that the BEM is always superior to the FEM, and to be fair have included two applications where a combination of the BEM and the FEM leads to the best results. Indeed, we believe that the analyst should be given a choice by the program used and make a case for more commercial software which allows the use of either method independently or in combination.

15.7   REFERENCES

1. Beer, G. and Watson, J.O. (1991) *Introduction to finite and boundary element methods for engineers*. Wiley, Chichester.
2. Payer, H.J. and Mang, H.A. (1997) Iterative strategies for solving systems of linear algebraic equations arising in 3D BE-FE Analyses of tunnel driving. *Numerical Linear Algebra with Applications*, 4(3), 239-268.
3. Smith, I.M., Wong, S.W., Gladwell, I. and Gilvary, B. (1989) PCG methods in transient FE analysis, Part I: First order problems. *Int. J. Numer. Methods Eng.*, **28** (7), 1557-1566.
4. Beer, G. (2000) BEFE Users Manual, CSS, Geidorfgürtel 46, Graz, Austria.
5. Tonon, F. (2000) PhD dissertation, University of Colorado, Boulder.
6. Beer, G., Sigl, O. and Brandl, J. (1997) Recent developments and application of the boundary element method. *Numerical Models in Geomechanics*. (S. Pietruszczak and G.N. Pande, eds.) Balkema, Rotterdam, 461-467.
7. Beer, G., Golser, H., Jedlitschka, G. and Zacher, P. (1999) Coupled finite element/boundary element analysis in rock mechanics - industrial applications. *Rock Mechanics for Industry*, (B. Amadei, R.L. Kranz, G. A. Scott and P. Smeallie, eds.) Balkema, Rotterdam. 133-140.
8. Smith, I.M. and Griffiths, D.V. (1998) *Programming the Finite Element Method* (3rd ed.). Wiley, Chichester.

9. Sloan, A., Moy, D. and Kidger, D. (1996) 3D modelling for underground excavation at point 1, CERN. EUROCK '98. Balkema, Rotterdam.
10. Beer, G. and Poulsen, B.A. (1994) Efficient numerical modelling of faulted rock using the boundary element method. *Int. J. Rock. Mech. Min. Sci. & Geomech. Abstr.*, **31** (5),485-506.

Appendix A
Program libraries

A.1. UTILITY_LIB

```
MODULE Utility_lib
!    Utility programs
CONTAINS
  SUBROUTINE Error_Message(TEXT)
  !-----------------------------------
  ! Writes an error message onto an error file
  ! and the console and terminates the program
  !-----------------------------------
  implicit none
  CHARACTER (LEN=*) TEXT
  LOGICAL :: EXST
  INQUIRE(FILE='ERR.DAT', EXIST= EXST)
  IF(EXST) THEN
   OPEN (UNIT=99,FILE='ERR.DAT',STATUS='OLD',FORM='FORMATTED',POSITION='APPEND')
  ELSE
   OPEN (UNIT=99,FILE='ERR.DAT',STATUS='NEW',FORM='FORMATTED')
  END IF
   WRITE(99,'(A)') TEXT
  CALL PERROR('Fatal Error, see file ERR.DAT')
  CALL EXIT(1)
  END SUBROUTINE Error_Message

  SUBROUTINE Solve(Lhs,Rhs,F)
  !-----------------------------------
  !    Solution of system of equations
  !    by Gauss Elimination
  !-----------------------------------
  IMPLICIT NONE
  REAL(KIND=8)::   Lhs(:,:)   !   Equation left hand side
  REAL(KIND=8)::   Rhs(:)     !   Equation right hand side
  REAL(KIND=8)::   F(:)       !   Unknowns
  REAL(KIND=8)::   FAC
  INTEGER::        M          !   Size of system
  INTEGER::        i,n
  M= UBOUND(Lhs,1)
  ! Reduction
  Equation_n:DO n=1,M-1
          IF(Lhs(n,n) < 1.0E-10 .and. Lhs(n,n) > -1.0E-10) THEN
    CALL Error_Message('Singular Matrix')
          END IF
   Equation_i: DO i=n+1,M
```

```
              FAC= Lhs(i,n)/Lhs(n,n)
              Lhs(i,n+1:M)= Lhs(i,n+1:M) - Lhs(n,n+1:M)*FAC
              Rhs(i)= Rhs(i) - Rhs(n)*FAC
            END DO   Equation_i
         END DO Equation_n
   !    Backsubstitution
   Unknown_n: DO n= M,1,-1
            F(n)= -1.0/Lhs(n,n)*(SUM(Lhs(n,n+1:M)*F(n+1:M)) - Rhs(n))
      END DO Unknown_n
   RETURN
   END SUBROUTINE Solve


   SUBROUTINE Assembly(Lhs,Rhs,DTe,DUe,Ldest,BCode,Ncode &
         ,Elres_u,Elres_te,Diag,Ndofe,Ndof,Nodel,Fac)
   !--------------------------------------------
   ! Assembles Element contributions DTe, DUe
   ! into global matrix Lhs and vector Rhs
   ! Also sums off-diagonal coefficients
   ! for the computation of diagonal coefficients
   !--------------------------------------------
   REAL(KIND=8) :: Lhs(:,:),Rhs(:)                  !   Global arrays
   REAL(KIND=8), INTENT(IN) :: DTe(:,:),DUe(:,:)    !   Element arrays
   INTEGER , INTENT(IN) :: LDest(:)                 !   Element destination vector
   INTEGER , INTENT(IN) :: BCode(:)                 !   Boundary code (local)
   INTEGER , INTENT(IN) :: NCode(:)                 !   Boundary code (global)
   INTEGER , INTENT(IN) :: Ndofe                    !   D.o.F´s / Elem
   INTEGER , INTENT(IN) :: Ndof                     !   D.o.F´s / Node
   INTEGER , INTENT(IN) :: Nodel                    !   Nodes/Element
   REAL , INTENT(IN) :: Elres_u(:)                  !   Vector u for element
   REAL , INTENT(IN) :: Elres_te(:)                 !   Vector t for element
   REAL , INTENT(IN) :: Fac(:)                      !   Mult. factors for symmetry
   REAL(KIND=8) :: Diag(:,:)                        !   Array containing diagonal coeff of DT
   INTEGER :: n,Ncol
   DoF_per_Element:&
   DO m=1,Ndofe
      Ncol=Ldest(m)    !  Column number
      IF(BCode(m) == 0) THEN    !  Neumann BC
        Rhs(:) = Rhs(:) + DUe(:,m)*Elres_te(m)*Fac(m)
   !     The assembly of dTe depends on the global BC
        IF (NCode(Ldest(m)) == 0 .and. Ncol /= 0) THEN
          Lhs(:,Ncol)= Lhs(:,Ncol) + DTe(:,m)*Fac(m)
        ELSE
          Rhs(:) = Rhs(:) - DTe(:,m) * Elres_u(m)*Fac(m)
        END IF
      END IF
      IF(BCode(m) == 1) THEN    !  Dirichlet BC
        Lhs(:,Ncol) = Lhs(:,Ncol) - DUe(:,m)*Fac(m)
        Rhs(:)= Rhs(:) - DTe(:,m) * Elres_u(m)*Fac(m)
      END IF
   END DO &
   DoF_per_Element
   !       Sum of off-diagonal coefficients
   DO n=1,Nodel
      DO k=1,Ndof
        l=(n-1)*Ndof+k
        Diag(:,k)= Diag(:,k) - DTe(:,l)
      END DO
   END DO
      RETURN
```

END SUBROUTINE Assembly

```
SUBROUTINE Mirror(Isym,nsy,Nodes,Elcor,Fac,Incie,Ldeste,Elres_te,Elres_ue &
             ,Nodel,Ndof,Cdim)
!-----------------------------------------
!   Creates mirror image of element
!-----------------------------------------
IMPLICIT NONE
INTEGER, INTENT(IN)       ::  Isym       !  Symmetry indicator
INTEGER, INTENT(IN) .      ::  nsy        !  Symmetry count
INTEGER, INTENT(IN)       ::  nodes      !  Highest node no
REAL, INTENT(IN OUT)      ::  Elcor(:,:) !  Coords (will be modified)
REAL, INTENT(IN OUT)      ::  Elres_te(:) !  Tractions of element
REAL, INTENT(IN OUT)      ::  Elres_ue(:) !  Displacements of element
REAL, INTENT(OUT)         ::  Fac(:)     !  Multiplication factors
INTEGER, INTENT(IN OUT)   ::  Incie(:)   !  Incidences   (will be
INTEGER, INTENT(IN OUT)   ::  Ldeste(:)  !  Destinations   modified)
INTEGER, INTENT(IN)       ::  Nodel      !  Nodes per element
INTEGER, INTENT(IN)       ::  Ndof       !  d.o.F. per Node
INTEGER, INTENT(IN)       ::  Cdim       !  Cartesian dimension
REAL  :: TD(3) ! Transformation vector (diagonal elements of T)
INTEGER :: n,m,Ison1,Ison2,Ison3,i
IF(nsy == 1)RETURN
!   Assign coefficients of TD
SELECT CASE (nsy-1)
  CASE(1)
    TD=(/-1.0,1.0,1.0/)
  CASE(2)
    TD=(/-1.0,-1.0,1.0/)
  CASE(3)
    TD=(/1.0,-1.0,1.0/)
  CASE(4)
    TD=(/1.0,1.0,-1.0/)
  CASE(5)
    TD=(/-1.0,1.0,-1.0/)
  CASE(6)
    TD=(/-1.0,-1.0,-1.0/)
  CASE(7)
    TD=(/1.0,-1.0,-1.0/)
END SELECT
!   generate coordinates and incidences
Nodes1: &
DO n=1,nodel
  Direction: &
  DO m=1,Cdim
    Elcor(m,n)= Elcor(m,n)*TD(m)
  END DO &
  Direction
  !   Check if point is on any symmetry plane
  Ison1= 0
  Ison2= 0
  Ison3= 0
  IF(Elcor(1,n)==0.0) Ison1=1
  IF(Elcor(2,n)==0.0) Ison2=1
  IF(Cdim > 2 .AND. Elcor(3,n)==0.0) Ison3=1
  !   only change incidences for unprimed nodes
  IF(ison1==1 .AND. nsy-1==1)CYCLE
  IF(ison2==1 .AND. nsy-1==3) CYCLE
  IF(ison1+ison2+ison3 > 1 .AND. nsy-1<4) CYCLE
```

```
      Incie(n)= Incie(n) + Nodes
END DO &
Nodes1
!    generate multiplication factors elast. Problems only
IF(Ndof > 1) THEN
   I=0
   Nodes2: &
   DO n=1,nodel
      Degrees_of_freedom1: &
      DO m=1,Ndof
         I=I+1
         Fac(I)= TD(m)    ! Multiplication factor for symmetry
      END DO &
      Degrees_of_freedom1
   END DO &
   Nodes2
END IF
!   Reverse destination vector for selected elem
SELECT CASE (nsy-1)
CASE (1,3,4,6)
   CALL Reverse(Incie,elcor,ldeste,Elres_te,Elres_ue,Ndof,Cdim,nodel)
CASE DEFAULT
END SELECT
RETURN
END SUBROUTINE Mirror

SUBROUTINE Reverse(Inci,elcor,ldest,Elres_te,Elres_ue,Ndof,Cdim,nodel)
!-------------------------------------
! reverses incidences, destination vector
! and co-ordinates
! so that outward normal is reversed
!-------------------------------------
IMPLICIT NONE
INTEGER, INTENT (INOUT)  ::   Inci(:)              !  Incidences
REAL, INTENT (INOUT)     ::   Elcor(:,:)           !  Coordinates
REAL, INTENT (INOUT)     ::   Elres_te(:)          !  Tractions of element
REAL, INTENT (INOUT)     ::   Elres_ue(:)          !  Displacements of element
INTEGER, INTENT (INOUT)  ::   Ldest(:)             !  Destination vector
INTEGER, INTENT (IN)     ::   Ndof                 !  No of degrees of freedom per node
INTEGER, INTENT (IN)     ::   Cdim                 !  Cartesian dimension
INTEGER, INTENT (IN)     ::   Nodel                !  No of nodes per element
REAL, ALLOCATABLE        ::   Elcort(:,:)          !  Temps
REAL, ALLOCATABLE        ::   Elres_tet(:)         !  Temps
REAL, ALLOCATABLE        ::   Elres_uet(:)         !  Temps
INTEGER, ALLOCATABLE     ::   Incit(:),Ldestt(:)   !  Temps
INTEGER                  ::   Node(8)              !  Reversing sequence
INTEGER                  ::   n,nc,Nchanges
ALLOCATE
(Incit(nodel),Elcort(Cdim,nodel),Ldestt(Nodel*ndof),Elres_tet(Nodel*ndof),Elres_uet(Nodel*ndof))
Incit= Inci
Elcort= Elcor
Ldestt= Ldest
Elres_tet= Elres_te
Elres_uet= Elres_ue
SELECT CASE (Cdim)
   CASE (2)   !  2-D problem
      Node(1:2)= (/2,1/) ; Nchanges= 2
   CASE (3)   !  3-D problem
      Node= (/1,4,3,2,8,7,6,5/) ; Nchanges= nodel !-1
```

```
END SELECT
Number_changes: &
DO n=1,Nchanges
   nc= Node(n)
   inci(n)= Incit(nc) ; Elcor(:,n)= Elcort(:,nc)
   Ldest(Ndof*(n-1)+1:Ndof*n)= Ldestt(Ndof*(nc-1)+1:Ndof*nc)
   Elres_te(Ndof*(n-1)+1:Ndof*n)=Elres_tet(Ndof*(nc-1)+1:Ndof*nc)
   Elres_ue(Ndof*(n-1)+1:Ndof*n)=Elres_uet(Ndof*(nc-1)+1:Ndof*nc)
END DO &
Number_changes
DEALLOCATE(Incit,Elcort,Ldestt,Elres_tet,Elres_uet)
RETURN
END SUBROUTINE Reverse

SUBROUTINE Jobin(Title,Cdim,Ndof,Toa,Nreg,Ltyp,Con,E,ny &
         ,Isym,nodel,nodes,maxe)
!-----------------------------------------------
!   Subroutine to read in basic job information
!-----------------------------------------------
CHARACTER(LEN=80), INTENT(OUT):: Title
INTEGER, INTENT(OUT) :: Cdim,Ndof,Toa,Nreg,Ltyp,Isym,nodel
INTEGER, INTENT(OUT) :: Nodes,Maxe
REAL, INTENT(OUT)    :: Con,E,ny
READ(1,'(A80)') Title
WRITE(2,*)'Project:',Title
READ(1,*) Cdim
WRITE(2,*)'Cartesian_dimension:',Cdim
READ(1,*) Ndof      !  Degrees of freedom per node
IF(NDof == 1) THEN
    WRITE(2,*)'Potential Problem'
ELSE
    WRITE(2,*)'Elasticity Problem'
END IF
IF(Ndof == 2)THEN
    READ(1,*) Toa        ! Toa ....Type of analysis (solid plane strain = 1,solid plane stress = 2)
    IF(Toa == 1)THEN
       WRITE(2,*)'Type of Analysis: Solid Plane Strain'
    ELSE
       WRITE(2,*)'Type of Analysis: Solid Plane Stress'
    END IF
END IF
READ(1,*) Nreg     !  Type of region
IF(NReg == 1) THEN
    WRITE(2,*)'Finite Region'
ELSE
    WRITE(2,*)'Infinite Region'
END IF
READ(1,*) Isym     !  Symmetry code
SELECT CASE (isym)
CASE(0)
WRITE(2,*)'No symmetry'
CASE(1)
WRITE(2,*)'Symmetry about y-z plane'
CASE(2)
WRITE(2,*)'Symmetry about y-z and x-z planes'
CASE(3)
WRITE(2,*)'Symmetry about all planes'
END SELECT
READ(1,*) Ltyp     !  Element type
```

```
IF(Ltyp == 1) THEN
WRITE(2,*)'Linear Elements'
ELSE
WRITE(2,*)'Quadratic Elements'
END IF
!   Determine number of nodes per element
IF(Cdim == 2) THEN   !   Line elements
 IF(Ltyp == 1) THEN
 Nodel= 2
 ELSE
 Nodel= 3
 END IF
ELSE                 !   Surface elements
 IF(Ltyp == 1) THEN
 Nodel= 4
 ELSE
 Nodel= 8
 END IF
END IF
!   Read properties
IF(Ndof == 1) THEN
   READ(1,*) Con
   WRITE(2,*)'Conductivity=',Con
ELSE
   READ(1,*) E,ny
   IF(ToA == 2) ny = ny/(1+ny)        !  Solid Plane Stress
   WRITE(2,*)'Modulus:',E
   WRITE(2,*)'Poissons ratio:',ny
END IF

READ(1,*) Nodes
WRITE(2,*)'Number of Nodes of System:',Nodes
READ(1,*) Maxe
WRITE(2,*)'Number of Elements of System:', Maxe
RETURN
END SUBROUTINE Jobin

SUBROUTINE JobinMR(Title,Cdim,Ndof,Toa,Ltyp,Isym,nodel,nodes,maxe)
!------------------------------------------------
!   Subroutine to read in basic job information
!------------------------------------------------
IMPLICIT NONE
INTEGER, INTENT(OUT)     :: Cdim      !  Cartesian dimension
INTEGER                  :: Idim      !  Dimension of Element
INTEGER, INTENT(OUT)     :: Ndof      !  No. of degeres of freedom per node
INTEGER, INTENT(OUT)     :: Toa       !  Type of analysis (plane strain = 1, plane stress = 2)
INTEGER                  :: Ltyp      !  Element type (linear = 1, quadratic = 2)
INTEGER, INTENT(OUT)     :: Nodel     !  No. of nodes per element
INTEGER, INTENT(OUT)     :: Nodes     !  No. of nodes of system
INTEGER, INTENT(OUT)     :: Maxe      !  Number of elements of system
INTEGER                  :: Isym      !  Symmetry code
CHARACTER(LEN=80)        :: Title     !  Title of calculation
INTEGER                  :: nr,nb

READ(1,'(A80)') Title
WRITE(2,*)'Project:',Title
READ(1,*) Cdim
WRITE(2,*)'Cartesian_dimension:',Cdim
Idim= Cdim - 1
```

```
READ(1,*) Ndof      !  Degrees of freedom per node
IF(NDof == 1) THEN
   WRITE(2,*)'Potential Problem'
ELSE
   WRITE(2,*)'Elasticity Problem'
END IF
IF(Ndof == 2)THEN
   READ(1,*) Toa        ! Toa ....Type of analysis (solid plane strain = 1, solid plane stress = 2)
   IF(Toa == 1)THEN
      WRITE(2,*)'Type of Analysis: Solid Plane Strain'
   ELSE
      WRITE(2,*)'Type of Analysis: Solid Plane Stress'
   END IF
END IF
READ(1,*) Ltyp      !  Element type
IF(Ltyp == 1) THEN
WRITE(2,*)'Linear Elements'
ELSE
WRITE(2,*)'Quadratic Elements'
END IF
!   Determine number of nodes per element
IF(Cdim == 2) THEN  !   Line elements
 IF(Ltyp == 1) THEN
   Nodel= 2
 ELSE
   Nodel= 3
 END IF
ELSE               ! Surface elements
 IF(Ltyp == 1) THEN
   Nodel= 4
 ELSE
   Nodel= 8
 END IF
END IF
READ(1,*) Nodes
WRITE(2,*)'Number of Nodes of System:',Nodes
READ(1,*) Maxe
WRITE(2,*)'Number of Elements of System:', Maxe
END SUBROUTINE JobinMR


SUBROUTINE Reg_Info(Nregs,ToA,Ndof,TypeR,ConR,ER,nyR,Nbel,ListR)
!-------------------------------------------------------------------
!  Subroutine to read in basic job information for each region
!-------------------------------------------------------------------
IMPLICIT NONE
INTEGER,INTENT(IN)     :: Nregs     ! Number of regions
INTEGER,INTENT(IN)     :: ToA       ! Type of analysis (solid plane strain = 1,solid plane stress
= 2)
INTEGER,INTENT(INOUT) :: TypeR(:)   ! Type of BE-regions (1 == finite, 2 == Infinite)
INTEGER,INTENT(INOUT) :: Nbel(:)    ! Number of boundary elements each region
INTEGER,INTENT(INOUT) :: ListR(:,:) ! List of element numbers each region
INTEGER,INTENT(IN)     :: Ndof      ! No. of degrees of freedom per node
REAL,INTENT(INOUT)     :: ConR(:)   ! Conductivity of each region
REAL,INTENT(INOUT)     :: ER(:)     ! Young's modulus of regions
REAL,INTENT(INOUT)     :: nyR(:)    ! Poisson's ratio of regions
INTEGER                :: Isym      ! Symmetry code
INTEGER                :: nr,nb
ListR= 0
Region_loop: &
```

```
DO nr=1,Nregs
  WRITE(2,*)' Region ',nr
  READ(1,*) TypeR(nr)        !  Type of region
  IF(TypeR(nr) == 1) THEN
    WRITE(2,*)'Finite Region'
  ELSE
    WRITE(2,*)'Infinite Region'
  END IF
  READ(1,*) Isym     !  Symmetry code
  SELECT CASE (Isym)
  CASE(0)
  WRITE(2,*)'No symmetry'
  CASE(1)
  WRITE(2,*)'Symmetry about y-z plane'
  CASE(2)
  WRITE(2,*)'Symmetry about y-z and x-z planes'
  CASE(3)
  WRITE(2,*)'Symmetry about all planes'
  END SELECT
  !  Read properties
  IF(Ndof == 1) THEN
    READ(1,*) ConR(nr)
    WRITE(2,*)'Conductivity=',ConR(nr)
  ELSE
    READ(1,*) ER(nr),nyR(nr)
    IF(ToA == 2) nyR(nr) = nyR(nr)/(1+nyR(nr))              !  Solid Plane Stress
    WRITE(2,*)'Young' smodulus:',ER(nr)
    WRITE(2,*)'Poisson's ratio:',nyR(nr)
  END IF
  READ(1,*)Nbel(nr)
! IF(Nbel(nr) > MaxeR)MaxeR= Nbel(nr)
  READ(1,*)(ListR(nr,nb),nb=1,Nbel(nr))
  WRITE(2,*) ' List of Boundary Elements: '
  WRITE(2,*)(ListR(nr,nb),nb=1,Nbel(nr))
END DO &
Region_loop
RETURN
END SUBROUTINE Reg_info

SUBROUTINE BCInput(Elres_u,Elres_t,Bcode,nodel,ndofe,ndof)
!-----------------------------------------
!       Reads boundary conditions
!-----------------------------------------
REAL,INTENT(OUT)      :: Elres_u(:,:) !  Element results , u
REAL,INTENT(OUT)      :: Elres_t(:,:) !  Element results , t
INTEGER,INTENT(OUT)   :: BCode(:,:)  !  Element BC's
INTEGER,INTENT(IN)    :: nodel       !  Nodes per element
INTEGER,INTENT(IN)    :: ndofe       !  D.o.F. per element
INTEGER,INTENT(IN)    :: ndof        !  D.o.F per node
INTEGER :: NE_u,NE_t
WRITE(2,*)''
WRITE(2,*)'Elements with Dirichlet BC's: '
WRITE(2,*)''
Elres_u(:,:)=0  ! Default prescribed values for u = 0.0
BCode = 0        ! Default BC= Neumann Condition
READ(1,*)NE_u
IF(NE_u > 0) THEN
Elem_presc_displ: &
DO n=1,NE_u
```

```
READ(1,*) Nel,(Elres_u(Nel,m),m=1,Ndofe)
READ(1,*) Nel,(BCode(Nel,m),m=1,Ndofe)
! BCode(Nel,:)=1
WRITE(2,*)'Element ',Nel,' Prescribed values: '
Na= 1
Nodes: &
DO M= 1,Nodel
   WRITE(2,*) Elres_u(Nel,na:na+ndof-1)
   Na= na+Ndof
END DO &
Nodes
END DO &
Elem_presc_displ
END IF
WRITE(2,*)"
WRITE(2,*)'Elements with Neuman BC´s: '
WRITE(2,*)"
Elres_t(:,:)=0   !  Default prescribed values = 0.0
READ(1,*)NE_t
Elem_presc_trac:  &
DO n=1,NE_t
   READ(1,*) Nel,(Elres_t(Nel,m),m=1,Ndofe)
   WRITE(2,*)'Element ',Nel,' Prescribed values: '
   Na= 1
   Nodes1: &
   DO M= 1,Nodel
      WRITE(2,*) Elres_t(Nel,na:na+ndof-1)
      Na= na+Ndof
   END DO &
   Nodes1
END DO &
Elem_presc_trac
RETURN
END SUBROUTINE BCInput

SUBROUTINE Geomin(Nodes,Maxe,xp,Inci,Nodel,Cdim)
!-----------------------------------
!  Inputs mesh geometry
!-----------------------------------
IMPLICIT NONE
INTEGER, INTENT(IN)    ::  Nodes   !  Number of nodes
INTEGER, INTENT(IN)    ::  Maxe    !  Number of elements
INTEGER, INTENT(IN)    ::  Nodel   !  Number of nodes of elements
INTEGER, INTENT(IN)    ::  Cdim    !  Cartesian dimension
REAL, INTENT(OUT)      ::  xP(:,:)  !  Node co-ordinates
REAL                   ::  xmax(Cdim),xmin(Cdim),delta_x(Cdim)
INTEGER, INTENT(OUT) ::  Inci(:,:) !  Element incidences
INTEGER                ::  Node,Nel,M,n
!------------------------------------------------------------
!    Read Node Co-ordinates from Input file
!------------------------------------------------------------
DO Node=1,Nodes
 READ(1,*) (xP(M,Node),M=1,Cdim)
 WRITE(2,'(A5,I5,A8,3F8.2)') 'Node ',Node,&
     ' Coor ',(xP(M,Node),M=1,Cdim)
END DO
```

```
!-------------------------------------------------
!    Read Incidences from Input file
!-------------------------------------------------
WRITE(2,*)"
WRITE(2,*)'Incidences: '
WRITE(2,*)"
Elements_1:&
   DO Nel=1,Maxe
READ(1,*) (Inci(Nel,n),n=1,Nodel)
WRITE(2,'(A3,I5,A8,24I5)')'EL ',Nel,'  Inci  ',Inci(Nel,:)
END DO &
Elements_1
RETURN
END SUBROUTINE Geomin

LOGICAL FUNCTION Match(Inci1,Inci2)
!----------------------------------
!    Returns a value of TRUE if the incidences
!    Inci1 and Inci2 match
!----------------------------------
IMPLICIT NONE
INTEGER, INTENT (IN) :: Inci1(:) !  1. incidence array
INTEGER, INTENT (IN) :: Inci2(:) !  2. incidence array
INTEGER :: Nodes,Node,N1,N2,Ncount
Nodes= UBOUND(Inci1,1)
Ncount= 0
Node_loop1: &
DO N1=1,Nodes
   Node= Inci1(n1)
   Node_loop2: &
   DO N2=1,Nodes
      IF(Node == Inci2(n2)) Ncount= Ncount+1
   END DO &
   Node_loop2
END DO &
Node_loop1
IF(Ncount == Nodes) THEN
 Match= .TRUE.
ELSE
 Match= .FALSE.
END IF
END FUNCTION Match


SUBROUTINE Destination(Isym,Ndest,Ldest,xP,Inci,Ndofs,nodes,Ndof,Nodel,Maxe)
!------------------------------------------------------------------
! Determine Node destination vector and Element destination vector
!------------------------------------------------------------------
IMPLICIT NONE
REAL, INTENT (IN)              :: xP(:,:)        !  Node co-ordinates
INTEGER, INTENT (IN OUT)       :: Ndest(:,:)     !  Node destination vector
INTEGER, INTENT (IN OUT)       :: Ldest(:,:)     !  Element destination vector
INTEGER, INTENT (IN OUT)       :: Ndofs          !  DoF's of system
INTEGER, INTENT (IN)           :: Inci(:,:)      !  Element incidences
INTEGER, INTENT (IN)           :: Isym,nodes,Ndof,Nodel,Maxe
INTEGER                        :: k,m,n,Nel,l
!-----------------------------------------------
!    Determine Node destination vector
!       Set Ndest == 0 if Point is on a symmetry plane
!-----------------------------------------------
```

```
!    no symmetry
IF(Isym == 0) THEN
   k=1
   Nodes0:  DO m=1, nodes
                DO n=1, Ndof
                   Ndest(m,n)= k
                   k=k+1
                END DO
            END DO Nodes0
!    y-z symmetry
ELSE IF(Isym == 1) THEN
   k=1
   Nodes1:  DO m=1, nodes
                IF(xP(1,m) == 0.0)THEN
                   Ndest(m,1)= 0
                ELSE
                   Ndest(m,1)= k
                   k=k+1
                END IF
                DO n=2, Ndof
                   Ndest(m,n)= k
                   k=k+1
                END DO
            END DO Nodes1
!    x-z and y-z symmetry
ELSE IF(Isym == 2) THEN
   k=1
   Nodes2:  DO m=1, nodes
                IF(xP(1,m) == 0.0)THEN
                   Ndest(m,1)= 0
                ELSE
                   Ndest(m,1)= k
                   k=k+1
                END IF
                IF(xP(2,m) == 0.0)THEN
                   Ndest(m,2)= 0
                ELSE
                   Ndest(m,2)= k
                   k=k+1
                END IF
                IF (Ndof == 3) THEN
                   Ndest(m,3)= k
                   k=k+1
                END IF
            END DO Nodes2
!    x-y, x-z and y-z symmetry
ELSE
   k=1
   Nodes3: DO m=1, nodes
                IF(xP(1,m) == 0.0)THEN
                   Ndest(m,1)= 0
                ELSE
                   Ndest(m,1)= k
                   k=k+1
                END IF
                IF(xP(2,m) == 0.0)THEN
                   Ndest(m,2)= 0
                ELSE
                   Ndest(m,2)= k
```

```
                    k=k+1
                 END IF
                 IF(xP(3,m) == 0.0)THEN
                    Ndest(m,3)= 0
                 ELSE
                    Ndest(m,3)= k
                    k=k+1
                 END IF
              END DO Nodes3
      END IF
      Ndofs= k-1                 ! DoF's of System
      !-------------------------------------------
      !   Determine Element destination vector
      !-------------------------------------------
      Elements:&
      DO Nel=1,Maxe
        DO n=1,Nodel
          k= (n-1)*Ndof+1
          l= n*Ndof
          Ldest(Nel,k:l)=Ndest(Inci(Nel,n),:)
        END DO
      END DO &
      Elements
      END SUBROUTINE Destination

      SUBROUTINE Scal(E,xP,Elres_u,Elres_t,Cdim,Scad,Scat)
      IMPLICIT NONE
      REAL, INTENT (INOUT)  :: E                 ! Young's modulus
      REAL, INTENT (INOUT)  :: xP(:,:)           ! Node co-ordinates
      REAL, INTENT (INOUT)  :: Elres_u(:,:)      ! Element results, u
      REAL, INTENT (INOUT)  :: Elres_t(:,:)      ! Element results, t
      REAL, INTENT (OUT)    :: Scad
      REAL, INTENT (OUT)    :: Scat
      INTEGER, INTENT(IN)   :: Cdim              ! Cartesian dimension
      REAL                  :: xmax(Cdim),xmin(Cdim),delta_x(Cdim)


      !-----------------------------------------------
      !   Determine Scalefactor for Tractions
      !   Scat ... 1/E
      !-----------------------------------------------
      Scat= 1./E                             ! Scalefactor for tractions
      E=1.0                                  ! Scaled Young's modulus
      Elres_t=Elres_t*Scat                   ! Scaled prescribed tractions by Scat
      !-----------------------------------------------------
      !   Determine Scalefactor for Displacements
      !   Scad ... max. Distance in any co-ordinate direction
      !-----------------------------------------------------
      xmax(1)= MAXVAL(xp(1,:))
      xmax(2)= MAXVAL(xp(2,:))
      IF(Cdim == 3)xmax(3)= MAXVAL(xp(3,:))
      xmin(1)= MINVAL(xp(1,:))
      xmin(2)= MINVAL(xp(2,:))
      IF(Cdim == 3)xmin(3)= MINVAL(xp(3,:))
      delta_x= xmax - xmin
      Scad= MAXVAL(delta_x)                  ! Scad ... Scale factor for displacements
      xP=xP/Scad                             ! Scaled Node co-ordinates
      Elres_u=Elres_u/Scad                   ! Scaled prescribed displacements by Scad
      END SUBROUTINE Scal
      END MODULE Utility_lib
```

A.2. GEOMETRY_LIB

```
!    Last change: CD   25 Mar 2000   8:10 pm
MODULE Geometry_lib
USE Utility_lib
CONTAINS
SUBROUTINE Normal_Jac(v3,Jac,xsi,eta,ldim,nodes,inci,coords)
!------------------------------------------------------
!  Computes normal vector and Jacobian
!  at point with local coordinates xsi,eta
!------------------------------------------------------
IMPLICIT NONE
REAL, INTENT(IN):: xsi,eta          !    Intrinsic co-ordinates of point
INTEGER,INTENT(IN):: ldim           !    Element dimension
INTEGER,INTENT(IN):: nodes          !    Number of nodes
INTEGER,INTENT(IN):: inci(:)        !    Element incidences
REAL, INTENT(IN)  :: coords(:,:)    !    Node coordinates
REAL,INTENT(OUT):: v3(:)            !    Vector normal to point
REAL,INTENT(OUT):: Jac              !    Jacobian
REAL,ALLOCATABLE  :: DNi(:,:)       !    Derivatives of shape function
REAL,ALLOCATABLE  :: v1(:),v2(:)    !    Vectors in xsi,eta directions
INTEGER :: Cdim ,i                  !    Cartesian dimension
!    Cartesian dimension:
 Cdim= ldim+1
!    Allocate temporary arrays
ALLOCATE (DNi(nodes,ldim),V1(Cdim),V2(Cdim))
!    Compute derivatives of shape function
Call Serendip_deriv(DNi,xsi,eta,ldim,nodes,inci)
!    Compute vectors in xsi (eta) direction(s)
DO I=1,Cdim
  V1(I)= DOT_PRODUCT(DNi(:,1),COORDS(I,:))
  IF(ldim == 2) THEN
    V2(I)= DOT_PRODUCT(DNi(:,2),COORDS(I,:))
  END IF
END DO
!    Compute normal vector
IF(ldim == 1) THEN
   v3(1)= V1(2)
   v3(2)= -V1(1)
   ELSE
  V3= Vector_ex(v1,v2)
END IF
!   Normalise
CAll Vector_norm(V3,Jac)
DEALLOCATE (DNi,V1,V2)
RETURN
END SUBROUTINE Normal_Jac


SUBROUTINE Tangent(v1,v2,xsi,eta,ldim,nodes,inci,coords)
!------------------------------------------------------
!  Computes vectors tangent to BE
!  at point with local coordinates xsi,eta
!------------------------------------------------------
IMPLICIT NONE
INTEGER,INTENT(IN):: ldim           ! element dimension
REAL, INTENT(IN)  :: xsi,eta        !   Intrinsic co-ordinates of point
INTEGER,INTENT(IN):: nodes          !   Number of nodes
INTEGER,INTENT(IN):: inci(:)        !   Element incidences
REAL, INTENT(IN)  :: coords(:,:)    !   Node coordinates
```

```
REAL,INTENT(OUT)  :: v1(ldim+1),v2(ldim+1) !  Vector normal to point
REAL,ALLOCATABLE  :: DNi(:,:)              !  Derivatives of shape function
!   REAL,ALLOCATABLE  :: v1(:),v2(:)        !  Vectors in xsi,eta directions
INTEGER :: Cdim ,i                         !  Cartesian dimension
!   Cartesian dimension:
Cdim= ldim+1
!   Allocate temporary arrays
ALLOCATE (DNi(nodes,ldim))
!   Compute derivatives of shape function
Call Serendip_deriv(DNi,xsi,eta,ldim,nodes,inci)
!   Compute vectors in xsi (eta) direction(s)
DO i=1,Cdim
   v1(i)= DOT_PRODUCT(DNi(:,1),COORDS(i,:))
   IF(ldim == 2) THEN
      v2(i)= DOT_PRODUCT(DNi(:,2),COORDS(i,:))
   END IF
END DO
DEALLOCATE (DNi)
RETURN
END SUBROUTINE Tangent

Subroutine Serendip_func(Ni,xsi,eta,ldim,nodes,inci)
!-------------------------------
!   Computes Serendipity shape functions  Ni(xsi,eta)
!   for one and two-dimensional (linear/parabolic) finite boundary elements
!-------------------------------
REAL,INTENT(OUT):: Ni(:)            !  Array with shape function values at xsi,eta
REAL, INTENT(IN):: xsi,eta          !  Intrinsic co-ordinates
INTEGER,INTENT(IN):: ldim           !  Element dimension
INTEGER,INTENT(IN):: nodes          !  Number of nodes
INTEGER,INTENT(IN):: inci(:)        !  Element incidences
REAL              :: mxs,pxs,met,pet !  Temporary variables
SELECT CASE (ldim)
   CASE (1)  !  one-dimensional element
      Ni(1)= 0.5*(1.0 - xsi) ;  Ni(2)= 0.5*(1.0 + xsi)
      IF(nodes == 2) RETURN !  linear element finished
      Ni(3)=  1.0 - xsi*xsi
      Ni(1)= Ni(1) - 0.5*Ni(3) ; Ni(2)= Ni(2) - 0.5*Ni(3)
   CASE(2)  !   two-dimensional element
      mxs= 1.0-xsi ; pxs= 1.0+xsi ; met= 1.0-eta ; pet= 1.0+eta
      Ni(1)= 0.25*mxs*met ; Ni(2)= 0.25*pxs*met
      Ni(3)= 0.25*pxs*pet ; Ni(4)= 0.25*mxs*pet
   IF(nodes == 4) RETURN   !  linear element finished
   IF(Inci(5) > 0) THEN        ! zero node number means node is missing
       Ni(5)= 0.5*(1.0 -xsi*xsi)*met
       Ni(1)= Ni(1) - 0.5*Ni(5) ; Ni(2)= Ni(2) - 0.5*Ni(5)
   END IF
   IF(Inci(6) > 0) THEN
       Ni(6)= 0.5*(1.0 -eta*eta)*pxs
       Ni(2)= Ni(2) - 0.5*Ni(6) ;  Ni(3)= Ni(3) - 0.5*Ni(6)
   END IF
   IF(Inci(7) > 0) THEN
       Ni(7)= 0.5*(1.0 -xsi*xsi)*pet
       Ni(3)= Ni(3) - 0.5*Ni(7) ; Ni(4)= Ni(4) - 0.5*Ni(7)
   END IF
   IF(Inci(8) > 0) THEN
       Ni(8)= 0.5*(1.0 -eta*eta)*mxs
       Ni(4)= Ni(4) - 0.5*Ni(8) ; Ni(1)= Ni(1) - 0.5*Ni(8)
   END IF
```

```
CASE DEFAULT    !  error message
CALL Error_message('Element dimension not 1 or 2' )
END SELECT
RETURN
END SUBROUTINE Serendip_func

SUBROUTINE Serendip_deriv(DNi,xsi,eta,ldim,nodes,inci)
!---------------------------------
!  Computes Derivatives of Serendipity shape functions Ni(xsi,eta)
!  for one and two-dimensional (linear/parabolic) finite boundary elements
!---------------------------------
IMPLICIT NONE
REAL,INTENT(OUT):: DNi(:,:)      !  Array with shape function derivatives at xsi,eta
REAL, INTENT(IN):: xsi,eta       !  Intrinsic co-ordinates
INTEGER,INTENT(IN):: ldim        !  Element dimension
INTEGER,INTENT(IN):: nodes       !  Number of nodes
INTEGER,INTENT(IN):: inci(:)     !  Element incidences
REAL         :: mxs,pxs,met,pet  !  Temporary variables
SELECT CASE (ldim)
   CASE (1)  !  one-dimensional element
      DNi(1,1)= -0.5
      DNi(2,1)= 0.5
      IF(nodes == 2) RETURN  ! linear element finished
      DNi(3,1)=  -2.0*xsi
      DNi(1,1)= DNi(1,1) - 0.5*DNi(3,1)
      DNi(2,1)= DNi(2,1) - 0.5*DNi(3,1)
   CASE (2)  !   two-dimensional element
      mxs= 1.0-xsi
      pxs= 1.0+xsi
      met= 1.0-eta
      pet= 1.0+eta
      DNi(1,1)= -0.25*met
      DNi(1,2)= -0.25*mxs
      DNi(2,1)=  0.25*met
      DNi(2,2)= -0.25*pxs
      DNi(3,1)=  0.25*pet
      DNi(3,2)=  0.25*pxs
      DNi(4,1)= -0.25*pet
      DNi(4,2)=  0.25*mxs
      IF(nodes == 4) RETURN  ! linear element finished
      IF(Inci(5) > 0) THEN   ! zero node number means node is missing
         DNi(5,1)= -xsi*met
         DNi(5,2)= -0.5*(1.0 -xsi*xsi)
         DNi(1,1)= DNi(1,1) - 0.5*DNi(5,1)
         DNi(1,2)= DNi(1,2) - 0.5*DNi(5,2)
         DNi(2,1)= DNi(2,1) - 0.5*DNi(5,1)
         DNi(2,2)= DNi(2,2) - 0.5*DNi(5,2)
      END IF
      IF(Inci(6) > 0) THEN
         DNi(6,1)= 0.5*(1.0 -eta*eta)
         DNi(6,2)= -eta*pxs
         DNi(2,1)= DNi(2,1) - 0.5*DNi(6,1)
         DNi(2,2)= DNi(2,2) - 0.5*DNi(6,2)
         DNi(3,1)= DNi(3,1) - 0.5*DNi(6,1)
         DNi(3,2)= DNi(3,2) - 0.5*DNi(6,2)
      END IF
      IF(Inci(7) > 0) THEN
         DNi(7,1)= -xsi*pet
         DNi(7,2)= 0.5*(1.0 -xsi*xsi)
```

```fortran
      DNi(3,1)= DNi(3,1) - 0.5*DNi(7,1)
      DNi(3,2)= DNi(3,2) - 0.5*DNi(7,2)
      DNi(4,1)= DNi(4,1) - 0.5*DNi(7,1)
      DNi(4,2)= DNi(4,2) - 0.5*DNi(7,2)
    END IF
    IF(Inci(8) > 0) THEN
      DNi(8,1)= -0.5*(1.0 -eta*eta)
      DNi(8,2)= -eta*mxs
      DNi(4,1)= DNi(4,1) - 0.5*DNi(8,1)
      DNi(4,2)= DNi(4,2) - 0.5*DNi(8,2)
      DNi(1,1)= DNi(1,1) - 0.5*DNi(8,1)
      DNi(1,2)= DNi(1,2) - 0.5*DNi(8,2)
    END IF
  CASE DEFAULT    !   error message
CALL Error_message('Element dimension not 1 or 2' )
END SELECT
RETURN
END SUBROUTINE Serendip_deriv

SUBROUTINE Cartesian(Ccor,Ni,Idim,elcor)
!-------------------------------------------------
!  Computes Cartesian coordinates
!  at point with local coordinates xsi,eta
!-------------------------------------------------
IMPLICIT NONE
REAL,INTENT(OUT) :: Ccor(:)        !   Cart. coords of point xsi,eta
REAL,INTENT(IN)  :: Ni(:)          !   Shape functions at xsi,eta
REAL, INTENT(IN) :: elcor(:,:)     !   Element coordinates
INTEGER          :: Idim, Cdim,I   !   Cartesian dimension
!   Cartesian dimension:
Cdim= Idim+1
!   Compute vectors in xsi (eta) direction(s)
DO I=1,Cdim
   Ccor(I)= DOT_PRODUCT(Ni(:),Elcor(I,:))
END DO
RETURN
END SUBROUTINE Cartesian

SUBROUTINE Vector_norm(v3,Vlen)
!-------------------------------------------------
!  Normalise vector
!-------------------------------------------------
IMPLICIT NONE
REAL, INTENT(INOUT) :: V3(:)       !  Vector to be normalised
REAL, INTENT(OUT)   :: Vlen        !  Length of vector
Vlen= SQRT( SUM(v3*v3))
IF(Vlen == 0.) RETURN
V3= V3/Vlen
RETURN
END SUBROUTINE Vector_norm

FUNCTION Vector_ex(v1,v2)
!-------------------------------------------------
!  Returns vector x-product v1xv2
!  Where v1 and v2 are dimension 3
!-------------------------------------------------
IMPLICIT NONE
REAL, INTENT(IN) :: V1(3),V2(3)    !   Input
REAL             :: Vector_ex(3)   !   Result
```

```
Vector_ex(1)=V1(2)*V2(3)-V2(2)*V1(3)
Vector_ex(2)=V1(3)*V2(1)-V1(1)*V2(3)
Vector_ex(3)=V1(1)*V2(2)-V1(2)*V2(1)
RETURN
END FUNCTION Vector_ex

REAL FUNCTION Dist(xa,xe,Cdim)
!---------------------------------------
!   Computes the distance between two points
!   with coordinates (xa,ya) and (xe,ye)
!---------------------------------------
IMPLICIT NONE
REAL, INTENT(IN)    :: xa(:)      !   Coords of point 1
REAL, INTENT(IN)    :: xe(:)      !   Coords of point 2
INTEGER, INTENT(IN) :: Cdim       !   Cartesian dimension
INTEGER             :: N
REAL                :: SUMS
SUMS= 0.0
DO N=1,Cdim
   SUMS= SUMS + (xa(n)-xe(n))**2
END DO
Dist= SQRT(SUMS)
RETURN
END FUNCTION Dist

REAL FUNCTION Direc(xA,xE)
!-----------------------------------------------------------
!  Computes the Direction-angle from point xA to point xE
!-----------------------------------------------------------
IMPLICIT NONE
REAL, INTENT(IN)    :: xA(2)
REAL, INTENT(IN)    :: xE(2)
REAL            :: pi=3.1415926536
Direc=ATAN2((xE(2)-xA(2)),(xE(1)-xA(1)))
IF (Direc < 0.00000000) THEN
   Direc= Direc + 2*pi
END IF
RETURN
END FUNCTION Direc

SUBROUTINE Ortho(v3,v1,v2)
!-----------------------------------------------------------
!   DETERMINES ORTHOGONAL VECTORS
!   V1, V2, V3
!   USING
!   V2 = V3 X VX
!   OR
!   V2 = V3 X VY     IF V3=VX
!   AND
!   V1 = V2 X V3
!-----------------------------------------------------------
REAL, INTENT (IN) :: v3(:)   !   Normal vector
REAL           :: V1(:), V2(:)  !   Orthogonal vectors
REAL           :: vx(3),vy(3)   !   Vectors in coordinate directions
vx= (/1.0,0.0,0.0/) ; vy= (/0.0,1.0,0.0/)
IF(ABS(V3(1)) + 0.005 .GE. 1.0) THEN
   v2= Vector_ex(v3,vy)
ELSE
   v2= Vector_ex(v3,vx)
```

```
END IF
V1= Vector_ex(v2,v3)
RETURN
END SUBROUTINE Ortho

EAL FUNCTION Min_dist (Elcor,xPi,Nodel,inci,ELengx,Elenge)
IMPLICIT NONE
REAL,INTENT(IN)           :: Elcor(:,:)      ! Coordinates of Element
REAL,INTENT(IN)           :: xPi(:)          ! Coordinates of Collocation point
REAL,INTENT(IN)           :: ELengx,ELenge    ! Elementlength xsi and eta
REAL           :: DET,A,B,C,D,F1,F2
REAL           :: xsi,eta,Dxsi,Deta
REAL           :: DistPS,DistPS_N
REAL           :: L
REAL           :: ERR
REAL,ALLOCATABLE      :: Ni(:)      ! Shape function
REAL,ALLOCATABLE      :: DNi(:,:)      ! Derivatives of shape function
REAL,ALLOCATABLE      :: r(:),xS(:),Dxs(:,:)
INTEGER,INTENT(IN)      :: Nodel
INTEGER,INTENT(IN)      :: inci(:)
INTEGER           :: n
ALLOCATE(Ni(Nodel),DNi(Nodel,Idim),r(Cdim),xS(Cdim),DxS(Cdim,Idim))
SELECT CASE(Cdim)
  CASE(2)
    xsi=0.0
    CALL Serendip_func(Ni,xsi,eta,Nodel,inci)
    CALL Cartesian(xS,Ni,Elcor)
    r= xS-xPi
    DistPS=Dist(xPi(:),xS(:))
    IF(((DistPS-ELengx/2)/Elengx) > 4.)THEN
      Min_dist1=DistPS-Elengx/2
      RETURN
    END IF
    DO n=1,40
      IF(n > 1)DistPS= DistPS_N
      CALL Serendip_deriv(DNi,xsi,eta,nodel,inci)
      DxS(1,1)= DOT_PRODUCT(DNi(:,1),Elcor(1,:))
      DxS(2,1)= DOT_PRODUCT(DNi(:,1),Elcor(2,:))
      DET= DxS(1,1)**2+DxS(2,1)**2
      Dxsi= -1/DET * DOT_PRODUCT(DxS(:,1),r(:))
      xsi= xsi+ Dxsi
      IF(ABS(xsi) > 1.) xsi= xsi/ABS(xsi)
      CALL Serendip_func(Ni,xsi,eta,Nodel,inci)
      CALL Cartesian(xS,Ni,Elcor)
      r= xS- xPi
      DistPS_N= Dist(xPi(:),xS(:))
      IF(DistPS_N > DistPS)THEN
        Min_dist1= DistPS
        RETURN
      END IF
      ERR= (DistPS- DistPS_N)/DistPS_N
      IF(ERR < 0.05)THEN
        Min_dist1= DistPS_N
        RETURN
      END IF
    END DO
  CASE(3)
    xsi=0.0
    eta=0.0
```

```
      CALL Serendip_func(Ni,xsi,eta,Nodel,inci)
      CALL Cartesian(xS,Ni,Elcor)
      r= xS-xPi
      DistPS=Dist(xPi(:),xS(:))
      IF(((DistPS-ELengx/2)/ELengx) > 4. .and.((DistPS-ELenge/2)/ELenge)> 4.)THEN
         Min_dist1=DistPS
         RETURN
      END IF
      DO n=1,40
         IF(n > 1)DistPS= DistPS_N
         CALL Serendip_deriv(DNi,xsi,eta,nodel,inci)
         DxS(1,1)= DOT_PRODUCT(DNi(:,1),Elcor(1,:))
         DxS(2,1)= DOT_PRODUCT(DNi(:,1),Elcor(2,:))
         DxS(3,1)= DOT_PRODUCT(DNi(:,1),Elcor(3,:))
         DxS(1,2)= DOT_PRODUCT(DNi(:,2),Elcor(1,:))
         DxS(2,2)= DOT_PRODUCT(DNi(:,2),Elcor(2,:))
         DxS(3,2)= DOT_PRODUCT(DNi(:,2),Elcor(3,:))
         A= DxS(1,1)**2+DxS(2,1)**2+DxS(3,1)**2
         B= DxS(1,1)*DxS(1,2)+DxS(2,1)*DxS(2,2)+DxS(3,1)*DxS(3,2)
         C=B
         D= DxS(1,2)**2+DxS(2,2)**2+DxS(3,2)**2
         DET= A*D - C*B
         F1= DOT_PRODUCT(DxS(:,1),r(:))
         F2= DOT_PRODUCT(DxS(:,2),r(:))
         Dxsi = -1/DET * (F1*D - F2*B)
         Deta = -1/DET * (F2*A - F1*C)
         xsi= xsi+ Dxsi
         eta= eta+ Deta
         IF(ABS(xsi) > 1.) xsi= xsi/ABS(xsi)
         IF(ABS(eta) > 1.) eta= eta/ABS(eta)
         CALL Serendip_func(Ni,xsi,eta,Nodel,inci)
         CALL Cartesian(xS,Ni,Elcor)
         r= xS- xPi
         DistPS_N=Dist(xPi(:),xS(:))
         IF(DistPS_N > DistPS)THEN
            xsi=xsi- Dxsi
            eta=eta- Deta
            Min_dist1= DistPS
            RETURN
         END IF
         ERR= (DistPS- DistPS_N)/DistPS_N
         IF(ERR < 0.05)THEN
            Min_dist1= DistPS_N
            RETURN
         END IF
      END DO
END SELECT
END FUNCTION Min_dist

SUBROUTINE Elength(L,Elcor,nodes,ldim)
!-------------------------------------------------
!  Computes the length of a boundary element
!-------------------------------------------------
IMPLICIT NONE
REAL,INTENT (IN)      ::  Elcor(:,:)
INTEGER, INTENT (IN)  ::  Nodes, ldim
REAL,INTENT (OUT)     ::  L
REAL                  ::  B(ldim+1), distB3, distB2, p, a, c
INTEGER               ::  Cdim
```

```fortran
Cdim=ldim+1
SELECT CASE (Nodes)
  CASE (2)
    L=Dist(Elcor(:,1),Elcor(:,2),Cdim)
    RETURN
  CASE (3)
    B=(Elcor(:,1)+Elcor(:,2))/2.0
    distB3=Dist(B(:),Elcor(:,3),Cdim)
    distB2=Dist(B,Elcor(:,2),Cdim)
    IF (distB3/distB2 < 0.01) THEN
      L=Dist(Elcor(:,1),Elcor(:,2),Cdim)
      RETURN
    END IF
    IF (distB3/distB2 < 0.1) THEN
      c=distB3/distB2
      L=2.0*distB2*(1.0+2.0/3.0*c**2-2.0/5.0*c**4)      !  Length Parabel linearisiert
      RETURN
    END IF
    p=distB2**2/(2*DistB3)                              !  Parabel Parameter p=y**2/(2*x)
    a=SQRT(p**2+distB2**2)
    L=2.0*(distB2/(2.0*p)*a + p/2.0*LOG((distB2 + a)/p))  !  Length Parabel exakt
    RETURN
  CASE DEFAULT
END SELECT
END SUBROUTINE Elength
END MODULE Geometry_lib
```

A.3. INTEGRATION_LIB

```fortran
MODULE Integration_lib
USE Geometry_lib; USE Laplace_lib; USE Elast_lib
CONTAINS
SUBROUTINE Gauss_coor(Cor,Wi,Intord)
!-----------------------------------
!   Returns Gauss coordinates and Weights for up to 8 Gauss points
!-----------------------------------
IMPLICIT NONE
REAL, INTENT(OUT)  :: Cor(8) !  Gauss point coordinates
REAL, INTENT(OUT)  :: Wi(8)  !  Weights
INTEGER,INTENT(IN) :: Intord !  Integration order
SELECT CASE (Intord)
  CASE (1)
    Cor(1)= 0.
    Wi(1) = 2.0
  CASE(2)
    Cor(1)= .577350269  ; Cor(2)= -Cor(1)
    Wi(1) = 1.0 ;  Wi(2) = Wi(1)
  CASE(3)
    Cor(1)= .774596669  ; Cor(2)= 0.0 ; Cor(3)= -Cor(1)
    Wi(1) = .555555555   ; Wi(2) = .888888888 ; Wi(3) = Wi(1)
  CASE(4)
    Cor(1)= .861136311 ; Cor(2)= .339981043 ; Cor(3)= -Cor(2) ; Cor(4)= -Cor(1)
    Wi(1) = .347854845 ; Wi(2) = .652145154 ; Wi(3) = Wi(2) ; Wi(4) = Wi(1)
  CASE(5)
    Cor(1)= .9061798459 ; Cor(2)= .5384693101 ; Cor(3)= .0 ; Cor(4)= -Cor(2)
    Cor(5)= -Cor(1)
    Wi(1) = .236926885 ; Wi(2) = .478628670 ; Wi(3) = .568888888 ; Wi(4) = Wi(2)
```

```fortran
      Wi(5) = Wi(1)
  CASE(6)
     Cor(1)= .932469514 ; Cor(2)= .661209386 ; Cor(3)= .238619186
     Cor(4)= -Cor(3) ;  Cor(5)= -Cor(2) ; Cor(6)= -Cor(1)
     Wi(1) = .171324492 ; Wi(2) = .360761573 ; Wi(3) = .467913934
     Wi(4) = Wi(3) ; Wi(5) = Wi(2) ; Wi(6) = Wi(1)
  CASE(7)
     Cor(1)= .949107912 ; Cor(2)= .741531185 ; Cor(3)= .405845151
     Cor(4)= 0.
     Cor(5)= -Cor(3) ;Cor(6)= -Cor(2) ;Cor(7)= -Cor(1)
     Wi(1) = .129484966 ; Wi(2) = .279705391 ; Wi(3) = .381830050
     Wi(4) = .417959183
     Wi(5) = Wi(3) ; Wi(6) = Wi(2) ; Wi(7) = Wi(1)
  CASE(8)
     Cor(1)= .960289856 ; Cor(2)= .796666477 ; Cor(3)= .525532409 ; Cor(4)= .183434642
     Cor(5)= -Cor(4) ; Cor(6)= -Cor(3) ; Cor(7)= -Cor(2) ; Cor(8)= -Cor(1)
     Wi(1) = .101228536 ; Wi(2) = .222381034 ; Wi(3) = .313706645 ;Wi(4) = .362683783
     Wi(5) = Wi(4) ; Wi(6) = Wi(3) ; Wi(7) = Wi(2) ; Wi(8) = Wi(1)
  CASE DEFAULT
CALL Error_Message('Gauss points not in range 1-8')
END SELECT
END SUBROUTINE Gauss_coor


SUBROUTINE Gauss_Laguerre_coor(Cor,Wi,Intord)
!-----------------------------------
! Returns Gauss_Laguerre coordinates and Weights for up to 8 Gauss points
!-----------------------------------
IMPLICIT NONE
REAL, INTENT(OUT) :: Cor(8) !  Gauss point coordinates
REAL, INTENT(OUT) :: Wi(8) !  weights
INTEGER,INTENT(IN) :: Intord !  integration order
SELECT CASE (Intord)
  CASE (1)
     Cor(1)= 0.5
     Wi(1) = 1.0
  CASE(2)
     Cor(1)= .112008806  ; Cor(2)=.602276908
     Wi(1) = .718539319 ;Wi(2) =.281460680
  CASE(3)
     Cor(1)= .063890793  ; Cor(2)= .368997063 ; Cor(3)= .766880303
     Wi(1) = .513404552  ; Wi(2) = .391980041 ; Wi(3) = .0946154065
  CASE(4)
     Cor(1)= .0414484801 ; Cor(2)=.245274914 ; Cor(3)=.556165453 ; Cor(4)= .848982394
     Wi(1) = .383464068   ; Wi(2) =.386875317 ; Wi(3) =.190435126 ; Wi(4) = .0392254871
  CASE(5)
     Cor(1)= .0291344721 ; Cor(2)= .173977213 ; Cor(3)= .411702520; Cor(4)=.677314174
     Cor(5)= .894771361
     Wi(1) = .297893471  ; Wi(2) = .349776226 ; Wi(3) =.234488290 ; Wi(4) = .0989304595
     Wi(5) = .0189115521
  CASE(6)
     Cor(1)= .0216340058 ; Cor(2)= .129583391 ; Cor(3)= .314020449
     Cor(4)= .538657217  ; Cor(5)= .756915337 ; Cor(6)=.922668851
     Wi(1) = .238763662 ; Wi(2) =.308286573 ; Wi(3) =.245317426
     Wi(4) = .142008756 ; Wi(5) =.0554546223 ; Wi(6) =.0101689586
  CASE(7)
     Cor(1)= .0167193554 ; Cor(2)= .100185677 ; Cor(3)= .246294246
     Cor(4)= .433463493
     Cor(5)= .632350988  ; Cor(6)= .811118626 ; Cor(7)= .940848166
     Wi(1) = .196169389  ; Wi(2) = .270302644 ; Wi(3) = .239681873
```

```
      Wi(4) = .165775774
      Wi(5) = .0889432271 ; Wi(6) =.0331943043 ; Wi(7) = .00593278701
   CASE(8)
      Cor(1)= .0133202441 ; Cor(2)=.0797504290 ; Cor(3)= .197871029 ; Cor(4)= .354153994
      Cor(5)= .529458575  ; Cor(6)= .701814529 ; Cor(7)= .849379320 ; Cor(8)= .953326450
      Wi(1) = .164416604 ; Wi(2) = .237525610 ; Wi(3) = .226841984 ;Wi(4) = .175754079
      Wi(5) = .112924030 ; Wi(6) =.0578722107 ; Wi(7) =.0209790737 ;Wi(8) =.00368640710
   CASE DEFAULT
   CALL Error_Message('Gauss points not in range 1-8')
END SELECT
END SUBROUTINE Gauss_Laguerre_coor


INTEGER FUNCTION Ngaus(RonL,ne)
!--------------------------------------------------------
!   Function returns number of Gauss points needed
!   to integrate a function 1/rn
!--------------------------------------------------------
IMPLICIT NONE
REAL , INTENT(IN)        ::  RonL    !   R/L
INTEGER , INTENT(IN)     ::  ne      !   Exponent (1,2,3)
REAL                     ::  Rlim(7) !   Array to store values of table
INTEGER                  ::  n
SELECT CASE(ne)
   CASE(1)
      Rlim= (/1.6382, 0.6461, 0.3550, 0.2230, 0.1490, 0.1021, 0.0698/)
   CASE(2)
      Rlim= (/2.6230, 1.0276, 0.5779, 0.3783, 0.2679, 0.1986, 0.1512/)
   CASE(3)
      Rlim= (/3.5627, 1.3857, 0.7846, 0.5212, 0.3767, 0.2864, 0.2249/)
   CASE DEFAULT
END SELECT
Ngaus=0
DO  N=1,7
   IF(RonL >= Rlim(N)) THEN
      Ngaus= N+1
      EXIT
   END IF
END DO
IF (Ngaus == 0)THEN  !   Point is to near the surface
   Ngaus=8
END IF
RETURN
END FUNCTION Ngaus


SUBROUTINE Integ2P (Elcor,Inci,Nodel,Ncol,xP,k,dUe,dTe,Ndest,Isym)
See text
END SUBROUTINE Integ2P


SUBROUTINE Integ2E(Elcor,Inci,Nodel,Ncol,xP,E,ny,dUe,dTe,Ndest,Isym)
see text
END SUBROUTINE Integ2E


SUBROUTINE Integ3(Elcor,Inci,Nodel,Ncol,xPi,Ndof,E,ny,ko,dUe,dTe,Ndest,Isym)
see text
END SUBROUTINE Integ3


SUBROUTINE Triangel_Coord(cor_tri,Inod,ntr)
```

```
!-------------------------------------
! Assigns local coordinates of triangular
! subelements
!-------------------------------------
IMPLICIT NONE
INTEGER, INTENT(IN) :: lnod,ntr        ! node, subelement no.
REAL , INTENT(OUT) :: cor_tri(2,3)     ! xsi,eta of triangle nodes
REAL              :: xsii(8),etai(8)
INTEGER           :: nod_tri(3),n
SELECT CASE (ntr)
  CASE (1)
  SELECT CASE(lnod)
    CASE (1)
      nod_tri=(/2,3,1/)
    CASE (2)
      nod_tri=(/3,4,2/)
    CASE (3)
      nod_tri=(/1,2,3/)
    CASE (4)
      nod_tri=(/1,2,4/)
    CASE (5)
      nod_tri=(/4,1,5/)
    CASE (6)
      nod_tri=(/1,2,6/)
    CASE (7)
      nod_tri=(/4,1,7/)
    CASE (8)
      nod_tri=(/1,2,8/)
    CASE DEFAULT
  END SELECT
  CASE (2)
  SELECT CASE(lnod)
    CASE (1)
      nod_tri=(/3,4,1/)
    CASE (2)
      nod_tri=(/4,1,2/)
    CASE (3)
      nod_tri=(/4,1,3/)
    CASE (4)
      nod_tri=(/2,3,4/)
    CASE (5)
      nod_tri=(/2,3,5/)
    CASE (6)
      nod_tri=(/3,4,6/)
    CASE (7)
      nod_tri=(/2,3,7/)
    CASE (8)
      nod_tri=(/3,4,8/)
    CASE DEFAULT
  END SELECT
  CASE (3)
  SELECT CASE(lnod)
    CASE (5)
      nod_tri=(/3,4,5/)
    CASE (6)
      nod_tri=(/4,1,6/)
    CASE (7)
      nod_tri=(/1,2,7/)
    CASE (8)
```

```
        nod_tri=(/2,3,8/)
      CASE DEFAULT
    END SELECT
    CASE DEFAULT
END SELECT
xsii=(/-1.0,1.0,1.0,-1.0,0.0,1.0,0.0,-1.0/)
etai=(/-1.0,-1.0,1.0,1.0,-1.0,0.0,1.0,0.0/)
cor_tri=0
DO n=1, 3
   cor_tri(1,n)= xsii(nod_tri(n))
   cor_tri(2,n)= etai(nod_tri(n))
END DO
END SUBROUTINE Triangel_Coord

SUBROUTINE Trans_Tri(ntr,lnod,xsib,etab,xsi,eta,Jacb)
!-------------------------------------------
! Transforms from local triangle coordinates
! to xsi,eta coordinates and computes
! the Jacobean of the transformation
!-------------------------------------------
IMPLICIT NONE
REAL, INTENT (IN)     ::  xsib,etab        !  Local coordinates
INTEGER, INTENT (IN)  ::  ntr,lnod         !  Subelement no, local node no.
REAL, INTENT (OUT)    ::  Jacb,xsi,eta     !  Jacobean and xsi,eta coords
REAL                  ::  Nb(3),dNbdxb(3),dNbdeb(3),dxdxb,dxdeb,dedxb,dedeb,cor_tri(2,3)
INTEGER               ::  n,i
CALL Triangel_Coord(cor_tri,lnod,ntr)
!
!   Transform xsi-bar and eta-bar to xsi and eta
!
Nb(1)=0.25*(1.0+xsib)*(1.0-etab)
Nb(2)=0.25*(1.0+xsib)*(1.0+etab)
Nb(3)=0.5*(1.0-xsib)
xsi = 0.0
eta = 0.0
DO n=1,3
   xsi = xsi+Nb(n)*cor_tri(1,n)
   eta = eta+Nb(n)*cor_tri(2,n)
END DO
!
!   Jacobian of Transformation xsi-bar and eta-bar to xsi and eta
!
dNbdxb(1)=0.25*(1.0-etab)
dNbdxb(2)=0.25*(1.0+etab)
dNbdxb(3)=-0.5
dNbdeb(1)=-0.25*(1.0+xsib)
dNbdeb(2)=0.25*(1.0+xsib)
dNbdeb(3)=0.0
dxdxb=0.0
dedxb=0.0
dxdeb=0.0
dedeb=0.0
DO i=1,3
   dxdxb=dxdxb+dNbdxb(i)*cor_tri(1,i)
   dedxb=dedxb+dNbdxb(i)*cor_tri(2,i)
   dedeb=dedeb+dNbdeb(i)*cor_tri(2,i)
   dxdeb=dxdeb+dNbdeb(i)*cor_tri(1,i)
END DO
Jacb=dxdxb*dedeb-dedxb*dxdeb
```

END SUBROUTINE Trans_Tri

```
SUBROUTINE Tri_RL(RLx,RLe,Elengx,Elenge,Inod,ntr)
!---------------------------------------
!  Computes ize of triangular sub-element
!  and Rmin
!---------------------------------------
IMPLICIT NONE
REAL, INTENT (IN) :: Elengx,Elenge      !   Length of element in xsi,eta dir
INTEGER, INTENT (IN) :: Inod,ntr        !   Local node, subelement no.
REAL, INTENT (OUT) :: RLx,RLe           !   Lengths of subelement
IF(Inod <= 4) THEN
   RLx=Elengx/Elenge
   RLe=Elenge/Elengx
END IF
IF (Inod == 5 .or. Inod == 7) THEN
   SELECT CASE (ntr)
     CASE (1)
        RLx=(Elengx/2.0)/Elenge
        RLe=Elenge/(Elengx/2.0)
     CASE (2)
        RLx=(Elengx/2.0)/Elenge
        RLe=Elenge/(Elengx/2.0)
     CASE (3)
        RLx=Elengx/Elenge
        RLe=Elenge/Elengx
   END SELECT
END IF
IF (Inod == 6 .or. Inod == 8) THEN
   SELECT CASE (ntr)
     CASE (1)
        RLx=Elengx/(Elenge/2)
        RLe=(Elenge/2)/Elengx
     CASE (2)
        RLx=Elengx/(Elenge/2)
        RLe=(Elenge/2)/Elengx
     CASE (3)
        RLx=Elengx/Elenge
        RLe=Elenge/Elengx
   END SELECT
END IF
END SUBROUTINE Tri_RL
END MODULE Integration_lib
```

A.4. ELAST_LIB

```
MODULE Elast_lib
USE Laplace_lib
!   Library for elasticity problems
IMPLICIT NONE
CONTAINS

FUNCTION UK(dxr,r,E,ny,Cdim)
!---------------------------------------
!  FUNDAMENTAL SOLUTION FOR DISPLACEMENTS
!  isotropic material (Kelvin solution)
!---------------------------------------
```

```fortran
IMPLICIT NONE
REAL,INTENT(IN)      :: dxr(:)              !  rx/r etc.
REAL,INTENT(IN)      :: r                   !  r
REAL,INTENT(IN)      :: E                   !  Young's modulus
REAL,INTENT(IN)      :: ny                  !  Poisson's ratio
INTEGER,INTENT(IN) :: Cdim                  !  Cartesian dimension
REAL                 :: UK(Cdim,Cdim)       !  Function returns array of same dim as dxr
REAL                 :: G,c,c1,onr,clog,conr !  Temps
G= E/(2.0*(1+ny))
c1= 3.0 - 4.0*ny
SELECT CASE (Cdim)
   CASE (2)                      !   Two-dimensional solution
      c= 1.0/(8.0*Pi*G*(1.0 - ny))
      clog= -c1*LOG(r)
      UK(1,1)= c*(clog + dxr(1)*dxr(1))
      UK(1,2)= c*dxr(1)*dxr(2)
      UK(2,2)= c*(clog + dxr(2)*dxr(2))
      UK(2,1)= UK(1,2)
   CASE(3)                       !   Three-dimensional solution
      c= 1.0/(16.0*Pi*G*(1.0 - ny))
      conr=c/r
      UK(1,1)= conr*(c1 + dxr(1)*dxr(1))
      UK(1,2)= conr*dxr(1)*dxr(2)
      UK(1,3)= conr*dxr(1)*dxr(3)
      UK(2,1)= UK(1,2)
      UK(2,2)= conr*(c1 + dxr(2)*dxr(2))
      UK(2,3)= conr*dxr(2)*dxr(3)
      UK(3,1)= UK(1,3)
      UK(3,2)= UK(2,3)
      UK(3,3)= conr*(c1 + dxr(3)*dxr(3))
   CASE DEFAULT
END SELECT
RETURN
END FUNCTION UK


FUNCTION TK(dxr,r,Vnor,ny,Cdim)
!-------------------------------------------
!  FUNDAMENTAL SOLUTION FOR TRACTIONS
!  isotropic material (Kelvin solution)
!-------------------------------------------
IMPLICIT NONE
REAL,INTENT(IN)      :: dxr(:)              !  rx/r etc.
REAL,INTENT(IN)      :: r                   !  r
REAL,INTENT(IN)      :: Vnor(:)             !  Normal vector
REAL,INTENT(IN)      :: ny                  !  Poisson's ratio
INTEGER,INTENT(IN) :: Cdim                  !  Cartesian dimension
REAL                 :: TK(Cdim,Cdim)       !  Function returns array of same dim as dxr
REAL                 :: c2,c3,costh,Conr    !  Temps
c3= 1.0 - 2.0*ny
Costh= DOT_PRODUCT (Vnor,dxr)
SELECT CASE (Cdim)
   CASE (2)
      c2= 1.0/(4.0*Pi*(1.0 - ny))
      Conr= c2/r
      TK(1,1)= -(Conr*(c3 + 2.0*dxr(1)*dxr(1))*Costh)
      TK(1,2)= -(Conr*(2.0*dxr(1)*dxr(2)*Costh + c3*(Vnor(1)*dxr(2) - Vnor(2)*dxr(1))))
      TK(2,2)= -(Conr*(c3 + 2.0*dxr(2)*dxr(2))*Costh)
      TK(2,1)= -(Conr*(2.0*dxr(1)*dxr(2)*Costh + c3*(Vnor(2)*dxr(1) - Vnor(1)*dxr(2))))
   CASE(3)           !   Three-dimensional
```

```fortran
      c2= 1.0/(8.0*Pi*(1.0 - ny))
      Conr= c2/r**2
      TK(1,1)= -Conr*(c3 + 3.0*dxr(1)*dxr(1))*Costh
      TK(1,2)= -Conr*(3.0*dxr(1)*dxr(2)*Costh - c3*(Vnor(2)*dxr(1) - Vnor(1)*dxr(2)))
      TK(1,3)= -Conr*(3.0*dxr(1)*dxr(3)*Costh - c3*(Vnor(3)*dxr(1) - Vnor(1)*dxr(3)))
      TK(2,1)= -Conr*(3.0*dxr(1)*dxr(2)*Costh - c3*(Vnor(1)*dxr(2) - Vnor(2)*dxr(1)))
      TK(2,2)= -Conr*(c3 + 3.0*dxr(2)*dxr(2))*Costh
      TK(2,3)= -Conr*(3.0*dxr(2)*dxr(3)*Costh - c3*(Vnor(3)*dxr(2) - Vnor(2)*dxr(3)))
      TK(3,1)= -Conr*(3.0*dxr(1)*dxr(3)*Costh - c3*(Vnor(1)*dxr(3) - Vnor(3)*dxr(1)))
      TK(3,2)= -Conr*(3.0*dxr(2)*dxr(3)*Costh - c3*(Vnor(2)*dxr(3) - Vnor(3)*dxr(2)))
      TK(3,3)= -Conr*(c3 + 3.0*dxr(3)*dxr(3))*Costh
   CASE DEFAULT
END SELECT
END FUNCTION TK


SUBROUTINE SK(TS,DXR,R,C2,C3)
!------------------------------------------------------
! KELVIN SOLUTION FOR STRESS
!  TO BE MULTIPLIED WITH T
!------------------------------------------------------
REAL, INTENT(OUT) :: TS(:,:)      !   Fundamental solution
REAL, INTENT(IN) :: DXR(:)        !   rx , ry, rz
REAL, INTENT(IN) :: R             !   r
REAL, INTENT(IN) :: C2,C3         !   Elastic constants
INTEGER :: Cdim                   !   Cartesian dimension
INTEGER :: NSTRES  ! No. of stress components
INTEGER :: II(6), JJ(6) ! sequence of stresses in pseudo-vector
REAL   :: A
INTEGER :: I,N,J,K
Cdim= UBOUND(DXR,1)
IF(CDIM == 2) THEN
   NSTRES= 3
   II(1:3)= (/1,2,1/)
   JJ(1:3)= (/1,2,2/)
ELSE
   NSTRES= 6
   II= (/1,2,3,1,2,3/)
   JJ= (/1,2,3,2,3,1/)
END IF
Coor_directions:&
DO K=1,Cdim
   Stress_components:&
   DO N=1,NSTRES
        I= II(N)
        J= JJ(N)
        A= 0.
        IF(K .EQ. J) A= A + DXR(I)
        IF(I .EQ. J) A= A - DXR(K)
        IF(K .EQ. I) A= A + DXR(J)
        A= A*C3
        TS(N,K)= C2/R*(A + Cdim*DXR(I)*DXR(J)*DXR(K))
        IF(Cdim .EQ. 3) TS(N,K)= TS(N,K)/R
   END DO &
   Stress_components
END DO &
Coor_directions
RETURN
END SUBROUTINE SK
```

```fortran
SUBROUTINE RK(US,DXR,R,VNORM,C3,C5,C6,C7,ny)
!-------------------------------------------------------
!   KELVIN SOLUTION FOR STRESS COMPUTATION
!   TO BE MULTIPLIED WITH U
!-------------------------------------------------------
REAL, INTENT(OUT) :: US(:,:)        !   Fundamental solution
REAL, INTENT(IN)  :: DXR(:)         !   rx , ry, rz
REAL, INTENT(IN)  :: R              !   r
REAL, INTENT(IN)  :: VNORM(:)       !   nx , ny , nz
REAL, INTENT(IN)  :: C3,C5,C6,C7,ny !   Elastic constants
INTEGER ::  Cdim                    !   Cartesian dimension
INTEGER :: NSTRES                   !   No. of stress components
INTEGER :: II(6), JJ(6)             !   Sequence of stresses in pseudo-vector
REAL    :: costh, B,C,cny
INTEGER :: I,N,J,K
Cdim= UBOUND(DXR,1)
IF(CDIM == 2) THEN
  NSTRES= 3
  II(1:3)= (/1,2,1/)
  JJ(1:3)= (/1,2,2/)
ELSE
  NSTRES= 6
  II= (/1,2,3,1,2,3/)
  JJ= (/1,2,3,2,3,1/)
END IF
COSTH= DOT_Product(dxr,vnorm)
Cny= Cdim*ny
Coor_directions:&
DO K=1,Cdim
  Stress_components:&
  DO N=1,NSTRES
      I= II(N)
      J= JJ(N)
      B= 0.
      C= 0.
      IF(I .EQ. J) B= Cdim*C3*DXR(K)
      IF(I .EQ. K) B= B + Cny*DXR(J)
      IF(J .EQ. K) B= B + Cny*DXR(I)
      B= COSTH *(B - C6*DXR(I)*DXR(J)*DXR(K))
      C= DXR(J)*DXR(K)*Cny
      IF(J .EQ.K) C= C + C3
      C= C*VNORM(I)
      B= B+C
      C= DXR(I)*DXR(K)*Cny
      IF(I .EQ. K) C=C + C3
      C= C*VNORM(J)
      B= B+C
      C= DXR(I)*DXR(J)*Cdim*C3
      IF(I .EQ. J) C= C - C7
      C= C*VNORM(K)
      US(N,K)= (B + C)*C5/R/R
      IF(Cdim .EQ. 3) US(N,K)= US(N,K)/R
  END DO &
  Stress_components
END DO &
Coor_directions
RETURN
END SUBROUTINE RK
```

```fortran
SUBROUTINE Trans_mat(v1,v2,v3, T)
!----------------------------------------------
! Computes Stress Transformation Matrix in 3-D
!----------------------------------------------
IMPLICIT NONE
REAL, INTENT(IN)    :: v1(3),v2(3),v3(3)    !   Unit vectors in orthogonal directions
REAL, INTENT(OUT)   :: T(6,6)               !   Transformation matrix
REAL                :: v1x,v1y,v1z,v2x,v2y,v2z,v3x,v3y,v3z  ! temps
v1x= v1(1) ; v1y= v1(2) ; v1z= v1(3)
v2x= v2(1) ; v2y= v2(2) ; v2z= v2(3)
v3x= v3(1) ; v3y= v3(2) ; v3z= V3(3)
!  T?11
T(1,1)= v1x**2 ; T(1,2)= v2x**2 ; T(1,3)= v3x**2
T(2,1)= v1y**2 ; T(2,2)= v2y**2 ; T(2,3)= v3y**2
T(3,1)= v1z**2 ; T(3,2)= v2z**2 ; T(3,3)= v3z**2
!  T?21
T(1,4)= 2.0*v1y*v1x ; T(1,5)= 2.0*v2y*v2x ; T(1,6)= 2.0*v3y*v3x
T(2,4)= 2.0*v1y*v1z ; T(2,5)= 2.0*v2y*v2z ; T(2,6)= 2.0*v3y*v3z
T(3,4)= 2.0*v1x*v1z ; T(3,5)= 2.0*v2x*v2z ; T(3,6)= 2.0*v3x*v3z
!  T?12
T(4,1)= v1x*v2x ; T(4,2)= v2x*v3x ; T(4,3)= v1x*v3x
T(5,1)= v1y*v2y ; T(5,2)= v2y*v3y ; T(5,3)= v1y*v3y
T(6,1)= v1z*v2z ; T(6,2)= v2z*v3z ; T(6,3)= v1z*v3z
!  T?22
T(4,4)= v1x*v2y+v1y*v2x ; T(4,5)= v2x*v3y+v2y*v3x ; T(4,6)= v1x*v3y+v1y*v3x
T(5,4)= v1y*v2z+v1z*v2y ; T(5,5)= v2y*v3z+v2z*v3y ; T(5,6)= v1y*v3z+v1z*v3y
T(6,4)= v1x*v2z+v1z*v2x ; T(6,5)= v2x*v3z+v2z*v3x ; T(6,6)= v1x*v3z+v1z*v3x
RETURN
END  SUBROUTINE Trans_mat


SUBROUTINE D_mat(E,ny,D,Cdim)
!----------------------------------
! Computes isotropic D-matrix
! Plane-strain (Cdim= 2)
! or 3-D      (Cdim= 3)
!----------------------------------
IMPLICIT NONE
REAL, INTENT(IN) :: E          !  Young's modulus
REAL, INTENT(IN) :: ny         !  Poisson's ratio
INTEGER,INTENT(IN) :: Cdim     !  Cartesian dimension
REAL, INTENT(OUT) :: D(:,:)    !  D-matrix
REAL              :: c1,c2,G
c1= E*(1.0-ny)/( (1.0+ny)*(1.0-2.0*ny) )
c2= ny/(1.0-ny)
G = E/(2.0*(1.0+ny))
D = 0.0
SELECT CASE (Cdim)
   CASE (2)
     D(1,1)= 1.0  ; D(2,2)= 1.0
     D(2,1)= c2 ; D(1,2)= c2
     D(3,3)= G/c1
   CASE (3)        !  3-D
     D(1,1)= 1.0  ; D(2,2)= 1.0  ; D(3,3)= 1.0
     D(2,1)= c2   ; D(1,3)= c2   ; D(2,3)= c2
     D(1,2)= c2   ; D(3,1)= c2   ; D(3,2)= c2
     D(4,4)= G/c1 ; D(5,5)= G/c1 ; D(6,6)= G/c1
   CASE DEFAULT
END SELECT
D= c1*D
```

```
RETURN
END SUBROUTINE D_mat

SUBROUTINE D_mat_anis(D,E1,G1,E2,G2,ny2,Cdim)
!---------------------------------
!  Computes an-isotropic D-matrix
!  Plane-strain (Cdim= 2)
!  or 3-D     (Cdim= 3)
!---------------------------------
IMPLICIT NONE
REAL, INTENT(OUT) :: D(:,:)      !  D-matrix
REAL, INTENT(IN) :: E1           !  Young's modulus, dir 1
REAL, INTENT(IN) :: G1           !  Shear modulus , dir 1
REAL, INTENT(IN) :: E2           !  Young's modulus, dir 2
REAL, INTENT(IN) :: G2           !  Shear Modulus , dir 2
REAL, INTENT(IN) :: ny2          !  Poisson's ratio, dir 2
INTEGER,INTENT(IN):: Cdim        !  Cartesian Dimension
REAL :: n                        !  Ratio E1/E2
REAL :: cc,c1,c2,c3,c4,ny1  !  temps
ny1= 0.5*E1/G1 -1.0
n= E1/E2
cc= E2/(1.+ny1)/(1.-ny1-2.*n*ny2**2)
c1= n*(1.-n*ny2**2)*cc
c3= n*ny2*(1.0+ny1)*cc
c4= (1 - ny1**2)*cc
D= 0. !  only nonzero components of D are assigned
SELECT CASE (Cdim)
   CASE (2)       !  plane strain
      D(1,1)= c1 ; D(2,2)= c4
      D(1,2)= c3 ; D(2,1)= c3
      D(3,3)= G2
   CASE (3)       !  3-D
      c2= n*(ny1+n*ny2**2)*cc
      D(1,1)= C1 ; D(2,2)= c1 ; D(3,3)= c4
      D(1,2)= C2 ; D(1,3)= c3 ; D(2,3)= C3
      D(2,1)= C2 ; D(3,1)= c3 ; D(3,2)= C3
      D(4,4)= G1 ; D(5,5)= G2 ; D(6,6)= G2
   CASE DEFAULT
END SELECT
RETURN
END SUBROUTINE D_mat_anis

END MODULE Elast_lib
```

A.5. LAPLACE_LIB

```
MODULE Laplace_lib
REAL :: Pi= 3.14159265359
CONTAINS
  REAL FUNCTION U(r,k,Cdim)
  !---------------------------------
  !   Fundamental solution for Potential problems
  !   Temperature/Potential
  !---------------------------------
  REAL,INTENT(IN)  :: r      !  Distance between source and field point
  REAL,INTENT(IN)  :: k      !  Conducivity
  INTEGER,INTENT(IN) :: Cdim !  Cartesian dimension (2-D,3-D)
```

```
SELECT CASE (CDIM)
   CASE (2)        ! Two-dimensional solution
     U= 1.0/(2.0*Pi*k)*LOG(1/r)
   CASE (3)        ! Three-dimensional solution
     U= 1.0/(4.0*Pi*r*k)
   CASE DEFAULT
     U=0.0
     WRITE (11,*)'Cdim not equal 2 or 3 in Function U(...)'
END SELECT
END FUNCTION U
REAL FUNCTION T(r,dxr,Vnorm,Cdim)
!-------------------------------
!  Fundamental solution for Potential problems
!  Normal gradient
!-------------------------------
REAL,INTENT(IN)::      r          !  Distance between source and field point
REAL,INTENT(IN)::     dxr(:)      !  rx/r , ry/r , rz/r
REAL,INTENT(IN)::   Vnorm(:)      !  Normal vector
INTEGER,INTENT(IN) ::  Cdim       !  Cartesian dimension
SELECT CASE (Cdim)
   CASE (2)        ! Two-dimensional solution
    T= -DOT_PRODUCT (Vnorm,dxr)/(2.0*Pi*r)
   CASE (3)        ! Three-dimensional solution
    T= -DOT_PRODUCT (Vnorm,dxr)/(4.0*Pi*r*r)
   CASE DEFAULT
    T=0.0
     WRITE (11,*)'Cdim not equal 2 or 3 in Function U(...)'
END SELECT
END FUNCTION T
FUNCTION dU(r,dxr,Cdim)
!-------------------------------
!  Derivatives of Fundamental solution for Potential problems
!  Temperature/Potential
!-------------------------------
REAL,INTENT(IN)::    r         !  Distance between source and field point
REAL,INTENT(IN):: dxr(:)       !  Distances in Cartesian directions divided by r
REAL :: dU(UBOUND(dxr,1))      !  dU is array of same dim as dxr
INTEGER ,INTENT(IN) :: Cdim    !  Cartesian dimension (2-D,3-D)
REAL :: C
SELECT CASE (CDIM)
   CASE (2)        ! Two-dimensional solution
    C=1/(2.0*Pi*r)
    dU(1)= C*dxr(1)
    dU(2)= C*dxr(2)
   CASE (3)        ! Three-dimensional solution
    C=1/(4.0*Pi*r**2)
    dU(1)= C*dxr(1)
    dU(2)= C*dxr(2)
    dU(3)= C*dxr(3)
   CASE DEFAULT
END SELECT
END FUNCTION dU
FUNCTION dT(r,dxr,Vnorm,Cdim)
!-------------------------------
!  derivatives of the Fundamental solution for Potential problems
!  Normal gradient
!-------------------------------
INTEGER,INTENT(IN) :: Cdim     !  Cartesian dimension
REAL,INTENT(IN)::      r       !  Distance between source and field point
```

```fortran
REAL,INTENT(IN)::  dxr(:)        !  Distances in Cartesian directions divided by R
REAL,INTENT(IN):: Vnorm(:)      !  Normal vector
REAL :: dT(UBOUND(dxr,1))       !  dT is array of same dim as dxr
REAL :: C,COSTH
COSTH= DOT_PRODUCT (Vnorm,dxr)
SELECT CASE (Cdim)
   CASE (2)        ! Two-dimensional solution
   C= 1/(2.0*Pi*r**2)
   dT(1)= C*COSTH*dxr(1)
   dT(2)= C*COSTH*dxr(2)
   CASE (3)        ! Three-dimensional solution
   C= 3/(4.0*Pi*r**3)
   dT(1)= C*COSTH*dxr(1)
   dT(2)= C*COSTH*dxr(2)
   dT(3)= C*COSTH*dxr(3)
   CASE DEFAULT
END SELECT
END FUNCTION dT
END MODULE Laplace_lib
```

A.6. POSTPROC_LIB

```fortran
MODULE Postproc_lib
USE Geometry_lib ; USE Utility_lib
IMPLICIT NONE
CONTAINS
SUBROUTINE BFLOW(Flow,xsi,eta,u,Inci,Elcor,k)
!-------------------------------------------
!  Computes flow vectors in direction tangential to the
!  Boundary
!-------------------------------------------
REAL , INTENT(OUT) ::  Flow(:)     !  Flow vector
REAL , INTENT(IN) ::     xsi,eta    !  Intrinsic coordinates of point
REAL , INTENT(IN) ::      u(:,:)    !  Nodal temperatures/potentials
INTEGER, INTENT (IN) :: Inci(:)     !  Element Incidences
REAL, INTENT (IN) ::    Elcor(:,:)  !  Element coordinates
REAL, INTENT (IN) ::      k         !  Conductivity
REAL, ALLOCATABLE  :: Vxsi(:),Veta(:),DNi(:,:),V3(:)
INTEGER :: Nodes,Cdim,Ldim
REAL :: Jxsi,Jeta,Flows(2),v1(3),v2(3),CosA,CosB,CosG,CosT
Nodes= UBOUND(ELCOR,2) !  Number of nodes
Cdim= UBOUND(ELCOR,1)  !  Cartesian Dimension
Ldim= Cdim-1        !  Local (element) dimension
ALLOCATE (Vxsi(cdim),Dni(Nodes,Ldim),v3(cdim))
IF(ldim > 1) ALLOCATE (Veta(cdim))
!  Compute Vector(s) tangential to boundary surface
CALL Serendip_deriv(DNi,xsi,eta,ldim,nodes,inci)
Vxsi(1)= Dot_Product(Dni(:,1),Elcor(1,:))
Vxsi(2)= Dot_Product(Dni(:,1),Elcor(2,:))
IF(Cdim == 2) THEN
   CALL Vector_norm(Vxsi,Jxsi)
   Flow(1)= -k*Dot_product(Dni(:,1),u(:,1))/Jxsi
ELSE
   Vxsi(3)= Dot_Product(Dni(:,1),Elcor(3,:))
   CALL Vector_norm(Vxsi,Jxsi)
   Veta(1)= Dot_Product(Dni(:,2),Elcor(1,:))
```

```
  Veta(2)= Dot_Product(Dni(:,2),Elcor(2,:))
  Veta(3)= Dot_Product(Dni(:,2),Elcor(3,:))
  CALL Vector_norm(Veta,Jeta)
!   Flows in skew coordinate system
  Flows(1)= -k*Dot_product(Dni(:,1),u(:,1))/Jxsi
  Flows(2)= -k*Dot_product(Dni(:,2),u(:,1))/Jeta
!   Orthoginal system
  v3= Vector_ex(Vxsi,Veta)
  Call Ortho(v3,v1,v2)
  CosA= DOT_Product(Vxsi,v1)
  CosB= DOT_Product(Veta,v2)
  CosG= DOT_Product(Veta,v1)
  CosT= DOT_Product(Vxsi,v2)
  Flow(1)= Flows(1)*CosA**2 + Flows(2)* CosG**2
  Flow(2)= Flows(1)*CosT**2 + Flows(2)* CosB**2
END IF
RETURN
END SUBROUTINE BFLOW


SUBROUTINE BStress(Stress,xsi,eta,u,t,Inci,Elcor,E,ny,IPS)
!--------------------------------------------------
!   Computes stresses in a plane tangential to the
!   Boundary Element
!--------------------------------------------------
REAL , INTENT(OUT) ::   Stress(:) !   Stress vector
REAL , INTENT(IN) ::    xsi,eta   !   Intrinsic coordinates of point
REAL , INTENT(IN) ::    u(:,:)    !   Nodal displacements
REAL , INTENT(IN) ::    t(:,:)    !   Nodal tractions
INTEGER, INTENT (IN) :: Inci(:)   !   Element incidences
REAL, INTENT (IN) ::    Elcor(:,:) !  Element coordinates
REAL, INTENT (IN) ::    E,ny      !   Elastic constants
INTEGER, INTENT (IN) :: IPS ! IPS= 0 plane strain; =1 plane stress
REAL, ALLOCATABLE :: Vxsi(:),Veta(:),DNi(:,:),Ni(:),trac(:)
INTEGER :: Nodes, Cdim, Ldim
REAL :: Jxsi,Jeta,v1(3),v2(3),CosA, CosB, CosG, CosT,v3(3)
REAL :: C1,C2,G,tn
REAL , ALLOCATABLE :: Dudxsi(:),Dudeta(:),Strain(:),Strains(:)
Nodes= UBOUND(ELCOR,2) ! Number of nodes
Cdim = UBOUND(ELCOR,1) ! Cartesian Dimension
Ldim= Cdim-1          ! Local (element) dimension
ALLOCATE (Vxsi(cdim),Veta(cdim),Dni(Nodes,Ldim),Ni(Nodes))
ALLOCATE (Dudxsi(Cdim),Dudeta(Cdim),trac(Cdim))
!   Compute Vector(s) tangential to boundary surface
CALL Serendip_deriv(DNi,xsi,eta,ldim,nodes,inci)
CALL Serendip_func(Ni,xsi,eta,ldim,nodes,inci)
trac(1)= Dot_Product(Ni,t(:,1))
trac(2)= Dot_Product(Ni,t(:,2))
Vxsi(1)= Dot_Product(Dni(:,1),Elcor(1,:))
Vxsi(2)= Dot_Product(Dni(:,1),Elcor(2,:))
IF(Cdim == 2) THEN
    ALLOCATE (Strain(1))
    CALL Vector_norm(Vxsi,Jxsi)
    V3(1)=   Vxsi(2)
    V3(2)= - Vxsi(1)
    tn= Dot_Product(v3(1:2),trac)
    DuDxsi(1)= Dot_Product(Dni(:,1),u(:,1))
    DuDxsi(2)= Dot_Product(Dni(:,1),u(:,2))
    Strain(1)= Dot_Product(DuDxsi,Vxsi)/Jxsi
    IF(IPS == 2) THEN
```

```
      Stress(1)= E*Strain(1) + ny*tn      !     plane stress
      ELSE
       Stress(1)= 1/(1.0-ny)*(E/(1.0+ny)*Strain(1) + ny*tn)      ! Plane strain
      END IF
ELSE
      ALLOCATE (Strain(3),Strains(3))
      trac(3)= Dot_Product(Ni,t(:,3))
      Vxsi(3)= Dot_Product(Dni(:,1),Elcor(3,:))
      CALL Vector_norm(Vxsi,Jxsi)
      Veta(1)= Dot_Product(Dni(:,2),Elcor(1,:))
      Veta(2)= Dot_Product(Dni(:,2),Elcor(2,:))
      Veta(3)= Dot_Product(Dni(:,2),Elcor(3,:))
      CALL Vector_norm(Veta,Jeta)
      V3= Vector_ex(Vxsi,veta)
      tn= Dot_Product(v3,trac)
      DuDxsi(1)= Dot_Product(Dni(:,1),u(:,1))
      DuDxsi(2)= Dot_Product(Dni(:,1),u(:,2))
      DuDxsi(3)= Dot_Product(Dni(:,1),u(:,3))
      DuDeta(1)= Dot_Product(Dni(:,2),u(:,1))
      DuDeta(2)= Dot_Product(Dni(:,2),u(:,2))
      DuDeta(3)= Dot_Product(Dni(:,2),u(:,3))
!  Strains in skew coordinate system
      Strains(1)= Dot_product(DuDxsi,Vxsi)/Jxsi
      Strains(2)= Dot_product(DuDeta,Veta)/Jeta
      Strains(3)= Dot_product(DuDeta,Vxsi)/Jeta + &
                      Dot_product(DuDxsi,Veta)/Jxsi
!  Orthogonal system
      v3= Vector_ex(Vxsi,Veta)
      CALL Ortho(v3,v1,v2)
      CosA= DOT_Product(Vxsi,v1)
      CosB= DOT_Product(Veta,v2)
      CosG= DOT_Product(Veta,v1)
      CosT= DOT_Product(Vxsi,v2)
!  Compute Strains
      Strain(1)= Strains(1)*CosA**2 + Strains(2)*CosG**2 + &
                    Strains(3)*CosA*CosG
      Strain(2)= Strains(1)*CosT**2 + Strains(2)*CosB**2 + &
                    Strains(3)*CosT*CosB
      Strain(3)= Strains(1)*CosG*CosT + Strains(2)*CosG*CosB + &
                    Strains(3)*(CosA*CosB+CosG*CosT)
!  Compute stresses
      C1= E/(1.0-ny**2)  ;  C2= ny/(1.0-ny) ; G=E/(1.0-2*ny)
      Stress(1)= C1*(Strain(1)+ny*strain(2))+ C2*Tn
      Stress(2)= C1*(Strain(2)+ny*strain(1))+ C2*Tn
      Stress(3)= G*Strain(3)
END IF
RETURN
End SUBROUTINE BStress
END MODULE Postproc_lib
```

A.7. STIFFNESS_LIB

```
MODULE Stiffness_lib
USE Utility_lib ; USE Integration_lib ; USE Geometry_lib
CONTAINS
SUBROUTINE
Stiffness_BEM(nr,xP,Nodel,Ndof,Ndofe,NodeR,Ncode,NdofR,Ndofc,KBE,A,tc,Cdim,Elres_u,Elres_t,&
```

IncieR,LdesteR,Nbel,ListR,TypeR,Bcode,Con,E,ny,Ndest,Isym)

```
!------------------------------------------
!  Computes the stiffness matrix
!  of a boundary element region
!  no symmetry
!------------------------------------------
IMPLICIT NONE
REAL, INTENT(INOUT):: xP(:,:)         !   Array of node coordinates
INTEGER, INTENT(IN):: nr
INTEGER, INTENT(IN):: Ncode(:)        !   Global restraint code
INTEGER, INTENT(IN):: NdofR
INTEGER, INTENT(IN):: Ndofc           !   No of interface degrees of freedom
INTEGER, INTENT(IN):: NodeR(:)
INTEGER, INTENT(IN):: TypeR(:)
INTEGER, INTENT(IN):: Cdim
INTEGER, INTENT(IN):: IncieR(:,:)
INTEGER, INTENT(IN):: LdesteR(:,:)
INTEGER, INTENT(INOUT):: Bcode(:,:)
INTEGER, INTENT(IN):: Nbel(:)
INTEGER, INTENT(IN):: ListR(:,:)
INTEGER, INTENT(IN):: Nodel
INTEGER, INTENT(IN):: Ndof
INTEGER, INTENT(IN):: Ndofe
INTEGER, INTENT(IN):: Isym
INTEGER, INTENT(IN):: Ndest(:,:)
REAL, INTENT(INOUT):: Elres_u(:,:),Elres_t(:,:)
REAL, INTENT(INOUT) :: E,ny,Con
REAL(KIND=8), INTENT(OUT) :: KBE(:,:) ! Stiffness matrix
REAL(KIND=8), INTENT(OUT) :: A(:,:)   ! u due to únit values ui
REAL(KIND=8), INTENT(OUT) :: tc(:)    ! interface tractions
!  temporal arrays :
REAL(KIND=8), ALLOCATABLE :: dUe(:,:),dTe(:,:),Diag(:,:)
REAL(KIND=8), ALLOCATABLE :: Lhs(:,:)
REAL(KIND=8), ALLOCATABLE :: Rhs(:),RhsM(:,:) ! right hand sides
REAL(KIND=8), ALLOCATABLE :: u1(:),u2(:,:)    ! results
REAL, ALLOCATABLE :: Elcor(:,:)
REAL      :: Scat,Scad
INTEGER            :: NdofF
INTEGER :: Dof,k,l,nel
INTEGER :: n,m,Pos,i,j,nd,ne
ALLOCATE(dTe(NdofR,Ndofe),dUe(NdofR,Ndofe))
ALLOCATE(Diag(NdofR,Ndof))
ALLOCATE(Lhs(NdofR,NdofR),Rhs(NdofR),RhsM(NdofR,NdofR))
ALLOCATE(u1(NdofR),u2(NdofR,NdofR))
ALLOCATE(Elcor(Cdim,Nodel))
!------------------------------------------
!  Scaling
!------------------------------------------
CALL Scal(E,xP,Elres_u,Elres_t,Cdim,Scad,Scat)
!------------------------------------------------------------
!  Compute and assemble element coefficient matrices
!------------------------------------------------------------
Lhs= 0.0
Diag= 0.0
Rhs= 0.0
RhsM= 0.0
Elements_1:&
DO Nel=1,Nbel(nr)
    ne= ListR(nr,Nel)
```

```
      Elcor(:,:)= xP(:,IncieR(ne,:))      !   gather element coords
      IF(Cdim == 2) THEN
        IF(Ndof == 1) THEN
          CALL Integ2P(Elcor,IncieR(ne,:),Nodel,NodeR(nr),xP,Con,dUe,dTe)
        ELSE
          CALL Integ2E(Elcor,IncieR(ne,:),Nodel,NodeR(nr),xP,E,ny,dUe,dTe,Ndest,Isym)
        END IF
      ELSE
        CALL Integ3(Elcor,IncieR(ne,:),Nodel,NodeR(nr),xP,Ndof,E,ny,Con,dUe,dTe,Ndest,Isym)
      END IF
      CALL
AssemblyMR(ne,Ndof,Ndofe,Nodel,Lhs,Rhs,RhsM,DTe,DUe,LdesteR(ne,:),Ncode,Bcode,Diag,Elres_
u,Elres_t,Scad)
END DO &
Elements_1
!-----------------------------------------------------
!  Add azimuthal integral for infinite regions
!-----------------------------------------------------
IF(TypeR(nr) == 2) THEN
   DO m=1, NodeR(nr)
     DO n=1, Ndof
       k=Ndof*(m-1)+n
       Diag(k,n) = Diag(k,n) + 1.0
     END DO
   END DO
END IF
!-----------------------------------------------------
!  Add Diagonal coefficients
!-----------------------------------------------------
Nodes_global: &
DO m=1,NodeR(nr)
  Degrees_of_Freedoms_node: &
  DO n=1, Ndof
    DoF = (m-1)*Ndof + n   ! global degree of freedom no.
    k = (m-1)*Ndof + 1     ! address in coeff. matrix (row)
    l = k + Ndof - 1       ! address in coeff. matrix (column)
    IF (NCode(DoF) == 1 .or. NCode(DoF) == 2) THEN       ! Dirichlet - Add Diagonal to Rhs
      Pos = 0
      Nel = 0
      !  get local degree of freedom no corresponding to global one
      Elements_all: &
      DO i=1,Nbel(nr)
        ne= ListR(nr,i)
        Degrees_of_freedom_elem: &
        DO j=1,Ndofe
          IF (DoF == LdesteR(ne,j)) THEN
            Nel = ne
            Pos = j
            EXIT
          END IF
        END DO &
        Degrees_of_freedom_elem
      IF (Nel /= 0) EXIT
      END DO &
      Elements_all
      Rhs(k:l) = Rhs(k:l) - Diag(k:l,n)*Elres_u (Nel,Pos)
      IF(NCode(DoF) == 2)THEN
        RhsM(k:l,DoF) = RhsM(k:l,DoF) - Diag(k:l,n) / Scad
      END IF
```

```
      ELSE
         Lhs(k:l,Dof)= Lhs(k:l,Dof) + Diag(k:l,n)! Neuman - Add to Lhs
      END IF
   END DO &
   Degrees_of_Freedoms_node
END DO &
Nodes_global
!   Solve problem
CALL Solve_Multi(Lhs,Rhs,RhsM,u1,u2)
!-------------------------------------------
!   Back - Scaling
!-------------------------------------------
DO N=1,NdofC
   u1(N)= u1(N) / Scat
   u2(N,:)= u2(N,:) / Scat
END DO
M=NdofC
NdofF= NdofR-NdofC
DO N=1,NdofF
   M=M+1
   IF(NCode(M) == 0) THEN
      u1(M)= u1(M) * Scad
      u2(M,:)= u2(M,:) * Scad
   ELSE
      u1(M)= u1(M) / Scat
      u2(M,:)= u2(M,:) / Scat
   END IF
END DO
   Elres_u(:,:,:)= Elres_u(:,:,:) * Scad
   Elres_t(:,:,:)= Elres_t(:,:,:) / Scat

!-------------------------------------------
! Gather element results due to
! zero Dirichlet conditions at the interface
!-------------------------------------------
Elements2: &
DO nel=1,Nbel(nr)
   ne= ListR(nr,nel)
   D_o_F1:    &
   DO nd=1,Ndofe
      IF(Ncode(LdesteR(ne,nd)) == 0) THEN
         Elres_u(ne,nd) = u1(LdesteR(ne,nd))
      ELSE IF(Bcode(ne,nd) == 1 .or. Bcode(ne,nd) == 2) THEN
         Elres_t(ne,nd) = u1(LdesteR(ne,nd))
      END IF
   END DO &
   D_o_F1
END DO &
Elements2

!-------------------------------------------
! Gather stiffness matrix KBE and matrix A
!-------------------------------------------
Interface_DoFs: &
DO N=1,Ndofc
   KBE(N,:)= u2(N,:)
   tc(N)= u1(N)
END DO &
Interface_DoFs
```

```fortran
A= 0.0
M=NdofC
Free_DoFs: &
DO N=1,NdofF
  M= M+1
  A(N,1:NdofC)= u2(M,:)
END DO &
Free_DoFs
DEALLOCATE (dUe,dTe,Diag,Lhs,Rhs,RhsM,u1,u2,Elcor)
RETURN
END SUBROUTINE Stiffness_BEM

SUBROUTINE Solve_Multi(Lhs,Rhs,RhsM,u,uM)
!-------------------------------------------
!   Solution of system of equations
!   by Gauss Elimination
!   for multiple right hand sides
!-------------------------------------------
REAL(KIND=8) ::   Lhs(:,:)   !   Equation left hand side
REAL(KIND=8) ::   Rhs(:)     !   Equation right hand side 1
REAL(KIND=8) ::   RhsM(:,:)  !   Equation right hand sides 2
REAL(KIND=8) ::   u(:)       !   Unknowns 1
REAL(KIND=8) ::   uM(:,:)    !   Unknowns 2
REAL(KIND=8) ::   FAC
INTEGER  M,Nrhs        !   Size of system
INTEGER  i,n,nr
M= UBOUND(RhsM,1) ; Nrhs= UBOUND(RhsM,2)
! Reduction
Equation_n: &
DO n=1,M-1
  IF(ABS(Lhs(n,n)) < 1.0E-10) THEN
    CALL Error_Message('Singular Matrix')
  END IF
  Equation_i: &
  DO i=n+1,M
    FAC= Lhs(i,n)/Lhs(n,n)
    Lhs(i,n+1:M)= Lhs(i,n+1:M) - Lhs(n,n+1:M)*FAC
    Rhs(i)= Rhs(i) - Rhs(n)*FAC
      RhsM(i,:)= RhsM(i,:) - RhsM(n,:)*FAC
  END DO  &
  Equation_i
END DO &
Equation_n
!    Backsubstitution
Unknown_1: &
DO n= M,1,-1
  u(n)= -1.0/Lhs(n,n)*(SUM(Lhs(n , n+1:M)*u(n+1:M)) - Rhs(n))
END DO &
Unknown_1
Load_case: &
DO Nr=1,Nrhs
  Unknown_2: &
  DO n= M,1,-1
    uM(n,nr)= -1.0/Lhs(n,n)*(SUM(Lhs(n , n+1:M)*uM(n+1:M , nr)) - RhsM(n,nr))
  END DO &
  Unknown_2
END DO &
Load_case
RETURN
```

END SUBROUTINE Solve_Multi

```fortran
SUBROUTINE
AssemblyMR(Nel,Ndof,Ndofe,Nodel,Lhs,Rhs,RhsM,DTe,DUe,Ldest,Ncode,Bcode,Diag,Elres_u,Elres
_t,Scad)
!-------------------------------------------------
!  Assembles Element contributions DTe , DUe
!  into global matrix Lhs, vector Rhs
!  and matrix RhsM
!-------------------------------------------------
INTEGER,INTENT(IN) :: NEL
REAL(KIND=8):: Lhs(:,:)                        !  Eq. left hand side
REAL(KIND=8) :: Rhs(:)                         !  Right hand side
REAL(KIND=8)  :: RhsM(:,:)                     !  Matrix of right hand sides
REAL(KIND=8), INTENT(IN) :: DTe(:,:),DUe(:,:)  !  Element arrays
REAL, INTENT(INOUT) :: Elres_u(:,:),Elres_t(:,:)
INTEGER , INTENT(IN)  :: LDest(:)              !  Element destination vector
INTEGER , INTENT(IN) :: NCode(:)               !  Boundary code (global)
INTEGER , INTENT(IN) :: BCode(:,:)             !  Boundary code (global)
INTEGER , INTENT(IN) :: Ndof
INTEGER , INTENT(IN) :: Ndofe
INTEGER , INTENT(IN) :: Nodel
REAL(KIND=8) :: Diag(:,:) ! Array containing diagonal coeff of DT
INTEGER :: n,Ncol,m,k,l
DoF_per_Element:&
DO m=1,Ndofe
   Ncol=Ldest(m)    !  Column number
   IF(BCode(nel,m) == 0) THEN   !  Neumann BC
     Rhs(:) = Rhs(:) + DUe(:,m)*Elres_t(nel,m)
!    The assembly of dTe depends on the global BC
     IF (NCode(Ldest(m)) == 0) THEN
       Lhs(:,Ncol)= Lhs(:,Ncol) + DTe(:,m)
     ELSE
       Rhs(:) = Rhs(:) - DTe(:,m) * Elres_u(nel,m)
     END IF
   ELSE IF(BCode(nel,m) == 1) THEN   !  Dirichlet BC
     Lhs(:,Ncol) = Lhs(:,Ncol) - DUe(:,m)
     Rhs(:)= Rhs(:) - DTe(:,m) * Elres_u(nel,m)
   END IF
   IF(BCode(nel,m) == 2) THEN   !  Interface
     Lhs(:,Ncol) = Lhs(:,Ncol) - DUe(:,m)
   END IF
   IF(NCode(Ldest(m)) == 2) THEN   !  Interface
     RhsM(:,Ncol)= RhsM(:,Ncol) - DTe(:,m) / Scad
   END IF
END DO &
DoF_per_Element
!        Sum of off-diagonal coefficients
DO n=1,Nodel
   DO k=1,Ndof
     l=(n-1)*Ndof+k
     Diag(:,k)= Diag(:,k) - DTe(:,l)
   END DO
END DO
   RETURN
END SUBROUTINE AssemblyMR

END MODULE Stiffness_lib
```

Appendix B
Answers to exercises

Exercise 2.1
See Utility Library in Appendix A.

Exercise 2.2
Given a 2 x 2 matrix

$$\mathbf{A} = \begin{bmatrix} a_{11} & a_{12} \\ a_{21} & a_{22} \end{bmatrix} \tag{A.1}$$

the determinant is given by $det|\mathbf{A}| = a_{11}a_{22} - a_{12}a_{21}$
For a 3 x 3 matrix

$$\mathbf{A} = \begin{bmatrix} a_{11} & a_{12} & a_{13} \\ a_{21} & a_{22} & a_{23} \\ a_{31} & a_{32} & a_{33} \end{bmatrix} \tag{A.2}$$

The determinant is given by:

$$det|\mathbf{A}| = a_{11}(a_{22}a_{33} - a_{32}a_{23}) + a_{21}(a_{13}a_{32} - a_{12}a_{33}) + a_{31}(a_{12}a_{23} - a_{13}a_{22}) \tag{A.3}$$

The subroutine is listed below:

```
REAL FUNCTION DETERMINANT(A)
!---------------------------------
!    Function computes the determinant of A
!---------------------------------
REAL, INTENT(IN) :: A(:)
N= UBOUND(A,1)
IF(N == 2) THEN
 Determinant= a(1,1)*A(2,2)
ELSE
```

```
 Determinant= a(1,1)*(a(2,2)*a(3,3)-a(2,3)*a(3,2)) &
+a(2,1)*(a(1,3)*a(3,2)-a(1,2)*a(3,3)) &
+a(3,1)*(a(1,2)*a(2,3)-a(1,3)*a(2,2))
END IF
RETURN
END FUNCTION DETERMINANT
```

Exercise 2.3

```
SUBROUTINE ASSEMBLE(A,B,I,J)
!----------------------------------
!     Assembles sub-matrix A into matrix B
!     at location i,j
!----------------------------------
REAL, INTENT (IN)      :: A(2,2)
REAL, INTENT (IN OUT)  :: B(:,:)
INTEGER, INTENT (IN)   :: I,J
B(I:,J:)= A
RETURN
END SUBROUTINE Assemble
```

Exercise 3.1

The computed length dependent on the number of used elements is shown in Figure B.1. It can be seen that the accuracy of the parabolic element is even better than that of six linear elements.

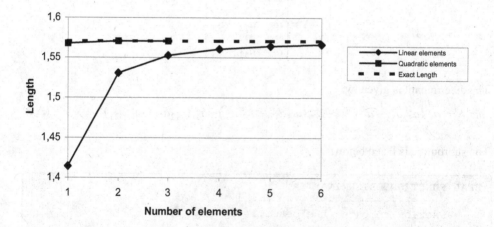

Figure B.1   Accuracy of length computation depending on discretisation used

Exercise 3.2
(a) Error = -0.075% b) Error is same as a).

Exercise 3.3
The input and output for program Compute_Area is presented in Figure B.2, which shows
the distribution of the Jacobian for Problem 3.3
(b) Error = -0.68 %

Inputfile

```
2  8  1  2
7  5  3  1  8  4  2  6
0  1  0
0.3827  0.9239  0
0.7071  0.7071  0
0.9239  0.3827  0
1  0  0
0  0.7071  0.7071
0  0  1
0.7071  0  0.7071
```

Output

```
Element dimension=          2
No. of element nodes=       8
Number of elements=         1
Integration order =         2
Area =              1.485628
```

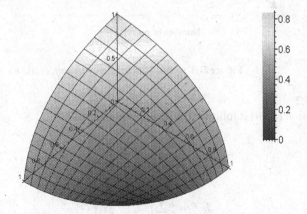

Figure B.2   Variation of Jacobian for Exercise 3.3 (a)

Exercise 5.1
(a)

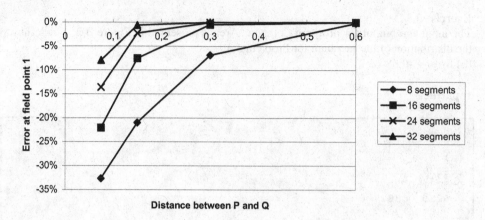

(b)

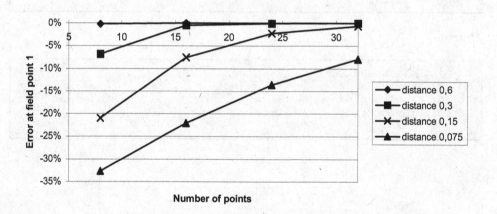

Figure B.3   Answers to exercises part a) and b)

Exercise 5.2
From equation (5.10) it follows that flow q is computed by:

$$q_x = \sum_{i=1}^{N} \left(-\frac{r_{xi}}{2 \cdot \pi \cdot r_i^2} \right) \cdot F_i$$

$$q_y = q_0 + \sum_{i=1}^{N} \left(-\frac{r_{yi}}{2 \cdot \pi \cdot r_i^2} \right) \cdot F_i$$

(A.4)

The DO LOOP *Int_points* of the source code has to be adapted in the following way (changes are printed bold):

```
Int_points: &
DO nin= 1,ninpts
 READ(10,*) xi,yi                    ! coordinates of interior points
 uq= 0.0
 qx= 0.0
 qy= 0.0
 Thetp= Pi/2.0                       ! angle to first source point P1
 Source_points2: &
 DO npp= 1,npnts
  xp= rp*COS(Thetp)                  ! x-coordinate of source point
  yp= rp*SIN(Thetp)                  ! y-coordinate of source point
  dxr(1)= xi-xp
  dxr(2)= yi-yp
  r= SQRT(dxr(1)**2 + dxr(2)**2)
  uq = uq+U(r,k,2)*F(npp)            ! function U from Laplace_lib
  qx = qx+dxr(1)/(r**2*2*Pi)*F(npp)  ! x-component of flow
  qy = qy+dxr(2)/(r**2*2*Pi)*F(npp)  ! y component of flow
  Thetp = Thetp+Delth                ! angle to next source point P
 END DO &
 Source_points2
 uq=uq-q/k*yi
 qy=qy+q
 WRITE(11,'(5(A,F10.3))') &
         'x=',xi,', y=',yi,', T=',uq,', q-x=',qx,', q-y=',qy
END DO &
Int_points
```

Exercise 5.3

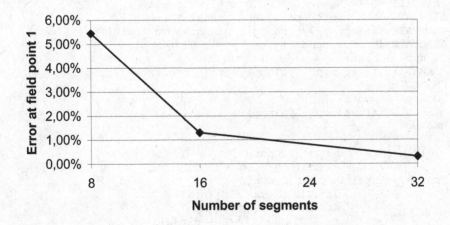

Figure B.4   Plot of error versus number of segments used

Exercise 5.4

Program 5.1 has to be changed in the following way, to be able to specify 'general' boundary shapes. Here the problem to be solved has to be defined by points on the boundary (the field points), the normal vector to the shape at that point and one source point for each field point.

```
PROGRAM Trefftz
!---------------------------------
!   Program to compute the heat flow past a cylindrical isolator
!   in a 2-D infinite domain using the Trefftz method
!---------------------------------
USE Laplace_lib ; USE Geometry_lib;
IMPLICIT NONE
REAL                       ::  q
REAL                       ::  k
INTEGER                    ::  npnts
REAL(KIND=8),ALLOCATABLE   ::  Lhs(:,:)
REAL(KIND=8),ALLOCATABLE   ::  Rhs(:)
REAL(KIND=8),ALLOCATABLE   ::  F(:)
REAL                       ::  dxr(2)
REAL                       ::  vnorm(2)
REAL,ALLOCATABLE           ::  xp(:),yp(:),xq(:),yq(:),norm(:,:)
REAL                       ::  Delth,Thetq,Thetp,xi,yi,r,uq,qx,qy
INTEGER                    ::  npq,npp,ninpts,nin,ios
OPEN(UNIT=10,FILE='INPUT.DAT',STATUS='OLD',ACTION='READ')
OPEN(UNIT=11,FILE='OUTPUT.DAT',STATUS='UNKNOWN',ACTION='WRITE')
READ(10,*) q,k,npnts
WRITE(11,*) ' Heat flow past a cylinder with Trefftz method'
WRITE(11,*) '   Heat inflow/outflow=  ',q
WRITE(11,*) '   Thermal conductivity= ',k
WRITE(11,*) '   Number of Points P,Q= ',npnts
ALLOCATE  (Lhs(npnts,npnts),Rhs(npnts),F(npnts))
ALLOCATE  (xp(npnts),yp(npnts),xq(npnts),yq(npnts))
ALLOCATE  (norm(2,npnts))
DO npp= 1,npnts
  READ(10,*) xq(npp), yq(npp),norm(1,npp),norm(2,npp)
END DO
DO npp= 1,npnts
  READ(10,*) xp(npp), yp(npp)
END DO

Field_points: DO npq= 1,npnts
                  Rhs(npq)= -q * norm(2,npq)
Source_points: DO npp= 1,npnts
                  dxr(1)= xp(npp)-xq(npq)
                  dxr(2)= yp(npp)-yq(npq)
                  r= SQRT(dxr(1)**2 + dxr(2)**2)
                  dxr= dxr/r
                  vnorm=norm(:,npq)
                  Lhs(npq,npp)= T(r,dxr,vnorm,2)
```

```
             END DO  Source_points
          END DO  Field_points
Lhs= - Lhs
Rhs= - Rhs
CALL Solve(Lhs,Rhs,F)
WRITE(11,*)  ''
WRITE(11,*)  'Temperatures at Boundary points:'
Field_points1: DO npq= 1,npnts
                  uq= 0.0
Source_points1:   DO npp= 1,npnts
                     dxr(1)= xp(npp)-xq(npq)
                     dxr(2)= yp(npp)-yq(npq)
                     r= SQRT(dxr(1)**2 + dxr(2)**2)
                     uq= uq + U(r,k,2)*F(npp)
                  END DO  Source_points1
                  uq=uq-q/k*yq(npq)
                  WRITE(11,*) &
                  'Temperature at field point',npq,' =',uq
               END DO  Field_points1
READ(10,*,IOSTAT=IOS) ninpts
IF(ninpts == 0 .OR. IOS /= 0) THEN
   STOP
END IF
WRITE(11,*)  ''
WRITE(11,*)  'Temperatures at interior points:'
Int_points:    DO nin= 1,ninpts
                  READ(10,*) xi,yi
                  uq= 0.0
               qx= 0.0
               qy= 0.0
Source_points2:   DO npp= 1,npnts
                   dxr(1)= xi-xp(npp)
                   dxr(2)= yi-yp(npp)
                   r= SQRT(dxr(1)**2 + dxr(2)**2)
                   uq = uq + U(r,k,2)*F(npp)
                   qx = qx + dxr(1)/(r**2*2*Pi)*F(npp)
                   qy = qy + dxr(2)/(r**2*2*Pi)*F(npp)
                  END DO  Source_points2
                  uq=uq-q/k*yi
                  qy=qy+q
                  WRITE(11,'(5(A,F10.3))') 'x=',xi,', &
                  y=',yi,', T=',uq,',  q-x=',qx,',  q-y=',qy
               END DO  Int_points
END PROGRAM Trefftz
```

The input file to compute the flow around an elliptical isolator with the dimension 4 in the x-direction and 2 in the y-direction. The first line specifies flow, conductivity and number of points, the following 24 lines show the field point coordinates and the components of the normal vector at each point. The following 24 lines specify the source point coordinates. The last 11 lines are used for computation of results inside the domain.

Input file

```
1.0 1.0 24
0  1  0 -1
-0.6840  0.9397  0.1790 -0.9838
-1.0000  0.8660  0.2774 -0.9608
-1.4142  0.7071  0.4472 -0.8944
-1.7321  0.5000  0.6547 -0.7559
-1.9319  0.2588  0.8814 -0.4723
-2  0  1  0
-1.9319 -0.2588  0.8814  0.4723
-1.7321 -0.5000  0.6547  0.7559
-1.4142 -0.7071  0.4472  0.8944
-1.0000 -0.8660  0.2774  0.9608
-0.6840 -0.9397  0.1790  0.9838
0 -1  0  1
0.6840 -0.9397 -0.1790  0.9838
1.0000 -0.8660 -0.2774  0.9608
1.4142 -0.7071 -0.4472  0.8944
1.7321 -0.5000 -0.6547  0.7559
1.9319 -0.2588 -0.8814  0.4723
2  0 -1  0
1.9319  0.2588 -0.8814 -0.4723
1.7321  0.5000 -0.6547 -0.7559
1.4142  0.7071 -0.4472 -0.8944
1.0000  0.8660 -0.2774 -0.9608
0.6840  0.9397 -0.1790 -0.9838
0  0.7
-0.6313  0.6499
-0.9199  0.5887
-1.2889  0.4564
-1.5489  0.2885
-1.6745  0.1209
-1.7  0
-1.6745 -0.1209
-1.5489 -0.2885
-1.2889 -0.4564
-0.9199 -0.5887
-0.6313 -0.6499

0 -0.7
0.6313 -0.6499
0.9199 -0.5887
1.2889 -0.4564
1.5489 -0.2885
1.6745 -0.1209
1.7  0
1.6745  0.1209
1.5489  0.2885
1.2889  0.4564
0.9199  0.5887
0.6313  0.6499
10
```

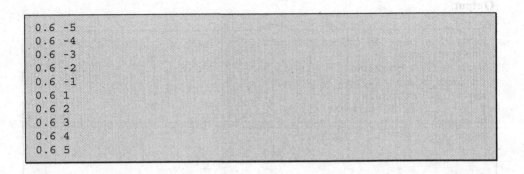

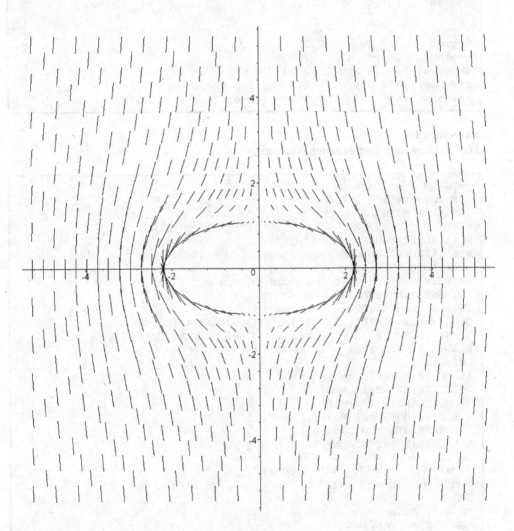

Figure B.5   Flow past an elliptical isolator: Flow vectors

Output

```
Heat flow past an object (direct BE method)
Input values:
Heat inflow/outflow=     1.000000
Thermal conductivity=    1.000000
Number of segments=               24
Temperatures at segment centers:
Segment     1  T=     -2.951
Segment     2  T=     -2.951



Segment    13  T=      2.951



Segment    23  T=     -2.383
Segment    24  T=     -2.717
Temperatures(T) and flow (q-x,q-y) at interior points:
x=0.600, y=-5.000, T=5.575,  q-x=0.020,  q-y=0.913
x=0.600, y=-4.000, T=4.704,  q-x=0.036,  q-y=0.872
```

Exercise 6.1

For this exercise a short program is created.

```
PROGRAM Test_Integ2P
USE Integration_lib
IMPLICIT NONE
REAL            :: ko
REAL,ALLOCATABLE      :: Elcor(:,:)
REAL(KIND=8),ALLOCATABLE:: dUe(:,:),dTe(:,:)
REAL,ALLOCATABLE      :: xPi(:,:)
INTEGER         :: Nodes,ldim,Cdim,n,m,i,Ncol
INTEGER,ALLOCATABLE    :: inci(:)
INTEGER,ALLOCATABLE    :: Ndest(:,:)
OPEN (UNIT=1,FILE='INPUT.dat',FORM='FORMATTED')
OPEN (UNIT=2,FILE='OUTPUT.dat',FORM='FORMATTED')
READ(1,*) ko, Nodes, ldim, Ncol
Cdim=ldim+1
ALLOCATE(Elcor(Cdim,Nodes))
ALLOCATE(xPi(Cdim,Ncol))
ALLOCATE(inci(Nodes))
ALLOCATE(dUe(Ncol,Nodes))
ALLOCATE(dTe(Ncol,Nodes))
ALLOCATE(Ndest(Nodes,1))
DO i=1, Ncol
 READ (1,*) (xPi(n,i),n=1,Cdim)
 WRITE (2,'(3(F10.3))') (xPi(n,i),n=1,Cdim)
END DO
DO n=1, Nodes
READ (1,*) inci(n)
 READ(1,*) (Elcor(m,n),m=1,Cdim)
```

```
END DO
WRITE (2,'(3(F10.3))') Elcor(1:Cdim,1:Nodes)
IF (Cdim==2) THEN
 CALL Integ2P (Elcor,Inci,Nodes,Ncol,xPi,ko,dUe,dTe)
ELSEIF (Cdim==3) THEN
 Ndest=1
 CALL Integ3(Elcor,Inci,Nodes,Ncol,xPi,1,ko,1.0,dUe,dTe,Ndest)
END IF
WRITE (2,*) 'dUe' ; WRITE (2,'(4(F15.8))') dUe
WRITE (2,*) 'dTe' ; WRITE (2,'(4(F15.8))') dTe
END PROGRAM Test_Integ2P
```

Using this program the following is obtained:

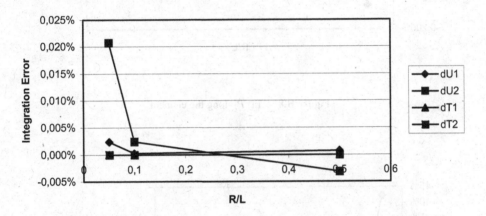

Figure B.6   a) P_i along the element.

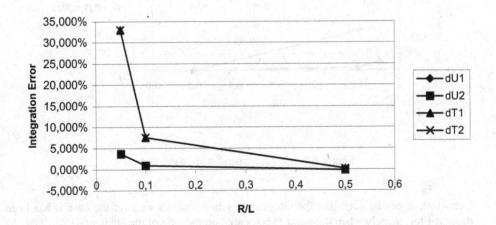

Figure B.7   b) P_i perpendicular to the element.

Exercise 6.2

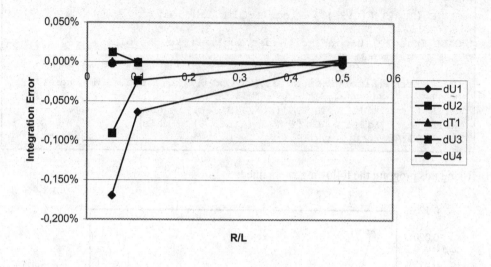

Figure B.8   (a) P_i along the element

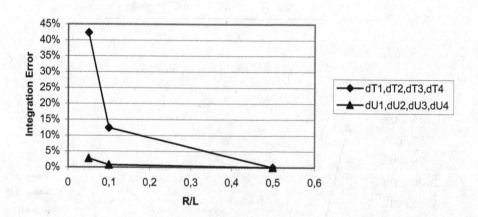

Figure B,9   (b) P_i perpendicular to the element

Comment: It can be seen that the integration scheme works well for the case it has been designed for, namely when the point P_i is located on the side of the element.

Exercise 7.1

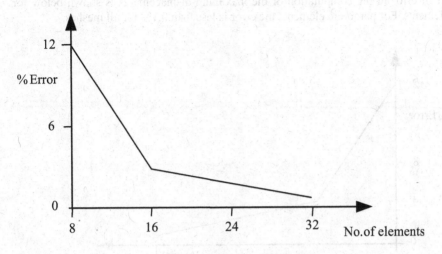

Figure B.10   a) Linear elements

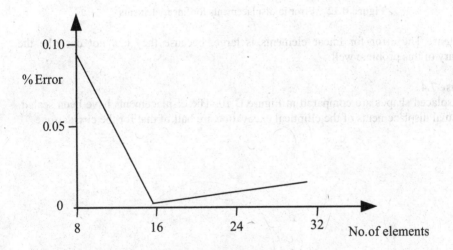

Figure B.11   b) Parabolic elements

Comment: We can see that parabolic elements give negligable error as compared with linear elements.

Exercise 7.2

The maximum temperature at the boundary is half the value of the circular isolator.

Exercise 7.3

The plot of error in the computation of the maximum displacements is shown below for linear elements. For parabolic elements the error is less than 0.1% for all meshes.

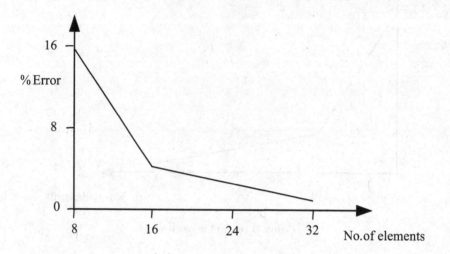

Figure B.12   Error in displacements for linear elements

Comment: The error for linear elements is large, because they can not describe the boundary of this problem well.

Exercise 7.4

The displaced shapes are compared in Figure B.10. The displacements have been scaled. The actual displacements of the elliptical excavation are half of that for the circular one.

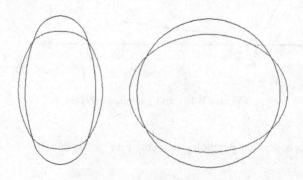

Figure B.13   Displaced shapes of elliptical and circular excavation

Exercise 7.5
The answer to this problem is trivial. The temperature on the right-hand side must be −1.0 so that a heat inflow at the left-hand side (= heat outflow at the right-hand side) is equal to 1.0. There is no heat flow in the vertical direction.

Exercise 7.6
See Chapter 9.

Exercise 7.7
This problem requires a small modification of the program. In BCInput it has been assumed that all degrees of freedom of an element have the same boundary code. In this example however there is an applied displacement in the x-direction, i.e. a *Dirichlet* boundary condition, whereas there is an applied traction of zero magnitude in the y-direction, i.e. a *Neumann* boundary condition.

In SUBROUTINE BCInput we simply replace the line

Bcode(Nel,:) = 1

with

READ(1,*)= Nel,(Bcode(Nel,m),m=1,Ndofe)

When this is done we get a rigid rotation node as a result (Figure B.14).

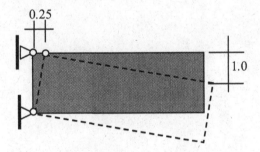

Figure B.14   Displaced shape

Exercise 8.1
See Figure B.15. Very good results can be achieved for all meshes.

Exercise 8.2
See Chapter 9.

Exercise 8.3
The answer to this problem is trivial. The flow is constant in the horizontal direction and zero in the vertical direction.

Exercise 8.4
See Chapter 9.

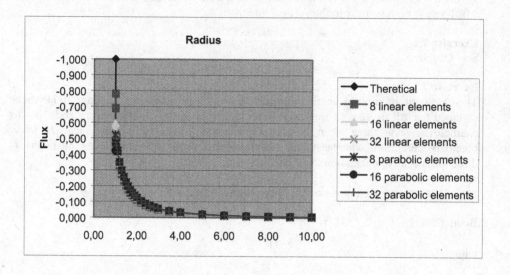

Figure B.15   Variation of flux along a horizontal line

INDEX